T0133036

Evolutionary Theory

Evolutionary Theory

A Hierarchical Perspective

Edited by NILES ELDREDGE, TELMO PIEVANI, EMANUELE SERRELLI, AND ILYA TËMKIN

THE UNIVERSITY OF CHICAGO PRESS   CHICAGO AND LONDON

The University of Chicago Press, Chicago 60637
The University of Chicago Press, Ltd., London
© 2016 by The University of Chicago

Printed in the United States of America

26 25 24 23 22 21 20 19 18 17   1 2 3 4 5

ISBN-13: 978-0-226-42605-1   (cloth)
ISBN-13: 978-0-226-42622-8   (paper)
ISBN-13: 978-0-226-42619-8   (e-book)

DOI: 10.7208/chicago/9780226426198.001.0001

This collective volume is the main outcome of the international research program "The Hierarchy Group: Approaching Complex Systems in Evolutionary Biology" (www.hierarchygroup.com) funded by the John Templeton Foundation (www.templeton .org) and held at the Department of Biology, University of Padua.

Library of Congress Cataloging-in-Publication Data

Names: Eldredge, Niles, editor. | Pievani, Telmo, editor. | Serrelli, Emanuele, editor. |
    Tëmkin, Ilya, editor.
Title: Evolutionary theory : a hierarchical perspective / edited by Niles Eldredge,
    Telmo Pievani, Emanuele Serrelli, and Ilya Tëmkin.
Description: Chicago ; London : The University of Chicago Press, 2016. |
    Includes bibliographical references and index.
Identifiers: LCCN 2016020354 | ISBN 9780226426051 (cloth : alk. paper) |
    ISBN 9780226426228 (pbk. : alk. paper) | ISBN 9780226426198 (e-book)
Subjects: LCSH: Evolution (Biology) | Evolution—Philosophy. | Hierarchies. |
    Biological systems. | Macroevolution.
Classification: LCC QH360.5.E97 2016 | DDC 576.8—dc23 LC record
    available at https://lccn.loc.gov/2016020354

Contents

INTRODUCTION The Checkered Career of Hierarchical Thinking in
Evolutionary Biology   1
Niles Eldredge

PART I **Hierarchy Theory of Evolution**

LINKING SECTION General Principles of Biological Hierarchical
Systems   19
Ilya Tëmkin and Emanuele Serrelli

CHAPTER 1 Pattern versus Process and Hierarchies: Revisiting
Eternal Metaphors in Macroevolutionary
Theory   29
Bruce S. Lieberman

CHAPTER 2 Lineages and Systems: A Conceptual Discontinuity in
Biological Hierarchies   47
Gustavo Caponi

CHAPTER 3 Biological Organization from a Hierarchical
Perspective: Articulation of Concepts and Interlevel
Relation   63
Jon Umerez

CHAPTER 4 Hierarchy: The Source of Teleology in Evolution   86
Daniel W. McShea

CHAPTER 5   Three Approaches to the Teleological and Normative
              Aspects of Ecological Functions   103
              *Gregory J. Cooper, Charbel N. El-Hani, and Nei F.
              Nunes-Neto*

PART 2   **Hierarchical Dynamics: Process Integration across
              Levels**

LINKING SECTION Information and Energy in Biological Hierarchical
              Systems   127
              Ilya Tëmkin and Emanuele Serrelli

CHAPTER 6   Why Genomics Needs Multilevel Evolutionary
              Theory   137
              T. Ryan Gregory, Tyler A. Elliott, and Stefan Linquist

CHAPTER 7   Revisiting the Phenotypic Hierarchy in Hierarchy
              Theory   151
              Silvia Caianiello

CHAPTER 8   Multilevel Selection in a Broader Hierarchical
              Perspective   174
              Telmo Pievani and Andrea Parravicini

CHAPTER 9   Systems Emergence: The Origin of Individuals in
              Biological and Biocultural Evolution   202
              *Mihaela Pavličev, Richard O. Prum, Gary Tomlinson,
              and Günter P. Wagner*

PART 3   **Biological Hierarchies and Macroevolutionary
              Patterns**

LINKING SECTION Ecology and Evolution: Neither Separate nor
              Merged   227
              Emanuele Serrelli and Ilya Tëmkin

CHAPTER 10   Unification of Macroevolutionary Theory: Biologic
              Hierarchies, Consonance, and the Possibility of
              Connecting the Dots   243
              William Miller III

CHAPTER 11   Coming to Terms with *Tempo and Mode*: Speciation,
             Anagenesis, and Assessing Relative Frequencies in
             Macroevolution   260
             Warren D. Allmon

CHAPTER 12   Niche Conservatism, Tracking, and Ecological Stasis:
             A Hierarchical Perspective   282
             Carlton E. Brett, Andrew Zaffos, and Arnold I. Miller

CHAPTER 13   The Stability of Ecological Communities as an
             Agent of Evolutionary Selection: Evidence from the
             Permian-Triassic Mass Extinction   307
             Peter D. Roopnarine and Kenneth D. Angielczyk

CHAPTER 14   Hierarchy Theory in the Anthropocene: Biocultural
             Homogenization, Urban Ecosystems, and Other
             Emerging Dynamics   334
             Michael L. McKinney

CONCLUSION   Hierarchy Theory and the Extended Synthesis
             Debate   351
             Telmo Pievani

List of Contributors   365
Index   377

Introduction

The Checkered Career of Hierarchical Thinking in Evolutionary Biology

Niles Eldredge

Hierarchy theory has played an important role in evolutionary theory since its inception in the early days of the nineteenth century. By "evolutionary theory," I mean the elaboration of causal mechanisms underlying a process of ancestry and descent that interlinks all organisms from the inception of life to the present. Crucial precursors to such an enterprise were the demonstration that all life, extinct and extant, is interconnected in network fashion: the long-familiar Linnaean Hierarchy. This was a necessary, but not sufficient, prerequisite to the formulation of non-miraculous theories on the history of life. Also critical was the acceptance of a Newtonian world view that hinged on the supposition that there is a causal explanation for all observed natural phenomena. Together, these are the two necessary precursors to the birth of causal analysis of the history of life. Both were in place by the late eighteenth century.

By "hierarchy," I simply mean that biological entities, be they molecules or species, are seen as parts of larger wholes—for example, populations are parts of species—and that this structural organization of biological entities is in itself germane to understanding the evolutionary process.

The quintessentially hierarchical observation that species are parts of larger collectivities (taxa, specifically genera) was there nearly from the start—which I trace to the work of Jean-Baptiste Lamarck in the section of fossils in his 1801 work *Animaux sans Vertèbres* (Lamarck 1801, 403–11). For greater detail on the works of Lamarck and all other texts cited in this short chapter, please see my *Eternal Ephemera* (Eldredge 2015).

The ontological status of species was the focal point of early evolutionary thinking, whether in hierarchical form or not. By 1801, and probably earlier, there were three general thoughts on the historical nature and ontological status of species. First, species are created by God, are stable and unchanging, and in the words of the philosopher William Whewell (1837, 626), "a transition from one to another does not exist." This of course was the standard creationist view of the nature and history of the species of the modern biota—held by educated savants in philosophy and the developing world of biological and geological science—as well as, of course, by the majority of ordinary citizens in the Western world.

But the nested pattern of resemblance suggested to several notable eighteenth-century savants (e.g., Darwin's grandfather Erasmus, the Frenchman Buffon, and others) that life must be connected historically, meaning that connections between species in a lineage-forming sense certainly do exist.

In this context, the second species concept in vogue early in the nineteenth century was simply that species empirically are indeed stable entities, as the creationist view insisted, and that a causal process of ancestry and descent forming lineages was in operation in the natural world. This view was first explicitly developed by another "John the Baptist," the Italian Giambattista Brocchi, in his 1814 monograph on the fossil shells of the subapennines of Tuscany. Brocchi saw his species as stable. But he also noticed that species are replaced in time by very similar species that were presumed descendants in a natural process, forming successions of species linked to form lineages.

Brocchi supposed that species are like individuals in that they have naturally caused births, histories (generally marked by morphological stability), and deaths (i.e., extinction). Historian Giuliano Pancaldi (1991) has aptly called this putative equivalence between species and individuals "Brocchi's analogy."

Crucial to Brocchi's thinking here is that individuals were acknowledged by laymen and savant alike to have naturally caused births, histories, and eventually deaths. This was known and acknowledged from time immemorial.

Thus the problem was not with individuals but with species. The reason species were nearly universally thought to be supernaturally created is that the Judeo-Christian Bible, especially in the two or three slightly different accounts of Creation in Genesis, said it was so. But the Bible, written piecemeal in the early ages of agriculturally based society, was written

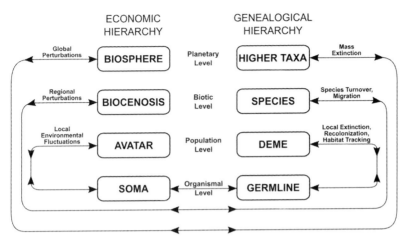

ECONOMIC   GENEALOGICAL
HIERARCHY   HIERARCHY

FIGURE 0.1 The sloshing bucket model relating the magnitude of environmental impact to the extent of evolutionary response. According to the model, the higher the level of external perturbation, the higher the level in the economic hierarchy at which its effects are expressed and, consequently, the higher the level of the genealogical hierarchy at which the evolutionary consequences are manifested.

by people steeped in the knowledge that individual humans, just like all the animals in the manger under their care, were born of sexual congress between a male and female. Insofar as I am aware, the only individual humans said in the Bible to have had supernatural births were Adam and Eve and, later, Jesus of Nazareth. The rest of us mere mortals from time immemorial were born of more prosaic natural causes.

As with births, so with deaths: individuals often die of accidents or disease, but failing such extrinsic causes is the certainty that individuals will end up dying of the intrinsic (innate) cause of simple old age. Brocchi posited that species age and eventually die (i.e., suffer extinction) of old age—unless they are cut off before their time by external environmental causes.

Thus Brocchi's gambit applied empirically acknowledged natural causes underlying the births, histories, and deaths of individuals to the entire species of which they were a part. This part/whole relationship between individuals and species, already present in the Linnaean hierarchy, is quintessentially hierarchical in Brocchi's extension in causal terms of processes underlying the births, histories, and deaths of individuals to species themselves. That was a bold and monumentally significant break from traditional, biblically based thinking—in the spirit of what might be called "Newtonian naturalism."

Finally, the third and most radical view of the ontological status and historical nature of species was first promulgated by Lamarck in the aforementioned work on fossil invertebrates in 1801. This view of species maintained that, in effect, species were not stable but were always gradually transforming as time went on. In a sense, then, species were constantly evolving themselves out of existence as they slowly, smoothly, and intergradationally evolved into their descendants. Early savants grappling with evolution admired Lamarck for his insistence that life is indeed causally interrelated (i.e., life has evolved) but were perplexed by the lack of any real evidence of Lamarck's insistence on gradualism. In his eulogy to Lamarck, Georges Cuvier wryly observed that for a man who described so many species, it was ironic that Lamarck didn't believe that species actually exist in nature.

Later on, of course, the "mature" Charles Darwin surpassed even Lamarck in his dedication to a gradualist perspective, with species slowly but inexorably changing through time, gradually evolving into their descendants. More on that below.

Hierarchy Blossoms: From Lamarck and Brocchi to Darwin

Lamarck and Brocchi became the first truly scientific investigators of the evolutionary process in no small measure because both of them developed quantitative measures in their assessment of the history of life, discernible in the fossils they studied. Indeed, there can be little doubt that their quantitative approach, involving the percentage of species in a fossil fauna that can be considered as still living in the modern biota, was the direct precursor to Lyell's percentage approach to recognizing the temporal divisions of the Tertiary Period. Their fossils were primarily marine invertebrates, most especially mollusks.

Both men had the intention of demonstrating that fossil species were very much connected in a causal sense with species in the modern biota—either through simple survival from past times into the current mélange of life or by having, in one way or another, given rise to descendant species now alive.

Lamarck had a running duel with Georges Cuvier, who maintained that no species found as fossils are still alive in the extant biota. To this, Lamarck retorted that among his marine mollusks extracted as fossils from the Eocene rocks of the Paris Basin, at least 3 percent were alive and well, living in the modern seas surrounding France. The rest, to be sure, were extinct, but many of these fossil species belong to the same genera as living species and are morphologically quite similar to their living analogues. To explain this, Lamarck simply postulated that the Eocene species had slowly transformed into the modern species—a form of extinction through transmutation (evolution) discussed and taken seriously in modern times by paleontologist George Gaylord Simpson and others in the twentieth century.

Brocchi, whose rocks were Mio-Pliocene in age—thus much younger than Lamarck's—accordingly dealt with much younger fossils. He estimated that roughly 50 percent of his fossil species were still extant in the modern fauna offshore from the Tuscan coast. Like Lamarck, he saw that now-extinct species were replaced by closely similar modern species attributable to the same genus. Like Lamarck, he saw these closely related fossil and recent species as forming skeins of ancestral-descendant species—lineages.

Lamarck is best remembered not for these pioneering empirical efforts and his initial declarations of transmutation but for his later (1809) thoughts on the causal processes underlying his putative patterns of gradual change through time—thoughts (e.g., on the inheritance of acquired characters) for which he is still routinely mocked in the modern literature. But his foundational contributions to causal evolutionary theory deserve our respect, not mockery.

Brocchi did not speculate in print on the causal process underlying the "births" of new species from older species. But he left no doubt that he thought that there was such a process, and that, in general, the paleontological approach to such problems must be "like physics." Both Lamarck and Brocchi were avowed Newtonian naturalists—as indeed were Cuvier and other contemporary naturalists who did not embrace an evolutionary perspective.

Finally, that both Lamarck and Brocchi saw causal connections (births of new species from old by natural processes) interlinking species within genera (and, by implication, linked into higher taxa in the Linnaean Hierarchy) makes their views intrinsically hierarchical. Brocchi, seeing species as discrete—with births, histories, and deaths within the ancestral-descendant lineages thus formed—came a lot closer to expounding a hierarchical view of evolutionary process recognizable in modern thinking than did Lamarck. If there were to be recognized a "father of evolutionary hierarchy theory," Brocchi would be my candidate.

Lamarck and Brocchi's work was well known and appreciated throughout Europe—perhaps especially so in Edinburgh. Edinburgh is critical, as that is where Darwin attended two terms of medical school in the mid-1820s, hearing debates and learning ambient ideas on transmutation filling the Edinburgh intellectual environment. Darwin applied these ideas to his own thinking, probably in Edinburgh, and later in Cambridge, but certainly during his five-year voyage to southern South America and eventually around the world (1831–36). Lamarck and Brocchi went on to play strong roles in Darwin's thinking as he wrote his sections on transmutation in the *Red Notebook* (1837) and in his four *Transmutation Notebooks* (1837–39). Their explicit influence dwindled sharply when Darwin developed his "mature" evolutionary ideas in the 1840s and 1850s.

Thus, before Darwin reached medical school in 1825, Lamarck's and Brocchi's ideas were already well known to Edinburgh's naturalists. From the standpoint of the early history of evolutionary theory as it developed in the 1820s and 1830s, Edinburghian James Hutton was one of the early savants, certainly in Great Britain, to demonstrate that the earth has had a history that can be deciphered scientifically, without resorting to supernatural causality. That lesson was soon to be extended to the knottier problem of explaining the causal factors underlying the history of life—which was beginning to appear to have had nearly as long a history as the earth itself.

In 1816, Edinburgh geologist John Horner (later to become the father-in-law of Charles Lyell) wrote an extensive and generally favorable review of Brocchi's 1814 monograph. Horner alluded to Brocchi's speculative "system" but confined most of his remarks to what he deemed Brocchi's praiseworthy treatments of the sediments and fossils of the subapennines. The main value of his review, from the standpoint of evolutionary history, is that Brocchi's name—his very existence—was brought to the attention of the English-speaking world. Thereafter, even though Brocchi's concepts (especially "Brocchi's analogy") were readily discussed in print, his name was rarely mentioned.

Not so with Lamarck, whose name was in fairly common parlance—especially in the two anonymous papers of 1826 and 1827. It is my belief that the founding editor of the *New Edinburgh Philosophical Journal*, Robert Jameson, wrote both of them (see Eldredge 2015 for discussion). Jameson taught at the medical school, teaching a course or curriculum segment called "On the Origin of the Animal Species," which Darwin attended.

Regardless of the author, both essays are interesting for their commingling of the ideas of Lamarck and Brocchi—evidently not viewed as necessarily antithetical alternatives, but rather both held in general esteem for their protransmutational positions.

It is in the second of these essays, as well as in one of Jameson's essays in Jameson's (1827) fifth edition of his book on Cuvier, that the term *replacement* also became common parlance in reference to ancestral species becoming replaced in time by descendant species. The most famous such usage in this era was undoubtedly by John Herschel, who wrote in a letter to Lyell, "Of course I allude to that mystery of mysteries, the replacement of extinct species by others" (Herschel 1836). Darwin (1859) cites this simple passage in the second sentence of the *Origin of Species*.

At the very end of Jameson's fifth edition of his Cuvier volume, there are two tables spread over four pages, with no accompanying explanation. These show the distributions of species within genera and families through time. They are a graphic representation of the sorts of data compiled by Lamarck and Brocchi and clearly show that species can undergo extinction while the genus to which it belongs can persist. Such patterns of extinction within persisting taxa (species within genera) became the code words for replacement of species through time through transmutational replacement, as first exemplified by Darwin when he was on the *Beagle* in 1832 in Argentina.

Darwin's other "Edinburgh Lamarckian" mentor, Robert Grant, is known (mostly through Darwin's autobiography) to have been an ardent admirer of Lamarck. But he was also steeped in Brocchi's analogy and the concept of replacement of species through time. Grant left few words explicitly pertaining to transmutation behind him, but this brief snippet of a remarkable passage in his Inaugural Address (1828, 11–12) as professor at the University College London beautifully illustrates both the Brocchian and the Lamarckian elements of his transmutational thinking that Darwin surely had heard during the long hours he spent with Grant in 1827: "In this vast host of living beings, which all start into existence, vanish, and are renewed, in swift succession, like the shadows of the clouds in a summer's day, each species has its peculiar form, structure, properties, and habits, adapted to its situation, which serve to distinguish it from every other species; and each individual has its destined purpose in the economy of nature. Individuals appear and disappear in rapid succession upon the earth, and entire species of animals have their limited duration, which is but a moment, compared with the antiquity of the globe." Especially the passage on the dynamics of species within genera of Grant's lyrical passage is a ringing statement of Brocchi's analogy and their relation to transmutation. These thoughts are purely and dramatically hierarchical.

Darwin learned more useful aspects of biology and geology at Cambridge—but none, insofar as I am aware, that directly pertained to a hierarchical viewpoint. It was only when Darwin started recording his thoughts while on the HMS *Beagle* that his hierarchical thinking in relation to transmutation became stunningly clear—within just over six months of stepping aboard that storied ship.

Darwin and the Development, Application, and Eventual Rejection of Hierarchical Thinking in Evolutionary Theory

As he recounts in his *Geological Diary* (1832–36), Darwin began comparing fossils with living species on the very first stop of the *Beagle*, on the Cape Verde Islands in early 1832. It was quickly to become a habit. Here, in the Cape Verdes, Darwin thought the marine invertebrates along the shores belonged to the same species that he observed as fossils in the limestone that he observed cropping out along the shore. He realized there was a nontrivial age to the rocks and the fossils they contained, yet those fossils belonged to still-living species.

The very next opportunity to make such comparisons occurred roughly six months later, at Bahia Blanca in Argentina in the fall of 1832. There, Darwin hit the evolutionary hierarchical jackpot. The marine invertebrates he found in the cliffs along the shoreline, as in the earlier Cape Verde experience, once again struck him as belonging to the same species inhabiting the shallow marine coastal waters.

But the fossil mammals told a different story. Many, such as the giant ground sloths (including *Megatherium*) and the giant and palpably armadillo-like glyptodonts, were extinct—showing Darwin that without any doubt, not all species living at the same time in the same place are destined to become extinct at the same time.

Most exciting of all, Darwin collected the remains of a small rodent that he thought represented a smaller species belonging to the same genus as the still-living "Patagonian cavy," or "mara." Here, Darwin thought, was an example of the extinction of a species belonging to the same genus of a similar living species. With these specimens, Darwin felt he had an example of replacement of an extinct species by its descendant, though he was cautious expressing such conclusions in his notes while on the *Beagle*. That's where the circumspection of expressing his observations in terms of replacement of species within genera became critical. It was not until Darwin returned safely home that his notes grew explicitly transmutational, and he was able to write straightforwardly about the putative ancestral-descendant relationships of species—the temporal replacement of extinct species by close relatives still living.

Earlier biologists (including Brocchi) were well aware of patterns of geographical replacement of closely related species as well. But when Darwin turned his sights on the living biota of the pampas and Patagonian wilds of southern South America, the documentation of actual examples of geographic replacement of congeneric species became something of an obsession. Most famous is the more or less abrupt replacement of the greater rhea of the Argentine pampas with the lesser rhea of Patagonia. The Rio Negro seemed to be the dividing line, where the ranges of the two species overlapped slightly, but no interbreeding between them seemed to be taking place.

Darwin became especially entranced with what he later referred to as the "halo" pattern, whereby now familiar patterns of geographic replacement of closely related species on the mainland became augmented with the observation that species on the islands off the mainland were close relatives of mainland species.

The icing on the cake, of course, were the examples of still further divergence of species living on separate islands within an archipelago—the first example being the different versions of the Falkland Fox living on the East and West Falkland Islands. Most famously, it was the five species of mockingbirds on various islands on the Galapagos that drove Darwin to write the famous words in his *Ornithological Notes*, where he says the following of the mockingbirds (throwing in the Galápagos tortoises and Falkland Foxes for good measure):

> In each Isld. each kind is exclusively found: habits of all are indistinguishable. When I recollect, the fact of the form of the body, shape of scales & general size, the Spaniards can at once pronounce from which Island any Tortoise may have been brought. When I see these Islands in sight of each other, & possessed of but a scanty stock of animals, tenanted by these birds, but slightly differing in structure & filling the same place in Nature, I must

suspect they are only varieties. The only fact of a similar kind of which I am aware, is the constant/asserted difference—between the wolf-like Fox of East and West Falkland Islds.—If there is the slightest foundation for these remarks the zoology of Archipelagoes—will be well worth examining; for such facts [would] undermine the stability of Species.

These were by far his most explicitly transmutational words known to have been written while still on the *Beagle*. And the "halo" effect was pure hierarchical imagery.

Six months prior to reaching the Galápagos in August of 1835, Darwin did write his first true evolutionary essay, entitled simply "February 1835." The writing is so cryptic that some historians still question whether the essay is indeed evolutionary in tone and content. Here Darwin writes about the births and deaths of species for the first time and defends Brocchi's notion of the aging and eventual deaths of species against Lyell's (1832) dismissal of those very ideas. Darwin kept returning to elements of Brocchi's analogy throughout the remainder of the 1830s and, in more muted terms, all the way up through the publication of his *Origin of Species* in 1859.

Darwin's Post-*Beagle* Evolution

Though Darwin arrived home in the fall of 1836, it was not until sometime in mid-1837 that he started writing his transmutational thoughts. Safe in his own home, with no one to peer at his notes, Darwin wrote in his *Red Notebook* (started on the *Beagle*) about the replacement patterns he saw among South American species. He made no bones about his conclusions that he saw ancestry and descent—that, for example, the greater rhea was the ancestor of the lesser rhea, though both are still living (and Darwin, in a separate passage, urged looking for a "common parent"). As far as temporal replacements were concerned, because Richard Owen told him his "mara" fossil actually belonged to a smaller species of rodent (a tocu-tocu), Darwin opted instead to cite the camel-like fossil he had discovered as the ancestor of the living guanacos. He also said he was tempted to believe the origins of the newer from the older species must have been sudden (per saltum). Species remained to Darwin as real entities, with births from ancestors just as individual organisms have from their parents.

In 1844, Darwin wrote Leonard Jenyns a letter containing his explanation of how he had come to favor the idea of transmutation in the first place:

> With respect to my far-distant work on species, I must have expressed my-self with singular inaccuracy, if I led you to suppose that I meant to say that my conclusions were inevitable. They have become so, after years of weigh-ing puzzles, to myself alone; but in my wildest day-dream, I never expect more than to be able to show that there are two sides to the question of the immutability of species, i.e. whether species are directly created, or by in-termediate laws, (as with the life & death of individuals). I did not approach the subject on the side of the difficulty in determining what are species & what are varieties, but (though, why I shd give you such a history of my do-ings, it wd be hard to say) from such facts, as the relationship between the living & extinct mammifers in S. America, & between those living on the continent & on adjoining islands, such as the Galápagos—It occurred to me, that a collection of all such analogous facts would throw light either for or against the view of related species, being co-descendants from a common stock.

Darwin claims here a purely Brocchian pedigree, rejecting the suppo-sition that he adopted a transmutational view because of the intergrada-tional similarities between taxa. This passage is all the more remarkable because, by 1844, Darwin had all but abandoned Brocchi to adopt a purely Lamarckian viewpoint. In his *Transmutational Notebooks B–E* (1837–39), Darwin debated the relative merits of two situations where he saw adap-tive change occurring via some "law" involving heredity and heritable var-iation but missing up to that point an additional ingredient. In *Notebook D*, he formulated natural selection as that law.

The two scenarios were (1) the adaptive change in geographic isola-tion leading to the emergence of new species (consonant with, but a step beyond, Brocchi; Darwin of course had data to support that with his geo-graphic replacement patterns, especially on islands vis-à-vis the mainland) and (2) the gradual transformation of species through time, so that species slowly and inexorably evolve themselves out of existence. Darwin came to see these two models as antithetical; he knew that many naturalists conversant with the fossil record acknowledged that species tended to be stable rather than demonstrate gradual change. In a striking and crucial single-sentence passage in *Transmutation Notebook E* in 1839, Darwin fi-nally made his choice, coming out as a gradualist, while dropping anything more than lip service to the importance of geographic isolation in evolution: "If separation in horizontal direction is far more important in making species, than time (as cause of change) which can hardly be believed, then, uniformity in geological formation intelligible" (*Notebook E*, 13). This meant that if gradual transformation of species is less important than geographic isolation (allopatric speciation) in the origin of species, stasis—the stability of species—is understandable. The rest, of course, is history. The evolution that came down to the twentieth century after 1859 was the thoroughgoing gradualism that Darwin left us with, thus leaving evolutionary biology, for a period of seventy-six years, with only a distinctly nonhierarchical Lamarckian/Darwinian vision of the evolutionary process—a process of continual, gradual adaptive modification of morphologies so inexorable that species were no longer seen as real, stable entities with their own births, histories, and deaths.

The Resurrection of a Hierarchical Perspective in Evolutionary Biology

There were, inevitably, some biologists who continued to urge the importance of geographic isolation in evolution. But it was not until geneticist Theodosius Dobzhansky (1935), who pointed out that throughout the seventy-five years of harping on continuity following Darwin's lead (seventy-six, actually, since the publication of the *Origin*, but near enough!), biologists were ignoring the reality of (morphological) gaps between closely related species. Dobzhansky's ontology held that species are indeed real and have births in geographic isolation from their ancestral species. The theme was picked up and embellished by ornithologist Ernst Mayr and others. No one knew that the ontology was Brocchian, and that allopatric speciation had clearly and cleanly been articulated by Charles Darwin in *Notebook B* (1837).

Thus began the resurrection—a reinvention, not a new discovery—of hierarchy theory in evolutionary biology. Neither Dobzhansky nor Mayr saw any antithetical relationship between allopatric speciation and gradual transformation of entire species (Dobzhansky naively defended the latter, saying paleontologists know that gradual change is true. Wrong!).

So it took a couple of young paleontologists to seal the deal: "punctuated equilibria" (Eldredge 1971; Eldredge and Gould 1972) started the revolution that blew gradualism out of the water. Gradualism, empirically,

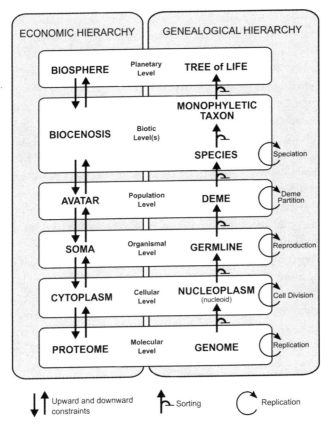

FIGURE 0.2 The dual hierarchical model of biological systems. The economic (ecological) hierarchy represents dynamics of matter and energy exchange, and, generally, corresponds to the spatial dimension of life. The genealogical (informational or evolutionary) hierarchy represents transmission of heritable information, corresponding to the temporal dimension of life.

is pretty much a fantasy. As in Darwin's day, there is little or no real empirical support for the concept. Stasis—the essential stability of species with little or no evolutionary change, often for millions of years—is the empirical norm. And in this case, we knew that our elders from Darwin's day were aware of that, hence Steve Gould's remark that stasis is "paleontology's trade secret."

So, as Darwin concluded in 1839, it really is an either/or situation. Gradualism versus stasis/allopatric speciation. Contrary to Darwin's passage from *Notebook E* quoted above, stasis is the empirical truth of the matter, not gradualism—which he favored with no supporting data but out of a sense that it would be difficult to achieve isolation over vast reaches of terrestrial environments, such as the Argentinian pampas and Patagonian wilds. He had no way of knowing of the effects of climate change on continental environments, and hence on evolution (Vrba 1985).

This time around, however, the effects on evolutionary theory have been far reaching in terms of explicitly recognizing the hierarchical structure of biological nature and its attendant implications. We have seen debates on whether species are individuals—as in the essay by Michael Ghiselin in 1974. We have asked questions about biases in the births and deaths of species—raising the possibility of "species selection" or at least "species sorting" (Eldredge and Gould 1972 for the initial proposition; Stanley 1975 for the name; and especially Vrba [e.g., 1984] for more sophisticated refinements).

We have come to see that biological nature is complexly organized into economic and genealogical hierarchies, and have come to ask how those twin hierarchies interact to produce evolutionary history (Eldredge and Salthe 1984). We see that the history of life resembles a "sloshing bucket" (Eldredge 2003), where large-scale environmental disruptions produce mass extinctions, with proportionally great concomitant evolutionary reactions—and with smaller environmental perturbations producing correspondingly lesser effects, and so forth.

As I look over the last few paragraphs above, I see my name cited with disconcerting monotony. I am hungry for change—for developments in hierarchy theory from the younger generation. That to me is the promise of this present endeavor, this book that holds great potential for extending and expanding on Brocchi's initial vision.

Hierarchical thinking has been present nearly from the get-go in evolutionary thinking. It went through a period of banishment. But it has enjoyed a renaissance, followed by a flowering into previously unexplored domains. I fully expect the work of our hierarchy group, as reflected in the pages of this book, to take this flowering into new and even more interesting domains.

References

Brocchi, Giambattista. 1814. *Conchiologia Fossile Subappennina*. Milan: Stamperia Reale.

Darwin, Charles. 1832–36. *Geological Diary/Geological Notes*. DAR 32–42. Cambridge University Library. [Includes the *Earthquake Portfolio* (DAR 42) containing two essays: "Reflection on Reading My Geological Notes" (DAR 42: 93–96) and "February 1835" (DAR 42: 97–99)]. See also http://darwin-online .org.uk.

———. 1835. "February 1835." In *Geological Diary/Geological Notes*. Cambridge University Library.

———. 1836. *Ornithological Notes*. Cambridge University Library, DAR 29.2. See Barlow, Nora, 1963 for a transcription. Also transcribed and online at http:// darwin-online.org.uk.

———. 1836–37. *The Red Notebook*. Down House. Original and transcription available online at http://darwin.amnh.org.

———. 1837–39. *Transmutation Notebooks B-E*. Cambridge University Library, DAR 121–24. For transcription, see Kohn, David, ed. 1987 and http://darwin .amnh.org; http://darwin-online.org.uk.

———. 1844. Letter to Jenyns. November 25, 1844. http://www.darwinproject.ac .uk/letter/entry-793.

———. 1859. *On the Origin of Species by Means of Natural Selection or, the Preservation of Favoured Races in the Struggle for Life*. London: John Murray. http:// darwin-online.org.uk and http://darwin.amnh.org.

Dobzhansky, Theodosius. 1935. "A Critique of the Species Concept in Biology." *Philosophy of Science* 2: 344–55.

Eldredge, Niles. 1971. "The Allopatric Model and Phylogeny in Paleozoic Invertebrates." *Evolution* 25: 156–67.

———. 2003. "The Sloshing Bucket: How the Physical Realm Controls Evolution." In *Evolutionary Dynamics: Exploring the Interplay of Selection, Accident, Neutrality, and Function*, edited by J. Crutchfield and P. Schuster, SFI Studies in the Sciences of Complexity Series, 3–32. New York: Oxford University Press.

———. 2015. *Eternal Ephemera: Adaptation and the Origin of Species from the Nineteenth Century through Punctuated Equilibria and Beyond*. New York: Columbia University Press.

Eldredge, Niles, and Stephen Jay Gould. 1972. "Punctuated Equilibria: An Alternative to Phyletic Gradualism." In *Models in Paleobiology*, edited by Thomas J. M. Schopf, 82–115. San Francisco: Freeman, Cooper. http://www.NilesEldredge .com, http://www.blackwellpublishing.com/ridley/classictexts/eldredge.pdf.

Eldredge, Niles, and Stanley N. Salthe. 1984. "Hierarchy and Evolution." *Oxford Reviews in Evolutionary Biology* 1: 182–206.

Ghiselin, Michael. 1974. "A Radical Solution to the Species Problem." *Systematic Zoology* 23: 536–44.

Herschel, John. 1836. Letter to Charles Lyell. http://darwin.amnh.org.

Horner, Leonard. 1816. "[Review of] G. B. Brocchi, *Conchiologia Fossile Subappennina*." *Edinburgh Review* 26: 156–80.

Jameson, Robert. (1813) 1827. *Essay on the Theory of the Earth: With Geological Illustrations by Professor Jameson*. 5th ed. Edinburgh: William Blackwood.

Jameson, Robert (Anon.). 1826. "Observations on the Nature and Importance of Geology." *Edinburgh New Philosophical Journal* 1: 293–302.

———. 1827. "Of the Changes Which Life Has Experienced on the Globe." *Edinburgh New Philosophical Journal* 3: 298–301.

Lamarck, Jean-Baptiste. 1801. *Systême des Animaux sans Vertèbres*. Paris.

———. 1809. *Philosophie Zoologique*. Paris.

Lyell, Charles. (1832) 1997. *Principles of Geology*, vol. 2. Edited by James Secord. Reprint, London: Penguin Books.

Pancaldi, Giuliano. 1983. *Darwin in Italy*. Bologna: Società editrice il Mulino. English translation 1991. Bloomington: Indiana University Press.

Stanley, Steven M. 1975. "A Theory of Evolution above the Species Level." *Proceedings of the National Academy of Sciences of the United States of America* 72: 646–50.

Vrba, Elisabeth S. 1984. "What Is Species Selection?" *Systematic Zoology* 33: 318–28.

———. 1985. "Environment and Evolution: Alternative Causes of the Temporal Distribution of Evolutionary Events." *South African Journal of Science* 81: 229–36.

Whewell, William. 1837. *History of the Inductive Sciences*. London: Parker.

PART I

Hierarchy Theory of Evolution

General Principles of Biological Hierarchical Systems

Ilya Tëmkin and Emanuele Serrelli

The hierarchy theory of evolution is ontologically committed to the existence of inherent nested hierarchies in nature and attempts to explain natural phenomena as a product of complex dynamics of biological systems in the context of scaling. Most generally, a *system* is a network of functionally interdependent and structurally interconnected parts of an integrated whole, where the complexity arises from nontrivial, nonlinear interactions among parts, so that the emergent global dynamics of the whole cannot be expressed as simply the sum of its parts.

Organizing Principles of a Hierarchy

A *hierarchy* is an arrangement of entities according to *levels*, or classes of entities of the same rank or significance. The meaning of levels and the relationship among them depend on which specific type of hierarchy is considered: order, inclusion, control, or level hierarchy. Despite the fact that multiple hierarchies can be recognized in living systems (ranging from purely epistemological constructs to specific ontological claims), the hierarchy theory of evolution focuses on a particular class of hierarchies—nested compositional hierarchies—as a fundamental physical organizational principle of real biological systems. A *nested compositional hierarchy* is an ordered organization in the context of scale based on the principle of increasing inclusiveness, so that entities at one level are composed of parts at lower levels, which themselves function as parts of more inclusive

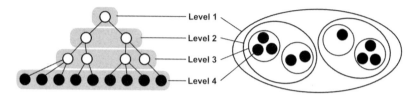

FIGURE PI.I General representations of hierarchical systems. In a rooted tree, or acyclical graph (left), links or edges (solid lines) designate different kinds of relationships (e.g., order or control) among entities (circles) at different levels. The Venn diagram (right) emphasizes a recursive organization of nested compositional hierarchies.

entities at a higher level. Levels are classes of such parts, and wholes and their ranks correspond to the scale of the entities that are their members. The term *focal level* refers to a level at which a particular phenomenon is observed, whereas the terms *higher* (or *upper*) and *lower* levels refer to more inclusive and less inclusive levels relative to the focal level, respectively. A nested hierarchy can be formalized mathematically as an ordered set and represented graphically as a rooted tree (an acyclical graph) or a Venn diagram (figure PI.I).

The discreteness of levels results from differences in the nature and rate of processes between entities at a given level and entities at different levels. However, it is not always easy to draw clear boundaries between levels when there are small differences in scale among entities at levels of adjacent ranks, particularly at the more weakly integrated higher levels of a hierarchy.

The nested part-whole relation in a hierarchy produces an important distinction between two kinds of attributes or traits of entities at any given level: aggregate and emergent. *Aggregate traits* are cumulative properties or combined attributes of entities at the lower level (for example, stenotopic-eurytopic characteristics of a species, which derive from the cumulative range of individual tolerances to environmental changes); *emergent traits* are properties that cannot be reduced to or be expressed in terms of properties of entities at the lower level (such as sex ratio, which characterizes a collective state of a population that cannot be applied to individual organisms). At a given level, all entities may either share the same traits or display variation in the traits.

Hierarchical Dynamics

The interactions of entities in a hierarchy fall into two major categories: within and between levels. Within-level, high-frequency dynamics involve direct and strong, typically time-independent (reversible) interactions of entities in the same set of processes at commensurate rates. Such interactions can be effectively represented as *complex networks*, or systems of interacting entities that are conventionally represented by a graph, a collection of nodes (vertices) connected by links (edges). Typically, directed links describe the flow of information, matter, or energy between a source and its target(s), whereas nondirected edges show mutual interactions, where information, matter, or energy are exchanged between a pair of nodes. Network dynamics allows for undirected interactions, such as cyclical relationships and feedback loops, so that the functional interactions within such a system are temporally restricted. Consequently, entities from different hierarchical levels cannot effectively be members of the same network. In biology, networks are present at all levels of organization, from metabolism and regulation of gene expression to ecological trophic webs and social networks within populations.

Despite the differences in the rules of interactions among entities at different levels, complex biological networks—from gene expression regulatory networks to ecosystem-wide food webs—share a set of common features: they tend to be highly modular and have a high clustering coefficient, a heavy tail in the degree distribution (the frequency distribution of the number of links per node), and a short mean path length (the mean number of nodes along the shortest path between two nodes). These fundamental architectural principles result in an astonishing network-wide property that is shared by real-world biological networks across levels of biological organization. Such networks display exceptional robustness and a high degree of tolerance against random failures and external perturbations.

Substantial differences in process rates prevent entities at different levels from interacting directly, allowing for relatively weak, generally time-dependent (irreversible) interactions in an aggregate fashion. This nontransitivity of direct effects across levels is what makes the levels discrete and quasi-independent to the extent that the details of within-level interactions can be ignored when considering between-level dynamics. Such property of hierarchical dynamics allows for decoupling and investigating processes at individual levels on their own right.

In biological systems, a pair of adjacent levels comprises a dual control system, where the interactions of entities at a lower level establish *initiating conditions* (*upward causation*) and the interactions of entities at a higher level exert constraints, or determine *boundary conditions* (*downward causation*). Noncontiguous levels may affect the dynamics at the focal level indirectly through cascading upward and downward effects across levels. However, there is a fundamental asymmetry in the effects of upward and downward causation: because downward causation affects all subsystems at all the less-inclusive levels contained within the system (level) where it originates, the dynamics at higher levels will always propagate downward, whereas the dynamics at lower levels might never manifest at higher levels.

Hierarchy and Complexity

A global architecture of a hierarchy is best approximated by a quasi-fractal representation, resulting from nesting of self-similar elements—that is, complex networks distributed across the levels—so that an entity at a focal level is composed of a network of its parts at a lower level and itself functions as an element of a network that makes up a higher level entity. Taking into account the isomorphism in the architectural features of biological networks across the hierarchy, a system of hierarchically nested, complex networks is predicted to display remarkable metastability, despite nonequilibrial dynamics at all levels of organization triggered by extrinsic disturbances, a prediction consistent with empirical data on living systems and the general tendency of complex systems to attain and remain in equilibrium. A change in such a system is a systemic, synergetic response to sufficiently strong perturbations—capable of disrupting networks at multiple levels of biological organization—that may have evolutionary consequences. Consequently, complexity in biological evolutionary phenomena arises from the synergetic effect of the network dynamics of entities at the focal level, the boundary and initiation conditions emerging from network dynamics of entities at adjacent levels, and cascading upward and downward effects from more remote levels.

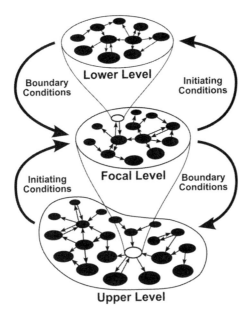

FIGURE PI.2 Quasi-fractal architecture and global dynamics in a nested compositional hierarchy. The intralevel direct interactions (straight arrows) connect individual entities (circles) within networks at each level and may include time-independent processes, such as cyclical relationships and feedback loops. The interlevel indirect interactions (curved arrows) establish constraints, or boundary and initiating conditions, for the processes at the focal level, exerted by the relative upper and lower levels, respectively.

The Ontological Paradigm of the Hierarchy Theory of Evolution

Living systems in general and, more specifically, evolving systems are defined by two classes of processes: *interaction*, or active exchange of matter and energy, and *replication*, or transmission of heritable information. The dynamic and informational aspects characterize two distinct kinds of biological entities: the *replicators*, or closed systems, that pass on their structure (i.e., information) to offspring, and the *interactors*, or open systems that require continuous matter/energy exchange with their environment to maintain their equilibrium. Interactors are related to replicators in such a way that the survival and the work of the former are causally responsible for the differential propagation of the latter. In evolutionary history, the tasks of replication (heritability) and dynamic interactions became progressively and irreversibly decoupled, producing differentiated and specialized sets of entities at each level of biological organization of greater functional efficiency. At the cellular level, the genetic nucleus in eukaryotes is separated from metabolic cytoplasm and other organelles; at the organismal level, the genetically transmitted germline is segregated from the mortal soma; and at the population level—for instance, in eusocial insects—nonreproductive castes are distinct from reproductive individuals. The degree of the decoupling and hence interdependency between the interactors and replicators varies substantially across organizational levels.

This fundamental dualism of biological systems and the lack of one-to-one correspondence of replicator and interactor entities across levels of biological organization prevented accommodating the entirety of the biotic realm by a single hierarchy. The hierarchy theory of evolution adopts a model of two interconnected systems, corresponding to the dynamic and the informational aspects of life: (1) the economic, or ecological, compositional nested hierarchy that captures for dynamic interactions of entities within and across levels through upward and downward causation and (2) the genealogical, or reproductive, nested compositional hierarchy, which reflects the historical nature of biological systems stemming from the unidirectional control of information flow. Most generally, the economic and genealogical hierarchies represent, respectively, the spatial and temporal dimensions of the organic realm. Importantly, drawing an explicit distinction between the two types of hierarchies allows for elucidating causal relationships between them.

References

Albert, Réka, and Albert L. Barabási. 2002. "Statistical Mechanics of Complex Networks." *Reviews of Modern Physics* 74: 47–97.

Brooks, Daniel R., and E. O. Wiley. 1988. *Evolution as Entropy: Toward a Unified Theory of Biology*, 2nd ed. Chicago: University of Chicago Press.

Eldredge, Niles, and Stanley N. Salthe. 1984. "Hierarchy and Evolution." *Oxford Surveys in Evolutionary Biology* 1: 184–208.

Hull, David. 1980. "Individuality and Selection." *Annual Review of Ecology and Systematics* 11: 311–22.

Newman, M. E. J. 2003. "The Structure and Function of Complex Networks." *SIAM Review* 45: 167–256.

Pattee, Howard H. 1970. "The Problem of Biological Hierarchy." In *Toward a Theoretical Biology 3, Drafts*, edited by Conrad H. Waddington, 117–36. Edinburgh: Edinburgh University Press.

————, ed. 1973. *Hierarchy Theory: The Challenge of Complex Systems*. New York: G. Braziller.

Pumain, Denise, ed. 2006. *Hierarchy in Natural and Social Sciences*. Dordrecht: Springer.

Salthe, Stanley N. 1985. *Evolving Hierarchical Systems: Their Structure and Representation*. New York: Columbia University Press.

Simon, Herbert A. 1962. "The Architecture of Complexity." *Proceedings of the American Philosophical Society* 106: 467–82.

————. 1973. "The Organization of Complex Systems." In *Hierarchy Theory: The Challenge of Complex Systems*, edited by Howard H. Pattee, 1–27. New York: G. Braziller.

Strogatz, Steven H. 2001. "Exploring Complex Networks." *Nature* 410: 268–76.

Tëmkin, Ilya, and Niles Eldredge. 2015. "Networks and Hierarchies: Approaching Complexity in Evolutionary Theory." In *Macroevolution: Explanation, Interpretation and Evidence*, edited by Emanuele Serrelli and Nathalie Gontier, 183–226. New York: Springer.

Wimsatt, William C. 1976. "Reductionism, Levels of Organization, and the Mind-Body Problem." In *Consciousness and the Brain: A Scientific and Philosophical Inquiry*, edited by Globus Gordon, G. Maxwell, and I. Savodnik, 205–67. New York: Plenum Press.

Woodger, J. H. 1929. *Biological Principles: A Critical Study*. London: Kegan Paul, Trench, Trubner.

Summaries for Part 1

The contributions in this part address conceptual and terminological questions that received little attention in evolutionary theory and philosophy of biology but are of critical importance for the unification of evolutionary biology in general and the hierarchy theory of evolution in particular.

Bruce Lieberman tackles an epistemological issue about evolutionary biology in hierarchical perspective: Is it a science discovering mechanisms (a nomothetic science) or is it a historical science (an idiographic one)? Lieberman reconstructs the story of this distinction in paleobiology, showing that an opposition between the two kinds is not necessary. Epistemological proposals to unify nomothetic and idiographic sciences have been convincing. In fact, macroevolutionary theories such as punctuated equilibria, the turnover pulse hypothesis, and the sloshing bucket theory embody both aspects. They invoke unique historical events, contingent factors, and deterministic mechanisms to generate and explain repeated patterns. Lieberman argues that "at least from the perspective of macroevolutionary theory, repeated analysis of contingent histories is the key to discovering whatever nomothetic principles exist in the history of life."

Gustavo Caponi addresses a fundamental distinction between two kinds of biological individuals: systems and lineages. This chapter attempts to overcome the disconnect in logic that is traditionally applied in the analysis of these two kinds of individuals, leveling the ground for a unified theoretical perspective. To this aim, the chapter focuses on the "modes of incorporation" that differentiate systems and lineages and on causal integration, emphasizing that systemic causality does not apply to lineages, where causal integration is due to genealogical relationships. Further discussion focuses on the levels of individual organism and population (both simultaneously characterized by properties of systems and lineages) to examine the interaction between the ecological and the genealogical aspects of biological hierarchies. In light of the system/lineage dichotomy, an important distinction is made between the notions of character and part.

Jon Umerez proposes an articulated view of hierarchical levels that distinguishes five kinds of progressively restrictive relationships among elements, leading to the emergence of a level: composition, integration, emergence, regulation, and organization. The chapter analyzes asymmetry and other characteristics of these relations. Umerez argues for a plurality of definitions of "level of organization" because the organization in biological systems is determined by a host of diverse processes and different kinds of relations, in all instances resulting in some kind of ordered arrangement. Consequently, neither composition nor mechanistic integrity—currently most popular philosophical views—are sufficient for defining relations among levels. The chapter also offers a discussion of the notion of (inter-level) constraint and explores its mode of action as the operational materialization of relations among organizational levels.

Daniel W. McShea brings into focus the relationship between nested compositional hierarchical structures and teleology. According to this view, an apparent teleology is a byproduct of an inherent hierarchical organization of living systems. A "teleological entity" is the one physically embedded in a larger system (the "next higher level") that directs or drives the movement at the focal level. Such control always operates by means of a "field," a spatially distributed physical structure that directs the contained entities. The explanation of seemingly teleological behavior is completed by a third level, the next lower level below the focal one, where the mechanisms explain why and how the entity is affected by the field. The chapter spells out further conditions for teleology to take place and argues in favor of a structural similarity between adaptive evolution and other seemingly goal-oriented processes, such as organismal development. McShea's hierarchical view of teleology also opens to a number of larger and smaller scales in which teleology could take place.

Teleology and hierarchical structure are also addressed by *Gregory Cooper, Charbel El-Hani*, and *Nei Nunes-Neto*. Their chapter lifts ecosystems from the biological hierarchies that are identified by the hierarchy theory of evolution. Since ecosystems contain abiotic elements, they are not part of the economic hierarchy—they are a different relevant hierarchy. The authors review the theoretical difficulties of identifying ecosystems, but most of all they evaluate three philosophical approaches to functions that could make sense of the normativity of ecosystems. In ecosystems, future utilities seem to explain present phenomena, and the system seems to define the proper way of functioning (and therefore, malfunctioning) for its parts. The concept of function in evolution and in ecology is equally complex. The chapter explains the limits of the causal role approach to function (relying on mechanistic relationships) and of the etiological approach to function (relying on history) and probes the possibility and limits of an organizational account, in which function depends on how the part contributes to the maintenance of the organization while, at the same time, is produced and maintained by the organization.

CHAPTER 1

Pattern versus Process and Hierarchies

Revisiting Eternal Metaphors in Macroevolutionary Theory

Bruce S. Lieberman

Introduction

Debates about the nature and value of certain types of research in the historical sciences, including paleontology, evolutionary biology, astronomy, and so on, are often marked by a tension or dichotomy that exists between approaches that emphasize simply documenting events that happened, or history, and approaches that emphasize discovering mechanisms, or ahistorical scientific processes and universal truths (Eldredge 1999). Stephen Jay Gould (1970) referred to these two approaches as, respectively, "idiographic" and "nomothetic," and a detailed definition of these terms was also provided in Raup et al. (1973). The idiographic approach later became largely subsumed under Gould's (1989) concept of contingency. This chapter focuses on elucidating these issues in the context of paleontological research in general and on the work of Stephen Jay Gould, Niles Eldredge, and Elisabeth Vrba, three of the principal paleontological architects of modern-day macroevolutionary theory, in particular. (Macroevolutionary theory focuses on the study of the patterns and processes pertaining to the birth, death, and persistence of species [Lieberman and Eldredge 2014].) Further, contributions of phylogeneticists and philosophers of science were also very important to macroevolutionary theory, and therefore relevant works of Ed Wiley (e.g., Wiley 1981, 1989) and others are also considered.

Finally, the significance of punctuated equilibria (Eldredge 1971; Eldredge and Gould 1972) and the turnover pulse hypothesis (Vrba 1985, 1992) to the contingent versus nomothetic debate vis-à-vis the study of the tempo and mode of evolution will be considered, as will some scientific analyses germane to this topic.

Gould's views on this topic differed from those of Eldredge, Vrba, and Wiley. Especially noteworthy is that Eldredge, Vrba, and Wiley consistently saw what Gould generally considered to be intransigent dualities as the means for unifying perspectives. Thus considering the works of each of these authors in turn provides important illumination on the topic and can help resolve long-standing arguments in evolutionary biology.

Gould's Early Views on Contingent and Nomothetic Approaches

Early in his career, there were several indications that Gould already saw this as an important issue. Further, his early writings seemed to suggest that a nomothetic approach might be seen by some as scientifically more worthwhile, although an approach that emphasized the study of history was not without value and could even be transformed into a nomothetic approach (Gould 1968).

Gould (1970) also discussed the issue of contingent versus nomothetic approaches specifically in the context of Dollo's law. Absolute convergence in morphology was impossible to Dollo (and Gould) because organisms were so complex and history was a sequence of largely unique phenomena. Here, Gould drew on Simpson (1964), who essentially saw Dollo's law arising from the fact that history cannot repeat itself in a probabilistic sense. To Gould (1970), aspects of evolutionary biology were nomothetic—for instance, the operation of natural selection. However, he also specifically acknowledged the historical nature of evolutionary events and indeed suggested that it was paramount. Further, Gould (1970) concluded that this largely precluded the formulation of true evolutionary laws with one exception to the rule comprising Dollo's law: that law was valid because it demonstrated the importance of history. The distinction between contingent and nomothetic approaches is also emphasized in Gould (1980).

The MBL Group

Sepkoski (2012) described in detail the workings and history of a group of scientists who used the so-called marine biology laboratory (MBL) model. The MBL model was basically an attempt to simulate the random evolution of clades, and the results of these random iterations of simulated evolution could be compared to actual evolutionary events preserved in the fossil record.

One of the important premises of the MBL group was that until we know the "degree of apparent order that can arise within random systems, we have no basis for asserting that a pattern has a conventional cause" (Gould et al. 1977, 24)—that is, patterns must be compared against a null hypothesis, which should be some form of a random walk. Cornette and Lieberman (2004) also endorsed the value of applying an approach that compared actual diversity patterns in the history of life to those that could be generated by a random walk. They concluded that throughout the Phanerozoic, in terms of its overall diversity and patterns of origination and extinction, marine animal life was undergoing a random walk.

Raup et al. (1973) argued that one of the significant impediments to the success of paleontology as a scientific discipline was its focus on why any individual taxon went extinct at any given time period. Thus they were suggesting that the contingency-based approach to paleontology was antithetical to achieving scientific status for the discipline. Still, randomness in the history of life was useful because it enabled predictability (Gould 1981).

Another major conclusion of Raup et al. (1973, 534) was that the evolutionary histories of simulated groups operating under identical constraints can differ dramatically. This is significant because it basically reveals the importance of taking into account contingency when using the nomothetic approach to the study of the history of life. Simply by historical accident (or contingency), groups can show dramatic differences in their evolutionary patterns over time even if there are no differences in the evolutionary processes generating those patterns. This ties in with Kant's views on manifest biological complexity as distinguishing life from a Cartesian-Newtonian mechanical arrangement (Eldredge and Grene 1992). The existence of complexity is one of the reasons the reductionist approach held by those such as Dawkins (1976, 1982), Dennett (1995), and others fails. In fact, life is hierarchically organized, with processes operating at several different emergent levels including the gene, organism, population, and species (Eldredge and Salthe 1984; Vrba and Eldredge 1984; Eldredge 1985, 1986, 1989a). This indicates that processes operating at one specific level, especially a lower hierarchical level, cannot be extrapolated to explain patterns in higher-level entities like species.

Time's Arrow and Time's Cycle

One aspect of Gould's work that is quite relevant to the general debate about nomothetic versus contingent viewpoints that has not generally been considered was broached in Gould (1977). This represented an important attempt to unite the seemingly trenchant nomothetic and contingent approaches: "The task of history is to explain the contexts so clearly that they can be separated and subtracted, thus permitting us to see the unchanging themes" (Gould 1977, 4). Indeed, this is where the term *contingency* (Gould 1977, 9) first appeared in his writings. It is also where he introduced the category "time's arrow," a later harbinger of Gould (1987), where it was conjoined with "time's cycle." These should be viewed as broader metaphors relating to contingent and nomothetic, respectively. Time's arrow was a synonym to Gould (1977, 1987) for an approach predicated on seeing history as an irreversible sequence of unrepeatable events that further indicated directionality in the history of life. He argued that this perspective, when applied to geology, was most consonant with that of the catastrophists, who held that the earth was continually changing and had only a limited duration. The time's cycle frame of reference matched the perspectives of uniformitarians such as James Hutton and Charles Lyell (early in his career), who both held that all was constancy and the earth unchanging—with the earth's duration effectively infinite—and the same processes that operated today were operating in the distant past. In this standpoint, "events have no meaning as distinct episodes with causal impact upon a contingent history" (Gould 1987, 11).

One of the underlying purposes of Gould (1987) was to find a means of uniting these dichotomies vis-à-vis the study of evolution and geology. He argued lucidly that a time's arrow–only viewpoint makes the history of life (and Earth) unintelligible because nothing ever repeats and thus no scientific processes can be adduced. By contrast, a time's cycle viewpoint implies scientific processes that effectively operate outside of history such that the "history" of life can be viewed largely in an ahistorical context.

However, Gould's views did evolve. By the time Gould (1987) had been published, he was retreating from a singular focus on time's cycle. While praising the profound insights of Hutton that led to the discovery of deep time, he did somewhat chastise him for not grasping "the power, worth, and distinction of history" (Gould 1987, 97). In essence, Hutton strived to make geology like Newtonian physics, but it could and should never be like Newtonian physics (see also Eldredge and Grene 1992; for that matter, contemporary physics is not like Newtonian physics, either). Too much emphasis on time's cycle, to the detriment of time's arrow, could not give us an adequate theory of the earth (Gould 1987, 97) or of life, and "time's arrow and time's cycle both capture important aspects of reality" (Gould 1987, 178). Thus Gould (1987) implored accommodation for both views as equally valid and important.

Gould in "Wonderful Life" and Beyond

Gould's views on the importance of time's arrow and contingency for our understanding of evolution reached their apogee in Gould (1989), and they never receded from this high water mark until the very end of Gould (2002; discussed more fully below). It was important that Gould championed reinstating the importance of history or time's arrow into the history of life. He felt it might not lead to generalities, but Gould (1994, 3) argued that we "should treasure the intricate story of our planet and its life." He even asserted, "The discovery of timeless and universal laws and the prediction of all occurrences under their guidance cannot be an expectation or even a desideratum" (Gould 1994, 4). But one does wonder if Gould almost went too far in the realm of biological evolution, arguing that everything was about history and that there were no general unifying principles or mechanisms governing the history of life. Certainly in toto Gould did not believe this, as he continued to argue in various places and venues for the importance of punctuated equilibria (and thus also allopatric speciation) and the blanket power of mass extinctions, which do represent general unifying principles (e.g., Gould 2002, 1339–40). However, at least a certain aspect of Gould's (1989, 1994) work started to give up on such unifying principles and instead came to emphasize individualistic particulars. Part of his validation of contingency or time's arrow—to the exclusion of views that would have been more mechanistic and predictable—represented Gould's (1989, 35) reaction against "our hopes for a universe of intrinsic meaning defined in our terms."

This was codified in Gould's (1989) notion that if we were able to re-play the tape of life again, we would witness a very different result. It had become paramount for Gould to demonstrate the appositeness of paleon-tology to evolutionary biology, and whereas earlier he might have drawn on other theoretical threads, here he accentuated that "historical science is not worse, more restricted, or less capable of achieving firm conclusions because experiment, prediction, and subsumption under invariant laws of nature do not represent its usual working methods. The sciences of his-tory use a different mode of explanation, rooted in the comparative and observational richness of our data" (Gould 1989, 274).

Gould (2002, 1337) repeated these points, contending that "our in-creasing willingness to take narrative explanations seriously has sparked a great potential gain," but he (Gould 2002, 1337) did add that not "all *con-ceivable* evolutionary questions must invoke enough historical particulars to require a large contingent component in their full explanation" (em-phasis in original). By the near concluding paragraphs of Gould (2002, 1338), contingency became the leftovers of what cannot be explained by general laws. Gould did contend that to him, and he did embrace the power of the nomothetic approach and dignified it as the one that most motivated him as a scientist.

Wiley and Brooks's (1982) and Brooks and Wiley's (1986) discussion of evolution as entropy also viewed macroevolution as firmly rooted, and best viewed, as a matter of contingency (although there was an overarch-ing, mechanistic explanation for this: information transfer and entropy). In this case, information transmission is similar to energy transmission and must follow the Second Law of Thermodynamics. Therefore evolu-tionary change should be expected, and it is actually the conservation of information (or stasis) that requires special explanation (Brooks and Wi-ley 1986).

It is worth distinguishing between the accuracy of Gould's (e.g., 1989, 1994) vision of contingency when considered on the microevolutionary scale and on the macroevolutionary scale. The smaller the hierarchical level in the genealogical hierarchy considered (Eldredge and Salthe 1984; Vrba and Eldredge 1984; Eldredge 1985, 1986, 1989a; e.g., genes, cells, populations), and the shorter the evolutionary time scale involved, the more likely true episodes of actual convergence will be, the less impor-tant history and time's arrow will be, and the more dominant time's cycle will be. The longer the evolutionary time scale considered and the larger the hierarchical level in the genealogical level considered (e.g., species, clades), the more important history and time's arrow will be, and the less dominant time's cycle will be.

A key issue in the entire discussion about the importance of nomothetic versus contingent views of macroevolution has to do with the mode of speciation. The greater the predilection for speciation to occur via the allopatric mode (as predicted in Mayr [1963] and used in the punctuated equilibria model *sensu* Eldredge [1971] and Eldredge and Gould [1972]), the more important contingency will be in the history of life because it entails that speciation (and thus macroevolution) will be motivated by unique climatic or geologic factors that cause geographic isolation. By contrast, if speciation is usually sympatric and involves competition for various resources, it will be more predictable, and there will be a more muted role for chance and contingency. For now, allopatry seems to be the most common style of animal speciation (Brooks and McLennan 2002; Coyne and Orr 2009; Wiley and Lieberman 2011), suggesting that contingency must be factored into any treatment of macroevolution.

Eldredge's Views on Historical and Functional Sciences

Eldredge cast the contingent versus nomothetic debate in the context of the distinction he frequently made between the historical sciences (like paleontology and evolutionary biology) and functional sciences (like physics and chemistry)—a distinction he felt was ultimately neither valid nor productive (Eldredge 1989a). Early on, Eldredge had reached a more nuanced view, perhaps somewhat equivalent to that in Gould (1987). Thus Eldredge surmounted this issue with much greater facility than Gould. Part of the explanation may lie in recognizing that Eldredge was one of the earliest adopters of phylogenetics (e.g., Eldredge 1979; Eldredge and Cracraft 1980) and thus always appreciated the study of evolutionary history. Further, he held that the chief reason for studying that history was for gaining insight into evolution. (This view was shared by Wiley [1981], Vrba [1985], Brooks and McLennan [1991], and many others.) For instance, Eldredge argued that the key to testing evolutionary theories was to study patterns in the history of life: "We believe that the most important connection between the two areas (evolutionary pattern and process) . . . involves the comparison of the patterns of both intrinsic and extrinsic features of organisms predicted from theories of process, with those actually 'found' in nature" (Eldredge and Cracraft 1980, 4). That is, the only way to derive nomothetic principles was to study contingent realities. In particular, one needed to find repetitive historical patterns, and these provided tests of theory (Eldredge 1999, 2000). It is probably not a stretch to suggest that Eldredge saw time's arrow and time's cycle as ultimately conjoined in a duality. Notably he, like Gould (1987), identified Hutton as a scientist fixated on viewing the world as undergoing a series of endless cycles that made it hard to have a concept of evolution (Eldredge 1999). He also argued, "History is a natural experiment, but also it is a connected sequence of unique events" (Eldredge 1989a, 8). In essence, to solve the contingent versus nomothetic duality that Gould perceived, Eldredge argued that by being able to look at the dynamics of large-scale patterns, paleontologists became not just chroniclers of evolution; they were able to deduce processes as well (Eldredge 1989b). Eldredge also recognized very early on that an important means of unifying the historical and functional sciences (or contingent and nomothetic viewpoints) was to view species, and also higher taxa, as individuals rather than classes. If species (and clades) were individuals, they were imbued with reality and were true actors in the evolutionary process, rather than nominalistic, artificial constructs (Eldredge 1982; 1985; 1989a,b; Vrba and Eldredge 1984; see also further discussion on this topic in the next section). Another important duality that Eldredge focused on—which helped resolve this issue and thus is pertinent here—was the distinction between the nature of entities or their ontology and the methodological theory by which we identify them or their epistemology (Eldredge 1985, 1989a). Eldredge's (1989a, 9) ultimate solution for resolving the seemingly trenchant historical/functional (and thus contingent/nomothetic) divide in evolutionary science came by recognizing that "large scale interacting entities would allow the historical branch of biology to contribute in a 'functional' manner to describing entities that exist in general in biological nature, and to understanding the processes underlying their interactions and producing their histories." Eldredge explicated how "paleontologists, just like physicists and population geneticists, look for repeated phenomena, those classes of events that call for generalization and explanation in causal terms" (1993, 36).

Eldredge (1999, 7) referenced the writings of the philosopher and physicist Ernst Mach when he put forth the notion that science was the description of the material things of the universe and the interactions among them. In such a perspective, the description of various biological entities was patently what evolutionary biology should be about. Eldredge (1999,

11–12) also lucidly demonstrated how all science, be it physics or biology, was about documenting patterns, be they the arrangement of subatomic particles flying through a cloud chamber or the appearance of fossil shells found walking up a cliff face. It was these repeated patterns that tell us how the evolutionary process works (Eldredge 1993). Echoing this, Eldredge (2011, 366) argued that adaptation through natural selection "is one of the closest things to a natural 'law' in evolutionary biology (geographic speciation is an example of another such 'statistical law')." Indeed, when it came to geographic or allopatric speciation, it too was another such law (Ghiselin 1987; Eldredge 1993).

Pattern versus Process: Recognizing the Important Role of Wiley and the Phylogeneticists in Advancing Understanding of This Topic

In the realm of phylogenetics, the contingent versus nomothetic and historical versus functional divides took on the countenance of the pattern versus process divergence.

In phylogenetics, "historical groups function significantly in science because they are the result of the operation of natural processes on their parts" (Wiley and Lieberman 2011, 113). Treating species as individuals meant that hypotheses about species were singular historical statements (Wiley and Lieberman 2011). In addition, at their very core, species are built up from reproducing lineages. Thus any phylogeny or pattern is imbued with a process: genealogy (Wiley 1978, 1981, 1989; Sober 1988, 2008). In this respect, phylogeneticists like Wiley did not see a trenchant divide between what Gould termed nomothetic and contingent approaches because in phylogenetics, pattern and process were inextricably linked. Ultimately the fact that Gould was not a phylogeneticist may partly explain why he was so troubled by the issue of the nomothetic versus contingent distinction. It is also fair to say that because of Wiley's position as a phylogeneticist (Wiley 1981) and his work documenting the relationship between evolution and entropy (Wiley and Brooks 1982; Brooks and Wiley 1986), he saw evolution as largely a historical process and therefore contingent.

Pattern and process are not solely linked together in phylogenetic trees and the species they comprise. The very means for building phylogenetic trees—which convey patterns of relationship—is homology, which is a concept that again inextricably links pattern and process. In particular, homology is conceptually tied directly to ancestry and process (Wiley 1975).

Phylogenetic biology, and especially concepts pertinent to it like the nature of species and homologies, provides some of the best means for unifying pattern and process, and thus contingent and nomothetic and functional and historical approaches. For this reason, Gould's concern about solving the nomothetic versus contingent divide had little impact on phylogeneticists (Wiley, pers. comm. with author, September 8, 2014).

Vrba's Views Pertaining to the Issue of Nomothetic versus Contingent in Macroevolution

Again, unlike Gould, Elisabeth Vrba never became absorbed with the philosophy behind the nomothetic/contingent distinction. The reasons for this likely lie in the fact that she, like Eldredge, Wiley, and others, was a phylogeneticist. Further, although she did not consider the historical/functional discordance in the sciences that Eldredge did, ultimately her work does have some bearing on this topic. In particular, she argued cogently that evolution did not happen without physical environmental change (Vrba 1980, 1985, 1989). Further, "to the extent that punctuated equilibria characterizes the history of life we should expect evolutionary changes to be a function of physical environmental changes and to occur synchronously when they occur at all" (Vrba 1989, 36). She developed these ideas first in Vrba (1980) and then in the context of her work with the turnover pulse hypothesis (Vrba 1985, 1992, 1996), which extended the single lineage concept of punctuated equilibria to many lineages. On the one hand, this is a theory firmly in the realm of contingency and time's arrow (or history) because it implies that changes are driven by particular phenomena that impinge upon life at distinct time intervals. On the other hand, one would "predict regularities of pattern" (Vrba 1989, 30) as different groups are anticipated to respond similarly any time the physical environment changes. Even further, her work on the turnover pulse was originally framed specifically in the context of the Neogene time interval, when there were repeated oscillations between much colder ice ages and warmer greenhouse intervals. However, the turnover pulse does not only hold during the Neogene, and there are very similar patterns documented from marine invertebrate fossil faunas from much of the Paleozoic (spanning more than one hundred million years) of the Appalachian basin (Brett and Baird 1995; Lieberman et al. 2007). There are also very similar patterns documented in Devonian trilobites (Lieberman 1999). Thus Vrba had identified a nomothetic, lawlike pattern, beautifully epitomizing the time's cycle metaphor and matching Eldredge's dictum that the aim in making macroevolution a functional science should be to identify repeated patterns in higher-level entities. Of course, Eldredge's (1971) and Eldredge and Gould's (1972) punctuated equilibria is thus another lawlike phenomenon that makes general predictions about evolutionary patterns and processes derived from understanding the nature of the fossil record and the significance of allopatric speciation. Finally, Eldredge's (2000, 2001) sloshing bucket (see also Pievani and Serrelli 2013; Pievani 2015) is another macroevolutionary theory that united time's arrow and time's cycle. In particular, the sloshing bucket theory argues that the more profound a perturbation in the economic hierarchy, the more profound its effects on entities in the genealogical hierarchy. It links evolutionary patterns, such as smaller- or larger-scale mass extinctions in genealogical entities like species in clades, to processes playing out in ecological entities (Pievani and Serrelli 2013; Pievani 2015).

Paleontological Analyses Bearing on the Nomothetic versus Contingent and Functional versus Historical Debates

Additional analyses that addressed the importance of time's arrow versus time's cycle viewpoints included those that have documented evidence for cyclicity in patterns of extinction in Raup and Sepkoski (1986) and also total diversity and origination (Rohde and Muller 2005; Lieberman and Melott 2007, 2012) over much of the Phanerozoic. In particular, Raup and Sepkoski, Rohde and Muller, Lieberman and Melott, and other paleontologists documented evidence for external affronts to our planet's biota, operating over repeating cycles of many tens of millions of years, which powerfully influenced evolution and the history of life.

Any mass extinction, or indeed a smaller extinction event of the type viewed in the framework of the turnover pulse hypothesis, can be viewed as a manifestation of contingency, as those species that go on to survive will provide the ancestral wellsprings of future species, whereas those species that are gone are gone for good. Especially if the species that went extinct were manifestly successful prior to their demise, it would introduce an important element of chance into the history of life, abjuring predictability when it came to inferring which lineages would survive and which would not. However, the manifestation of repeatability in such a pattern also suggests that predictability could be embraced, at least at the larger scale. In particular, these overarching cycles might explain as much as 10 percent of the overall variance in diversity (Lieberman and Melott 2012). In such a perspective, time's arrow may dominate, but there is still an important role for time's cycle.

Lieberman and Melott (2013) provided another result bearing on this debate where, building on the work of Gilinsky (1994), they found that through time, the overall volatility of taxa—their propensity to speciate or go extinct—had declined during the Phanerozoic. This pattern materialized because those groups that had higher rates of origination were also prone to have higher rates of extinction. Because of their higher rates of extinction, they were more predisposed to go extinct and never return again (Lieberman and Melott 2013). The net result was the survival of the blandest taxa, because over many hundreds of millions of years of evolution, groups with lower rates of extinction and origination were all that remained, while the more rapidly evolving groups (with high rates of extinction) had disappeared. Such high origination and extinction groups were found to be especially susceptible during times of mass extinction. Lieberman and Melott (2013) identified similar patterns in the behavior of stocks in the stock market and stars in the universe. The commonalities across these disparate systems emerged because taxa, stocks, and stars were all historical entities, and once they disappear (go extinct), they are gone for good. In a sense, what they found is nomothetic behavior of contingent systems, and indeed, the behavior of such systems was predictable precisely because of the contingent nature of the entities being considered.

Caveats need to be raised for all these aforementioned studies, and indeed for all the analyses conducted by what Sepkoski (2012) and others have referred to as the "Chicago school" of paleontology. Unfortunately, species were not considered in any of these studies—only paleontological genera or families. (The various problems associated with this aspect of the Chicago school approach have been discussed in detail by Hendricks et al. [2014]). Further, many of these genera and families were not monophyletic and thus did not comprise true evolutionary individuals (see discussion above) but rather classes. For macroevolutionary studies to have validity, there is a requisite need that the data bear on individuals. Any studies that analyze datasets largely comprising entities of questionable reality, where the datasets and modeled evolutionary groups are not in-dividuals but instead simply artificial constructs, will be problematic and their results will be unable to reach complete fruition. In the particular case of paleontological studies, those that include actual or modeled (in the case of the MBL group) paraphyletic genera or families cannot claim that their results bear directly on the real world. Still, such studies com-prise what is available thus far in the realm of relevant paleontological tests addressing this issue, and it might be reasonable to suppose that they display a modicum of accuracy, at least in a general sense.

Conclusions

Scientists have debated whether sciences that focused on historical entities (e.g., paleontology, natural history, astronomy, etc.) were somehow funda-mentally different from, and inferior to, those sciences that were not con-cerned with the history of the entities they studied (e.g., Newtonian physics, physical chemistry, Mendelian genetics, etc.). This duality has been usefully treated in the context of the variance between contingent and nomothetic or historical and functional or pattern and process based scientific view-points. The debate has special relevance for macroevolution because this discipline perforce might be viewed as focusing more on contingency and history, and therefore might be construed by some as possibly less rigorous or scientific than microevolution. Some of the leaders in macroevolutionary theory have considered this issue.

For instance, Gould (1987) endorsed both nomothetic and contingent approaches, and the concepts of time's arrow and time's cycle are quin-tessential for describing the history of life and studying macroevolution. Gould's views on this topic did evolve and shift throughout his scientific career, and he was not always entirely in favor of the approach he advo-cated for in Gould (1987). Still, his ultimate musings on this topic at the very end of Gould (2002) did endorse Gould (1987). Niles Eldredge (e.g., 1986, 1989a, 1993, 1999) and Eldredge and Grene (1992) also considered this topic, using the terms *historical* and *functional* sciences; Eldredge use-fully introduced the perspective that the contrast sometimes made be-tween these two viewpoints was far less than it seemed. Eldredge (1999) argued that all scientific study is about the analysis of pattern, and that whether the entities being studied in a particular branch of science were individuals or classes provided the background for understanding why the apparent divide need not be troublesome. By its very nature, Newtonian physics must consider class-like entities, whereas paleontology must be grounded in the analysis of evolutionary individuals. Both were concerned with discovering relevant mechanisms, however.

The context provided by phylogenetic biology (Wiley and Lieberman 2011) to the debate is also relevant. In particular, contingency and nomothetic or historical and functional can be ultimately wedded in phylogenetics. There, the apparent pattern and process distinction can be resolved by recognizing that among the important entities phylogeneticists study (species and taxa) and the characters used to deduce the relations of species (homologs), there are concepts that ultimately involve genealogy and inheritance (Wiley 1975, 1978, 1981, 1989). There is thus no need to see pattern and process (or contingency and nomothesis) as separate.

In addition, ultimately key macroevolutionary theories such as punctuated equilibria (Eldredge 1971; Eldredge and Gould 1972), the turnover pulse hypothesis (Vrba 1985, 1992), and the sloshing bucket theory (Eldredge 2000, 2001; Pievani and Serrelli 2013; Pievani 2015) beautifully embody both aspects of what have been at times seen as dualities. This is because they invoked unique, historical events and contingent factors—through the touchstone of allopatric speciation triggered by environmental change—to generate and explain repeated patterns. Further, these theories dealt with real entities or individuals, like species and clades, which have a fundamental place in the evolutionary process writ large. The debate about the appositeness of contingent/historical/pattern approaches as opposed to nomothetic/functional/process ones in the sciences will likely continue. Yet, at least from the perspective of macroevolutionary theory, repeated analysis of contingent histories is the key to discovering whatever nomothetic principles exist in the history of life.

Acknowledgments

Thanks to Emanuele Serrelli and Ilya Tëmkin for inviting me to participate in this volume. Thanks to them, Ed Wiley, Niles Eldredge, and two anonymous reviewers for comments on an earlier version of this paper. This research was supported by NSF DEB-1256993.

References

Brett, Carlton E., and Gordon C. Baird. 1995. "Coordinated Stasis and Evolutionary Ecology of Silurian to Middle Devonian Faunas in the Appalachian Basin." In *New Approaches to Speciation in the Fossil Record*, edited by Doug H. Erwin and Robert L. Anstey, 285–315. New York: Columbia University Press.

Brooks, Daniel R., and Deborah A. McLennan. 1991. *Phylogeny, Ecology, and Behavior: A Research Program in Comparative Biology*. Chicago: University of Chicago Press.

———. 2002. *The Nature of Diversity*. Chicago: University of Chicago Press.

Brooks, Daniel R., and Edward O. Wiley. 1986. *Evolution as Entropy: Toward a Unified Theory of Biology*. Chicago: University of Chicago Press.

Cornette, James L., and Bruce S. Lieberman. 2004. "Random Walks in the History of Life." *Proceedings of the National Academy of Sciences of the United States of America* 101: 187–91.

Coyne, Jerry A., and Allen Orr. 2009. *Speciation*. Sunderland, MA: Sinauer.

Dawkins, Richard. 1976. *The Selfish Gene*. New York: Oxford University Press.

———. 1982. *The Extended Phenotype*. San Francisco: W. H. Freeman.

Dennett, Daniel. 1995. *Darwin's Dangerous Idea*. New York: Simon and Schuster.

Eldredge, Niles. 1971. "The Allopatric Model and Phylogeny in Paleozoic Invertebrates." *Evolution* 25: 156–67.

———. 1979. "Alternative Approaches to Evolutionary Theory." *Bulletin of the Carnegie Museum of Natural History* 13: 7–19.

———. 1982. "Phenomenological Levels and Evolutionary Rates." *Systematic Zoology* 31: 338–47.

———. 1985. *Unfinished Synthesis: Biological Hierarchies and Modern Evolutionary Thought*. New York: Oxford University Press.

———. 1986. "Information, Economics, and Evolution." *Annual Review of Ecology and Systematics* 17: 351–69.

———. 1989a. *Macroevolutionary Dynamics*. New York: McGraw-Hill.

———. 1989b. "Punctuated Equilibria, Rates of Change and Large-Scale Entities in Evolutionary Systems." *Journal of Social and Biological Structures* 12: 173–84.

———. 1993. "History, Function, and Evolutionary Biology." *Evolutionary Biology* 27: 33–50.

———. 1995. *Reinventing Darwin: The Great Debate at the High Table of Evolutionary Theory*. New York: John Wiley & Sons.

———. 1999. *The Pattern of Evolution*. New York: W. H. Freeman.

———. 2000. "Biological and Material Cultural Evolution: Are There Any True Parallels?" In *Perspectives in Ethology*, edited by Francois Tonneau and Nicholas S. Thompson, 113–53. Berlin: Springer.

———. 2001. "The Nature and Origin of Supraspecific Taxa Revisited: With Special Reference to Trilobita." In *Fossils, Phylogeny, and Form: An Analytical Approach*, edited by Jonathan M. Adrain, Gregory D. Edgecombe, and Bruce S. Lieberman, 341–75. New York: Kluwer Academic Publishing.

———. 2011. "Paleontology and Cornets: Thoughts on Material Cultural Evolution." *Evolution: Education and Outreach* 4: 364–73.

Eldredge, Niles, and Joel Cracraft. 1980. *Phylogenetic Patterns and the Evolutionary Process. Method and Theory in Comparative Biology.* New York: Columbia University Press.

Eldredge, Niles, and Stephen Jay Gould. 1972. "Punctuated Equilibria: An Alternative to Phyletic Gradualism." In *Models in Paleobiology*, edited by Thomas J. M. Schopf, 82–115. San Francisco: W. H. Freeman.

Eldredge, Niles, and Marjorie Grene. 1992. *Interactions: The Biological Context of Social Systems.* New York: Columbia University Press.

Eldredge, Niles, and Stanley N. Salthe. 1984. "Hierarchy and Evolution." *Oxford Surveys in Evolutionary Biology* 1: 184–208.

Ghiselin, Michael T. 1987. "Species Concepts, Individuality, and Objectivity." *Biology and Philosophy* 2: 127–43.

Gilinsky, Norman L. 1994. "Volatility and the Phanerozoic Decline of Background Extinction Intensity." *Paleobiology* 20: 445–58.

Gould, Stephen J. 1968. "The Molluscan Fauna of an Unusual Bermudian Pond: A Natural Experiment in Form and Composition." *Breviora* 308: 1–13.

———. 1970. "Dollo on Dollo's Law: Irreversibility and the Status of Evolutionary Laws." *Journal of the History of Biology* 3: 189–212.

———. 1977. "Eternal Metaphors of Palaeontology." *Patterns of Evolution*, edited by Anthony Hallam, 1–26. Amsterdam: Elsevier.

———. 1980. "The Promise of Paleobiology as a Nomothetic, Evolutionary Discipline." *Paleobiology* 6: 96–118.

———. 1981. "Chance That Shapes Our Ends." *New Scientist* 89: 347–49.

———. 1987. *Time's Arrow, Time's Cycle: Myth and Metaphor in the Discovery of Geological Time.* Cambridge, MA: Harvard University Press.

———. 1989. *Wonderful Life: The Burgess Shale and the Nature of History.* New York: W. W. Norton.

———. 1994. "Introduction: The Coherence of History." In *Early Life on Earth: Nobel Symposium*, no. 84, edited by Stefan Bengtson, 1–8. New York: Columbia University Press.

———. 2002. *The Structure of Evolutionary Theory.* Cambridge, MA: Harvard University Press.

Gould, Stephen J. David M. Raup, J. John Sepkoski Jr., Thomas J. M. Schopf, and Daniel S. Simberloff. 1977. "The Shape of Evolution: A Comparison of Real and Random Clades." *Paleobiology* 3: 23–40.

Hendricks, Jonathan R., Erin E. Saupe, Corinne E. Myers, Elizabeth J. Hermsen, and Warren D. Allmon. 2014. "The Generification of the Fossil Record." *Paleobiology* 40(4): 511–28.

Lieberman, Bruce S. 1999. "Turnover Pulse in Trilobites During the Acadian Orogeny." *Virginia Museum of Natural History Special Publications* 8: 99–108.

Lieberman, Bruce S., and Niles Eldredge. 2014. "What Is Punctuated Equilibrium? What Is Macroevolution?" *Trends in Ecology and Evolution* 29: 185–86.

Lieberman, Bruce S., and Adrian L. Melott. 2007. "Considering the Case for Biodiversity Cycles: Reexamining the Evidence for Periodicity in the Fossil Record." *PLoS ONE* 2(8) e759: 1–9.

———. 2012. "While This Planet Goes Cycling on: What Role for Periodic Astronomical Phenomena in Large Scale Patterns in the History of Life." In *Earth and Life: Global Biodiversity, Extinction Intervals, and Biogeographic Perturbations through Time, International Year of Planet Earth*, edited by John Talent, 37–50. Berlin: Springer.

———. 2013. "Declining Volatility, a General Property of Disparate Systems: From Fossils, to Stocks, to the Stars." *Palaeontology* 56: 1297–1304.

Lieberman, Bruce S., William Miller III, and Niles Eldredge. 2007. "Paleontological Patterns, Macroecological Dynamics and the Evolutionary Process." *Evolutionary Biology* 34: 28–48.

Mayr, Ernst. 1963. *Animal Species and Evolution*. Cambridge: Belknap Press of Harvard University Press.

Pievani, Telmo. 2015. "How to Rethink Evolutionary Theory: A Plurality of Evolutionary Patterns." *Evolutionary Biology* 42. doi:10.1007/s11692-015-9338-3.

Pievani, Telmo, and Emanuele Serrelli. 2013. "Bucket Thinking: The Future Framework for Evolutionary Explanation." *Contrastes. Revista Internacional de Filosofía—Suplementos* 18: 389–405.

Raup, David M., and J. John Sepkoski Jr. 1984. "Periodicity of Extinctions in the Geologic Past." *Proceedings of the National Academy of Sciences of the United States of America* 81: 801–5.

———. 1986. "Periodic Extinctions of Families and Genera." *Science* 231: 833–36.

Raup, David M., Stephen Jay Gould, Thomas J. M. Schopf, and Daniel S. Simberloff. 1973. "Stochastic Models of Phylogeny and the Evolution of Diversity." *Journal of Geology* 81: 525–42.

Rohde, Robert A., and Richard A. Muller. 2005. "Cycles in Fossil Diversity." *Nature* 434: 208–10.

Sepkoski, David. 2012. *Rereading the Fossil Record: The Growth of Paleobiology as an Evolutionary Discipline*. Chicago: University of Chicago Press.

Simpson, George G. 1944. *Tempo and Mode in Evolution*. New York: Columbia University Press.

———. 1964. *Evolution, and This View of Life*. New York: Harcourt, Brace & World.

Sober, Elliott. 1988. *Reconstructing the Past: Parsimony, Evolution and Inference*. Cambridge, MA: MIT Press.

———. 2008. *Evidence and Evolution: The Logic behind the Science*. New York: Cambridge University Press.

Vrba, Elisabeth S. 1980. "Evolution, Species, and Fossils: How Does Life Evolve?" *South African Journal of Science* 76: 61–84.

———. 1985. "Environment and Evolution: Alternative Causes of the Temporal Distribution of Evolutionary Events." *South African Journal of Science* 81: 229–36.

———. 1989. "The Environmental Context of the Evolution of Early Hominids and Their Culture." In *Bone Modification: Publication of the Center for the Study of the First Americans*, edited by Robson Bonnichsen and Marcella H. Sorg, 27–42. Orono: University of Maine Press.

———. 1992. "Mammals as a Key to Evolutionary Theory." *Journal of Mammalogy* 73: 1–28.

————. 1996. "On the Connection between Paleoclimate and Evolution." In *Paleoclimate and Evolution with Emphasis on Human Origins*, edited by Elisabeth S. Vrba, George H. Denton, Timothy C. Partridge, and Lloyd H. Burckle, 24–45. New Haven: Yale University Press.

Vrba, Elisabeth S., and Niles Eldredge. 1984. "Individuals, Hierarchies and Processes: Toward a More Complete Evolutionary Theory." *Paleobiology* 10: 146–71.

Wiley, Edward O. 1975. "Karl R. Popper, Systematics, and Classification: A Reply to Walter Bock and Other Evolutionary Taxonomists." *Systematic Biology* 24: 233–43.

————. 1978. "The Evolutionary Species Concept Reconsidered." *Systematic Biology* 27: 17–26.

————. 1981. *Phylogenetics*. New York: John Wiley & Sons.

————. 1988. "Parsimony Analysis and Vicariance Biogeography." *Systematic Zoology* 37(3): 271–90.

————. 1989. "Kinds, Individuals, and Theories." In *What the Philosophy of Biology Is*, edited by Michael Ruse, 289–300. Dordrecht: Kluwer Academic.

Wiley, Edward O., and Daniel R. Brooks. 1982. "Victims of History: A Non-Equilibrium Approach to Evolution." *Systematic Zoology* 31: 1–24.

Wiley, Edward O., and Bruce S. Lieberman. 2011. *Phylogenetics: Theory and Practice of Phylogenetic Systematics*, 2nd ed. New York: John Wiley & Sons.

CHAPTER 2

Lineages and Systems

A Conceptual Discontinuity in Biological Hierarchies

Gustavo Caponi

In a philosophical tradition, biological taxa were generally construed as natural kinds, or types, such as gold or homoeothermic animal (Mayr 1976). Therefore, the names of taxa (e.g., *Panthera* or Vertebrata) were regarded as common names—for instance, volcano or island. The development of phylogenetic systematics in the second half of the twentieth century prompted a revision of the notion of taxon in philosophy of biology (Ghiselin 1974; Griffiths 1974; Hull 1976), leading to the reconceptualization of taxa as individuals, or unique products of the evolutionary processes (Sober 1993, 149). In this perspective, taxonomic names are regarded as proper names, so that the phrase *Panthera leo* is alike to the name Vesuvius, rather than to the generic term *volcano*.

As individual entities, taxa can incorporate taxa of lower rank as their parts (Ghiselin 1974; Hull 1976). It follows that if *Panthera* is the name of an individual entity, then the terms *Panthera pardus* and *Panthera leo* designate parts of *Panthera*. Consequently, in relation to *Panthera leo*, the subspecies *Panthera leo persica* should be understood as part of it species. Furthermore, if we assume that species are lineages of populations, we also must accept that the individual living beings that comprise a population are its parts, and not instances of the taxa to which we ascribe them (Eldredge 1995, 174).[1] Alexander the Great's horse, Bucephalus, was a part, not an instance, of *Equus caballus*; the relationship between the horse and the species to which it belongs is analogous to that between the heart and the circulatory system, rather than between the horse and the class of homoeothermic animals it can be ascribed to.

However, it is important to emphasize that the relationship between Bucephalus's heart and its circulatory system significantly differs from that between *Equus caballus* and Bucephalus, or between *Panthera leo* and *Panthera*. These differences are the expression of a more fundamental distinction that exists between two general types of individual entities: *systems* and *lineages*. It is critical to distinguish between these two conceptually different types of individuals, as this distinction underscores the recognition of two major biological hierarchies: the *genealogical hierarchy* (a hierarchy of lineages) and the *ecological hierarchy* (a hierarchy of systems; Eldredge and Salthe 1984; Eldredge 1985; Salthe 1985; Pievani and Serrelli 2013).

Without considering the distinction between these two types of individuals, the relationships among the hierarchies, especially the epistemic articulations in pertinent discourses, cannot be properly understood. The failure to address the nature of biological individuals is likely to result in a proliferation of epistemological issues stemming from the conflation of two different epistemological orders—one that rules the discourse about lineages and the other about systems. In this chapter, I set out to explore the distinction between lineages and systems and to highlight its importance to the philosophy of biology and to the development of biology as a whole, particularly evolutionary biology. In light of recent application of the hierarchy theory to evolutionary biology, defining essential concepts, such as lineages and systems, is paramount. It should be clear that I do not attempt to propose major conceptual or theoretical innovations. My main goal is to show the importance of this distinction in the contemporary discourse in evolutionary biology in order to steer away from conceptual confusion that can definitely obstruct further development of the hierarchical perspective on biological processes in research and teaching.

Modes of Incorporating

By stating "Bucephalus was part of *Equus caballus*," we are predicating the relationship *part of* in a way that is not the same as when we say "the mane was a part of Bucephalus." In the first case, being "part of" is equal to "being a specimen of *Equus caballus*," but not so in the second case— the parts of a body are not specimens of that body. The parts of an engine are not specimens of the system that incorporates them. There are individuals, the lineages, to whom we can attribute parts that are recognizable as their specimens. There are also other individuals—that is, systems—whose parts we do not recognize as being their specimens.

Lineages can be structured hierarchically. They may contain sublineages as their parts and be sublineages of yet more encompassing lineages: *Panthera leo persica* is a sublineage of *Panthera leo*, whereas *Panthera leo* is a sublineage of *Panthera*. Systems, on the other hand, are always regarded as being composed of parts, or subsystems, and can themselves be subsystems of more encompassing, complex systems: the esophagus is a subsystem of the digestive system, which is, in turn, a subsystem of an individual organism. The esophagus, however, is not a sublineage of an organism, nor is *Panthera leo* a subsystem of *Panthera*: systems are integrated by and into systems, whereas lineages are integrated by and into lineages.

Furthermore, only lineages may contain variants. This also distinguishes them from systems. When it is said that there is a mottled variant and a dark variant of the peppered moth, *Biston betularia*, both variants are recognized as being parts of a single species. In this case, the predicate "being part of" is used in a way that has no analogy in cases where we allude to systems such as organisms, machines, or galaxies: The esophagus is not a variant of the digestive system and the solar system is not a variant of the Milky Way galaxy. These individuals, the esophagus and the solar system, are not lineages because, being systems, they have no variants of themselves. They change, of course, but it makes no sense to say that their phases are their variants. Unfortunately, I am not a variant of who I was at twenty.

Lineages evolve; systems develop, grow, and age. Evolving implies participating in microevolutionary (e.g., natural and sexual selection, geographic isolation, migration and genetic drift) or macroevolutionary (e.g., extinction, speciation, species sorting, and species selection) processes. Those individuals that do not engage in these processes—that is, they do not have a relatively autonomous evolutionary history—cannot be lineages in the sense that is adopted here. The growth and differentiation of lines of cells during ontogeny can be best thought as the construction of components of an organic system. Whereas linguistic uses seem to show that this discrimination is artificial and arbitrary, biologists tacitly recognize that difference. They never consider these lines of cells as being a taxon.

There is another element that not only makes more obvious the difference between lineages and systems but also explains the classical and lingering tendency to confuse lineages, as biological taxa, with kinds. The relationship between a sublineage and the more encompassing lineage does not have a parallel in the relationship between systems and subsystems. This former relationship is similar, but not identical, to that between an example and its class, or type. As the existence of just a single element is already sufficient to establish the nonvacuous character of a class, the persistence of just a sublineage, and even the survival of a single specimen of that sublineage, are capable of maintaining, by themselves, the existence of all the lineages to which this sublineage or this specimen belongs. As long as *Panthera leo persica* subsists, *Panthera leo* remains extant, even though all the other subspecies of *P. leo* have already gone extinct. Clearly, this is not the case of a subsystem and the system to which that subsystem belongs. The few pieces of a device that may remain intact after a fire turned all the others pieces into ashes will never be, by themselves, the device that burned.

If one day, a heterosexual, loving couple of astronauts are the only vertebrates to escape from a planetary catastrophe, at least for a while, they would keep *Vertebrata*—as well as *Mammalia*, *Homo*, and *Homo sapiens*—within the set of extant taxa. But if the Spirit of Saint Louis airplane was destroyed by fire, we could not point to its remaining parts and say: "This is Lindbergh's aircraft." Although some of the parts may be intact, that old and noble machine would no longer exist. Its remains would not preserve its existence. If the heart of a dead person is transplanted into another body, it would not become the man who has died. The subsystem can never be the whole system, whereas the sublineage can always become the entire lineage. Possibly, this capacity of the sublineage and the specimen to sustain the persistence of the more inclusive individuals is the most important reason to maintain and emphasize the difference between lineages and systems.

Such a rather superficial similarity between lineages and classes—the fact that the mere existence of just a single element is sufficient to establish a more inclusive entity, whether a lineage or a class—should not undermine the fundamental assumption that lineages are individuals and not classes or kinds. This condition seemed to be disregarded by Stanley Salthe (2012, 356), who considered the taxonomy generated by the phylogenetic systematics as simple subsumption hierarchy based on relations of similarity, rather than on ancestry. This would imply that taxa are typologically defined by relations of similarity and not of ancestry, which is equivalent to the claim that taxa are natural kinds and not individual monophyletic groups (cf. Sober 1993, 148). Of course, shared character states are important for inferring the genealogical relationships as well as the delineation of taxa, but those shared character states are just diagnostic indicators of membership in a taxon and not the defining features of this taxon (Panchen 1992, 343; Ghiselin 1997, 200).

It is also true that, superficially and from a certain point of view, the relationships of inclusion between classes and of incorporation in lineages may behave in a similar manner. The relationship of incorporation between lineages obeys a kind of transitivity, which is analogous to that in class inclusion. If the class A is included in the class B, and B is included in the class C, then A is included in C; and if the lineage A is a sublineage of B, and B is a sublineage of the lineage C, then A is a sublineage of C. But in fact, they are very different things: the first is a relationship between classes, the second is a relationship of incorporation between parts of individuals.

The Problem of the *Causal Integration* of Taxa

Usually, the causal interaction among their different parts is considered to be a distinctive characteristic of individuals (Brandon and Mishler 1996, 108). It is said that in all individual entities, there is a causal connection between the whole and its parts, which does not exist between the classes and their elements. The problem is that this causal interaction appears to be absent in taxa (cf. Eldredge 1985, 144; Ghiselin 1997, 51). Taxa are not cohesive wholes of truly interacting parts, and therein lies one of the difficulties that the advocates of the individuality of taxa often confront: the dubious internal causal cohesion of lineages. The distinction between lineages and systems is critical to addressing this problem. We must accept that interaction between parts, considered *sensu* Eldredge (1985, 166–67) as an exchange of matter and energy, is what characterizes systems but is not intrinsic to lineages (see also Tëmkin and Eldredge 2015). Different local populations of a single species do not interact with each other, as do populations of different species within an ecological community (Eldredge 1995, 185). The local populations, or "avatars,"[2] of one species are incorporated into different communities and ecosystems (Damuth 1985, 1137; Eldredge 1995, 185). There, these avatars interact with avatars of other species. They usually do not interact with other avatars of the same species because these other avatars are integrated into other communities and ecosystems. This geographical and ecological dispersion is even clearer in the case of different species of the same genus. For example, the existence of *Tapirus indicus*, the Malayan tapir, does not depend on the fate of the three American species of its genus.

In fact, *Panthera leo persica*, the Indian lion, can become extinct without impacting the persistence of the African subspecies of *Panthera leo*. The same could be said about the impact of the American tapir species on the increase, decrease, or total extinction of *Tapirus indicus*. The parts that comprise *Tapirus* and *Panthera leo* appear to be independent of each other. This independence may be higher or lower in the case of the avatars of the same species and, sometimes, it may be virtually complete, as in the case of different species in a genus. Still, that genus will be recognized as a real taxonomic unit and, in that sense, as an individual, due to the genealogical relationships among the included species. It is also based on genealogical, not ecological, considerations that we regard the particular populations of the same species as being parts of it. Causal interactions among some species of the same genus, or among some avatars of the same species, may exist, of course, but they are not considered as the defining principle of either taxonomic identity or unity.

The principle of integration on lineages is clearly entirely different from that of systems, such as organisms or ecosystems. In systems, the destruction of any of their constituent parts always has repercussions, however minor or major, on other parts of the whole. That interdependence of constituent parts is what defines systems as genuine individuals. In systems, the states and function of each part are more or less sensitive to the states and the overall performance of the other parts, as well as the entire system. The temperature of any part of a rock depends on the temperature of the rest of it, and the temperature of the whole rock depends on the temperature of each part. It is the existence of such close connections between the states of parts and whole that allows for considering that rock as being a genuine system, different from a mere collection of unconnected elements (cf. Carnap 1967, §36).

In short, the problem of *parts cohesion*, repeatedly raised against the individuality of taxa, appears to arise from seeking causal connection between parts and wholes in lineages, the notion applicable only to systems. Therefore, the requirement for individuality based on the idea of integration of parts through causal interaction is misguided when applied to lineages. Similarly, a requirement for a single entity would be misguided—for a machine to be composed of components that are all modifications of a single original part or are all produced by a single manufacturer. We then would risk regarding systems as lineages, which would be no less erroneous than considering the individuality of lineages based on criteria relevant to systems. The identity of lineages is purely genealogical, but nevertheless, it is sufficient to consider them as individuals (Ghiselin 1997, 54).

Species are true individual entities because "they have beginnings, histories, and ends" (Eldredge 1995, 120). The first are the events, or processes, of speciation; the last are other events, or processes, of either speciation or extinction; and the sequences of evolutionary contingencies occurring between those beginnings and ends are the evolutionary histories of species. But there is another element that leads one to assume the individuality of species: selective pressures and other evolutionary forces that act on species, leaving distinctive marks on each of them that are preserved in successive generations (Caponi 2013). Those marks, when resulting from selective pressures, are known as adaptations, and in all cases they are manifested in character states. The identity between today's nightingale (*Luscina megarhynchos*) and that whose voice was heard by John Keats is recorded in the traces left in the nightingale—and only in the nightingale—by natural selection and other evolutionary factors. But such features of individuality are unique to species. The individuality of supraspecific taxa, such as genera *Luscina* and *Panthera*, rests only on their status as monophyletic groups. Monophyly is the integrating key of any supraspecific taxon, and that key is not causally null. Many characters of *Panthera leo* are not explained by selective pressures that have that occurred during the evolutionary history of that species. They are derived directly from an ancestral form that was also the ancestor of *Panthera pardus*, which possibly had those same characters, though, perhaps, somewhat modified. This is so because common descent is not causally null: conservation of shared ancestral features explains why both species share many characters. The marks that natural selection and other microevolutionary factors left on the common ancestor of *Panthera leo* and *Panthera pardus*, however modified in the descendant species, indicate their membership in the genus *Panthera*. The common descent causally explains many of the characteristics of the entities that comprise a lineage. Common descent, according to the original formulation by Darwin (1859, 206), explains the *unity of type*—the character shared by two lineages. But if we aim to explain the morphological divergence existing between these lineages, it is very likely that natural selection comes to be the best hypothesis: different selective pressures, derived from different conditions of existence, typically are a good explanation of divergence in the character states.

Populations and Individual Living Beings

Populations and individual living beings present a sort of ontological duplicity in that they are registered in both biological hierarchies. This dual enrollment is as inevitable, as necessary. Without this duality, we would not be able to understand the causal crossover between the order of systems and the order of lineages. In populations and individual living beings, the genealogical hierarchy that rises from individual specimen to demes to higher taxa intersects with the ecological hierarchy that rises from the organism and avatars to ecosystems. And in the individual living being, the genealogical hierarchy also intersects the anatomical and physiological hierarchy, descending from the organism, and reaches the molecules that this organism incorporates (cf. MacMahon et al. 1978, 701; Panchen 1992, 342; Pickett et al. 2007, 28).

Individual living beings and populations are the gearwheels that connect the *genealogical* and the *systemic hierarchies*. This sheds light on how ecological processes impact lineages. Persisting in their obstinate reproduction, individual living beings give rise to new lineages and themselves engage in interaction with other individuals within ecological systems (cf. Eldredge 1985, 166; 1995, 186). Struggling for life and reproduction inside populations, the individual living beings link the fate of taxa to ecological opportunities and threats (cf. Hull 1987, 170). That is what natural selection is—the impact, or resonance, of ecological vicissitudes in the genealogical order; an evolutionary effect of an ecological phenomenon named "struggle for life."

In the case of populations, this duality is clearly revealed in the distinction between demes and avatars. Demes are populations considered as lineages, and their ecological correlate, the avatars, are populations considered as systems that interact within an ecosystem (cf. Eldredge 1985, 166–72). Whereas it may seem obvious, in the case of an individual living being, such undeniable duplicity (cf. Eldredge 1985, 188; 1995, 193) is not so easy to represent. That, I think, is what justifies, or excuses, the use of an expression such as "individual living being" instead of the seemingly more appropriate, usual, and elegant "individual organism." The rationale behind such a statement is that the latter term is only appropriate for the individual living being if considered as an ecological and/or physiological system. When the individual living being is regarded as a level of the genealogical hierarchy, the only minimally adequate expression seems to be "specimen."

In other words, as specimen of a taxon, an individual living being participates in the realm of lineages, and as individual organism, it participates in the realm of systems. An individual horse, Bucephalus again, can be considered to be part of a lineage—that is, a specimen of *Equus caballus* or of *Equus*. At the same time, it can also be considered as a system. Bucephalus can be regarded as an organism—as a system, divisible into parts such as organs and cells, which are neither its variants nor its specimens— and it can also be considered as a subsystem integrated into larger systems, such as a herd, an avatar, or an ecosystem. Bucephalus, of course, is not a variant of these systems, nor is it their specimen.

At present, the concept of individual organism appears to have lost the clarity it once had (cf. Bouchard and Huneman 2013). Here it is used only for referring the systemic correlate of taxonomic specimen. The expression "individual organism" designates, then, the systemic counterpart of specimen in the ecological hierarchy. In many cases, however, this counterpart may not have all the characteristics of ecological and physiological autonomy typically associated with the classic concept of organism (cf. Clarke 2013). This calls for a precise definition of specimen's systemic correlate. Maybe, for the ease the explanation of the duality of individual living being, this difficulty could be circumstantially overlooked. In practice, this would mean taking vertebrates as paradigms of organic individuality, although it must be pointed out that there are taxa, where the boundary and the nature of organic individuality is difficult to establish (Clarke and Okasha 2013, 64). For instance, lichens represent a symbiotic association of fungi and algae, where these last are not autonomous bodies and where the former are systems that do not have a recognized taxonomic correlate.

There is yet another issue that might be a source of much confusion—an asymmetry in the roles that an individual living being plays in the two hierarchies. As a specimen, an individual living being is the last element of a lineage. For not being further divisible into sublineages that could have their own independent evolutionary histories, the specimen itself is not a lineage. However, as an individual organism, the individual living being is a subsystem incorporated into a larger ecological system, but it is also a physiological system that incorporates subsystems like organs and cells. This difference between organism and specimen is significant because it establishes a mereological discontinuity between lineages and systems, which is frequently overlooked. For instance, Michael Ghiselin (1997, 65; 2007, 283) asserts that the relation *be part of* ascends transitively from the organ, or from any other part of the organism, directly to the lineage to which that individual living being is attributed. Thus if we accept that Bucephalus was part of *Equus caballus* and its heart was part of the individual organism designated by that name, we must also accept that the heart was part of *Equus caballus*. However, such transitivity is a product of the confusion between lineages and systems—a mirage that arises from disregarding the discontinuity between the two biological hierarchies and the dual nature of the individual living being.

Species are composed of demes. Demes, in turn, are composed of individual living beings that are their parts and, consequently, are also parts of those species. It is also true that the organs of these individual living beings are their parts: the pope is part of *Homo sapiens* and his heart is part of him. It would not be correct, however, to assert that the heart of the pope, or any of our hearts, is part of *Homo sapiens*. Taxa do not have organs and their constituting parts do not include the components of individual organisms. Lineages are neither organic systems nor superorganisms. It is clear that the pope's heart is neither a specimen of *Homo sapiens* nor a sublineage, or a variety of that species. It is also clear that the potential parts of lineages are always sublineages and specimens. It follows that the heart of the pope—or of any other mammal, even though it could be described as a vertebrate heart—cannot be described as being itself a vertebrate. That makes the heart of the pope something quite different from the pope himself and from *Homo sapiens*. So despite the fact that the pope is a (specimen of) vertebrate and that *Homo sapiens* is a lineage within a higher lineage *Vertebrata*, analogous statements cannot be made regarding the pope's heart. We could point at the heart and refer to it as "a part of a vertebrate," but we would never refer to it as "a vertebrate." This observation further highlights the discontinuity between the hierarchy of systems and the hierarchy of lineages, bringing about another important distinction—that of parts and characters.

Parts and Characters

Individual vertebrates do have hearts. Let us consider the possession of that organ to be a character of the taxon. Even though the heart of the pope, or of any other vertebrate, is not a part of *Vertebrata*, having a heart is a specific character of that lineage, just as having a particular configuration of organelles can be considered as a character of a lineage of a unicellular organism. On the one hand, organs and cells organelles are not parts of lineages—they are components of systems. Their presence and absence, on the other hand, can be used as characters, allowing for recognizing and differentiating distinct lineages, although not defining them (Ghiselin 1997, 199). In other words, lineages do not incorporate organs but possess characters. Therefore, understanding the difference between *talking about lineages* and *talking about systems* requires acknowledging the distinction between parts and characters (cf. Ghiselin 1997, 303).

This difference is not immediately obvious, because "the term *character*," as Michael Ghiselin (2005, 98) correctly affirmed, "is equivocal: it conflates the parts with their properties." The multiplicity of uses that this term has had, and still has, in the constellation of biological discourses (Fristrup 2001) is due, largely, to this ambiguity. The confusion is aggravated by the fact that "many of the words that refer to parts are used *attributively*" (Ghiselin 1997, 201). Although it is common—and correct—to say, as Ghiselin observed (1997, 201), that this or that animal is "lunged" or "winged," it blurs the distinction between speaking about parts (i.e., lungs or wings) and speaking about characters (i.e., *being lunged* or *being winged*). The difference between the two becomes clearer if we consider that "one dissects a bilateral organism, not its *bilaterality*, its digestive system, not its *herbivorousness*" (Ghiselin 1997, 201).

Parts of living beings—that is, the subsystems of organisms—can be dissected, damaged, removed, and incinerated, but this is not true of their characters. One can eat a chicken leg, but the apomorphies of *Gallus gallus domesticus* cannot be eaten: apomorphies are untouchable (Caponi 2011). This indicates that we are talking about things that are intuitively recognizable as being different. This recognition becomes clearer when we consider that the very lack of an organ can be considered as a character. The lack of an organ, in fact, can be an apomorphic character if the taxon without the organ descends from an ancestor characterized by having that organ. An example of such situation is the absence of limbs in snakes.[3]

The absence of limbs—a derived, or apomorphic, state that characterizes *Ophidia*—was a product of transformation of a plesiomorphic character state, the possession of four limbs, of its tetrapod ancestor (Ghiselin 1997, 200). However, to state that the absence of limbs is part of snakes makes no sense. Thus when we examine an individual *yarará* (*Bothrops alternatus*), considering it a specimen of *Ophidia*, we can identify the apomorphic character—namely, the lack of limbs. But if we consider this *yarará* as an organism, as a system, it would be impossible to identify that absence as if it were a part, or a subsystem, which is integrated into the organic whole. In the physiology of *Ophidia*, the lack of limbs simply does not exist.

In current discourse of systematics, the concept of character is intrinsically genealogical: a character is always treated as a difference, or similarity, between lineages. Thus it only makes sense to attribute a character to an individual living being if this living being is considered to be a specimen of a sublineage attributed to a higher lineage. If we were to think of that living being as a system, characters would be invisible; the apomorphy, the derivate state, is only identifiable through a reference to the primitive state, the plesiomorphy. The concept of character does not have a meaning without the evolutionary polarity between primitive and derivate states.

Disregarding an exclusively genealogical meaning of the concept of character invites confusion. For instance, Günter Wagner (2001, 3) considers a character to "be thought of as a part of an organism that exhibits causal coherence to have a well-defined identity and that plays a (causal) role in some biological processes." Wagner was clearly concerned with the relationship between development and evolution and was not thinking in terms of concepts of character and character-states developed in systematics. He viewed characters as if they were morphological modules, amenable to be individualized in a nonarbitrary manner. That way of thinking conflates a character of a lineage with a part of a system—it confuses a character with a subsystem of the system called "organism." The lack of limbs in snakes would never meet the requirements proposed by Wagner. Yet it is correctly recognized as a distinctive character of *Ophidia*, a character that, being real, should be subject to specific evolutionary explanations but that cannot be subject to functional analysis. An absence of an organ, or structure, cannot be the subject of a functional analysis because such an analysis should specify the causal role that the operation of an existing subsystem actually has in the operation of the whole that incorporates that subsystem among its part (Cummins 1975, 765). The absence, however, should have an evolutionary explanation. Such would be the case, for example, of a selective explanation that could show under what conditions—under what sequence of selective pressures—the loss of this structure could be adaptive. Evolutionary explanations are always explanations of character states. For apomorphies, the explanations are based on natural selection; for plesiomorphies, on common descent. In the first case, what is explained is the alteration of a character; in the second, the more or less integral preservation of the primitive state of that character.

A Final Remark

Finally, it should be also understood that a character is not a unit specific to the level below the specimen in the genealogic category. "Character," and predicates such as "apomorphic," "plesiomorphic," or "homoplasic," are concepts that can be used to identify and to qualify features of all the levels of the genealogical hierarchy. Characters are not parts of organism; neither are they parts of specimens, nor of lineages. Of course, they are not parts of specimens in the manner in which a component is part of a whole, but neither are parts of specimens and lineages in the manner in which a sublineage, or a specimen, is part of a lineage. Genera incorporate species and species incorporate specimens, but specimens do not incorporate characters in the way that organisms incorporate organs. In this regard, specimens should be considered as the irreducible basis of the genealogical hierarchy. Because they are part of the hierarchy, the specimens exhibit characters. But this is not like the possession of a part, removable from the system that possesses it.

As was discussed earlier, the genealogical and ecological hierarchies converge at the level of the individual living being. Unlike the hierarchy of systems, which progress from the organismal level to the level of molecules, the hierarchy of lineages finds in the individual living being its simplest and final element. There is nothing in the genealogical hierarchy that could be to specimens what organs, tissues, cells, and molecules are to organisms, and this also pertains to chromosomes and genes. The former are cellular components and the latter are components of chromosomes, both being entities at suborganismal levels of the ecological hierarchy of systems.

The preceding statement, however, applies to genes only when they are considered as material objects. But genes can also be conceptualized as packages *of information* that are passed from generation to generation (Williams 1992, 11), and this perspective allows for considering genes from in light of the hierarchy of lineages. But it is important to remark that the informational gene does not enroll in the genealogical hierarchy in the same way that lineages and specimens do—its behavior and way of functioning, in this respect, resembles those of characters and character states. *Panthera leo persica*, being a sublineage of *Panthera leo*, could never be a sublineage of *Panthera pardus*. Neither is it conceivable that a specimen of *Panthera leo* could also be a specimen of *Panthera pardus*. On the other hand, it is true that some characters of *Panthera leo* occur in *Panthera pardus*, and it is also possible to say that some genes are found in both *Panthera leo* and *Panthera pardus*, or in singular specimens of both species. Maybe that similarity in the behavior of characters and informational genes reflects the fact that the latter are nothing more but alternatives between possible states of a character—an idea that the use of the concept of allele in population genetics seems to corroborate.

It is possible to trace genealogies of characters in the same way as the genealogies of taxa are inferred. However, one must remember that the genealogies of characters and genes are not independent of the genealogies of taxa. The genealogies of characters are partial and specifically focus on reconstructions of taxonomic genealogies. This is true even for such characters as the presence of a chromosome or a gene. Gene genealogies can be traced in the same way as the genealogy of the vertebrate limb, but these gene genealogies are just markers, more or less accurate, of the genealogy of particular taxa. In some cases, it may even be that the genealogy of a gene comes to be the only, or the most reliable, marker of a phylogeny. But, at least from a conceptual point of view—though perhaps not from a strictly methodological standpoint—genealogy of genes cannot be regarded independently from genealogy of taxa.

Notes

1. In this discussion, the term *instance* is preferred over the term *example* to avoid confusion with the concept of "exemplar" in phylogenetic systematics, which has an entirely different meaning (denoting a specimen of a lineage).

2. John Damuth (1985) proposed the term *avatar* (which means "incarnation") to designate each of the different populations of the same species that are integrated in multiple communities, interacting with avatars of other species. Unfortunately, the term has not been so widely adopted by the scientific community. Following the proposal by Eldredge (1995, 187), it is adopted here because it appears to be particularly suitable for the explanation of the differences between the

hierarchy of biological systems and the hierarchy of biological lineages. "Avatars" are the systemic correlate of demes in the genealogical hierarchy (cf. Eldredge 1995, 185).

3. Given two character-states, one primitive and the other derived, the last is described as "apomorphic," while the primitive will be described as "plesiomorphic" (Eldredge and Cracraft 1980, 31).

References

Bouchard, Frédéric, and Philippe Huneman. 2013. Introduction to *From Groups to Individuals*, edited by Frédéric Bouchard and Philippe Huneman, 1–14. Cambridge, MA: MIT Press.

Brandon, Robert N., and Brent D. Mishler. 1996. "Individuality, Pluralism, and Phylogenetic Species Concept." In *Concepts and Methods in Evolutionary Biology*, edited by Robert N. Brandon, 106–23. Cambridge: Cambridge University Press.

Caponi, Gustavo. 2011. "Las apomorfías no se comen: Diseño de caracteres y funciones de partes en Biología." *Filosofia & História da Biologia* 6(2): 251–66.

———. 2013. "Las especies son linajes de poblaciones microevolutivamente interconectadas: Una mejor delimitación del concepto evolucionario de especie." *Principia* 17(3): 395–418.

Carnap, Rudolf. 1967. *The Logical Structure of the World*. Los Angeles: University of California Press.

Clarke, Ellen. 2013. "The Multiple Realizability of Biological Individuals." *Journal of Philosophy* 110(8): 413–35.

Clarke, Ellen, and Samir Okasha. 2013. "Species and Organisms: What are the Problems?" In *From Groups to Individuals*, edited Frédéric Bouchard and Philippe Huneman, 551–76. Cambridge, MA: MIT Press.

Cummins, Robert. 1975. "Functional Analysis." *Journal of Philosophy* 72(20): 741–65.

Damuth, John. 1985. "Selection among Species: A Formulation in Terms of Natural Functional Units." *Evolution* 39(5): 1132–46.

Darwin, Charles R. 1859. *On the Origin of Species by Means of Natural Selection, or the Preservation of Favoured Races in the Struggle for Life*. London: Murray.

Eldredge, Niles. 1985. *Unfinished Synthesis: Biological Hierarchies and Modern Evolutionary Thought*. Oxford: Oxford University Press.

———. 1995. *Reinventing Darwin: The Great Debate at the High Table of Evolutionary Theory*. New York: John Wiley & Sons.

Eldredge, Niles, and Joel Cracraft. 1980. *Phylogenetic Patterns and the Evolutionary Process*. New York: Columbia University Press.

Eldredge, Niles, and Stanley N. Salthe. 1984. "Hierarchy and Evolution." *Oxford Surveys in Evolutionary Biology* 1: 182–206.

Fristrup, Kurt M. 2001. "A History of Character Concepts in Evolutionary Biology."

In *The Character Concept in Evolutionary Biology*, edited by Günther P. Wagner, 13–36. San Diego: Academic Press.

Ghiselin, Michael T. 1974. "A Radical Solution to the Species Problem." *Systematic Zoology* 23: 536–44.

———. 1997. *Metaphysics and the Origin of Species*. Albany, NY: SUNY Press.

———. 2005. "Homology as a Relation of Correspondence between Parts of Individuals." *Theory in Bioscience* 124: 91–103.

———. 2007. "Is the Pope a Catholic?" *Biology & Philosophy* 22: 283–91.

Griffiths, Graham. 1974. "On the Foundations of Biological Systematics." *Acta Biotheorica* 23: 85–131.

Hull, David L. 1976. "Are Species Really Individuals?" *Systematic Zoology* 25(2): 174–91.

———. 1987. "Genealogical Actors in Ecological Roles." *Biology & Philosophy* 2: 168–84.

MacMahon, James A., Donald L. Phillips, James V. Robinson, and David J. M. Schimpf. 1978. "Levels of Biological Organization: An Organism-Centered Approach." *Bioscience* 28(11): 700–704.

Mayr, Ernst. 1976. "Typological versus Population Thinking." In *Evolution and the Diversity of Life: Selected Essays*, 53–63. Cambridge, MA: Harvard University Press.

Panchen, Alec L. 1992. *Classification, Evolution and the Nature of Biology*. Cambridge: Cambridge University Press.

Pickett, Steward T. A., Jurek Kolasa, and Clive G. Jones. 2007. *Ecological Understanding*. Amsterdam: Elsevier.

Pievani, Telmo, and Emanuele Serrelli. 2013. "Bucket Thinking: The Future Framework for Evolutionary Explanation." *Contrastes*, suplemento 18: 389–405.

Salthe, Stanley N. 1985. *Evolving Hierarchical Systems*. New York: Columbia University Press.

———. 2012. "Hierarchical Structures." *Axiomathes* 22: 355–83.

Sober, Elliott. 1993. *Philosophy of Biology*. Chicago: Chicago University Press.

Tëmkin, Ilya, and Niles Eldredge. 2014. "Networks and Hierarchies: Approaching Complexity in Evolutionary Theory." In *Macroevolution: Explanation, Interpretation and Evidence*, edited by Emanuele Serrelli and Nathalie Gontier, 183–226. Dordrecht: Springer.

Wagner, Günther P. 2001. "Characters, Units and Natural Kinds: An Introduction." In *The Character Concept in Evolutionary Biology*, edited by Günther P. Wagner, 1–10. San Diego: Academic Press.

Williams, George G. 1992. *Natural Selection: Domains, Levels and Challenges*. Oxford: Oxford University Press.

CHAPTER 3

Biological Organization from a Hierarchical Perspective

Articulation of Concepts and Interlevel Relation

Jon Umerez

1. Introduction

Introductory biology books often refer to what Ahl and Allen (1996) call the "conventional levels of organization": a sequence from the cell to the biosphere through intermediate levels of organism, population, community, and ecosystem (p. 77, fig. 4.1). This scheme is frequently supplemented with additional subcellular or molecular levels. Even without the reference to social hierarchies, it is also common to represent the hierarchical levels of natural systems as lower or higher (bottom or top), denoting their position with the use of a scale or rank of some sort.

The contrast between reductionist and nonreductionist approaches regarding the understanding of the specificity of biological systems, along with their constitution and dynamics, has several lines of contention in the philosophical literature. One such key issue is the different appreciation of the hierarchical arrangement of biological systems into levels of organization and the corresponding interpretation of the nature of such levels and, particularly, about the relationships established among them.

When the relation among levels is considered in its bottom-up form, it does not usually become controversial in discussions about reductionism. On the contrary, some variants of hierarchical representation are commonly proposed to be more fitting for a reductionist view of biological systems, despite their variety and complexity (Oppenheim and Putnam 1958).

In contrast, when the relation among levels is considered in its top-down form, it is frequently criticized or even dismissed entirely. This is exemplified in the widespread opposition to the concept of downward causation (Campbell 1974), even in approaches that are not strictly reductionist. In very general terms, the occurrence of downward causation implies that higher-level entities or processes have a genuine causal influence on those at lower levels—that is, that higher-level events are the cause of identifiable effects on events at a lower level.[1]

Until recently, in the philosophy of science, the subject of levels and their relation has been discussed mainly in the areas of the philosophy of mind and of cognitive sciences, where corollaries of the so-called mind-body problem figure centrally in the debates, that aim to understand the relation, if any, between physical properties and mental phenomena. In the philosophy of biology, the subject has been more directly addressed in the debate over levels of selection and macroevolution (see from Brandon and Burian 1984 to Okasha 2006). Even in this context, the nature of the relationship among levels is an unsettled issue (beyond the more direct debates about whether to accept multiple loci of selection or on reaching a consensus regarding which ones to admit and under which conditions). As attested by Samir Okasha, the problem stems from a philosophers' concern: "In a multi-level scenario matters are less simple. Presumably, multi-level selection involves causality at more than one level of the hierarchy. But this raises a number of questions. Are the higher- and lower-level causal processes autonomous, or are they interdependent? Might selection at one level ever be 'reduced' to selection at a lower level? If higher-level selection has an impact on lower-level phenomena, does this mean that 'downward causation' is occurring? (The significance of this question is that some philosophers regard downward causation as a suspect notion)" (Okasha 2006, 77).

Recently, the philosophy of biology has seen a renewed interest in levels of organization. Some of those perspectives aspire to break the very dichotomy between reductionism and antireductionism by focusing on the specific dynamics of the component parts and their relations in order to explain the behavior of complex systems. They also accept the relevance of higher levels in natural systems and the explanatory need to take them into account. Among these views, the most representative are the various neo-mechanistic views (see, for instance, Bechtel and Richardson 1993; Machamer et al. 2000; Bechtel 2008; Craver 2007; Craver and Darden 2013).

However, a more detailed analysis of the concept of level generated within that approach (i.e., Craver and Bechtel 2007) brought into ques-

tion the basic intuitions regarding the causally specific significance of interlevel relations held by current hierarchical approaches in biology,
claiming that "things at different levels of organization . . . do not causally
interact" (Craver 2015, 23) and, therefore, leave no room for genuine interlevel relation.

This line of thinking has been followed by discussions that have led to
even more deflationary accounts of levels and hierarchies,[2] "where levels
of organization give way to more well-defined and fundamental notions,
such as scale and composition" (Eronen 2015, 39), or even to eliminativism
about levels:[3] "I argue that no distinct notion of levels is needed for analyzing explanations and causal issues in neuroscience: it is better to rely
on more well-defined notions such as composition and scale. One outcome of this is that apparent cases of downward causation can be analyzed away" (Eronen 2013, 1,042).

There are, of course, exceptions to this trend, and we can find other philosophical approaches (for instance, Dupré 2012 or Mitchell 2009) that endorse multiple causal interactions among levels of organization in biological
systems. It could therefore be worthwhile to speculate about the reasons
for this widespread suspicion in the philosophical literature regarding the
legitimacy of considering interlevel relations.

One reason for the different acceptance of those two kinds of relations
among levels derives from the natural and standard characterization of levels of organization in the context of "nested compositional hierarchies"
(Tëmkin and Eldredge 2015). However, explicitly hierarchical approaches
in ecology, evolutionary biology, or biological complexity do customar-
ily clarify and develop, in each case, the precise meaning of that general
description in terms of nested composition.[4] The compositional aspect in
these hierarchical perspectives involves a complex, rich set of more specific
relations among the levels that extends beyond mere spatial inclusion (being part of) or material constitution (being made of) (see section 3).

But such a nuanced view of nested compositional hierarchies is not
equally acknowledged by nonexplicitly hierarchical perspectives that regard the hierarchical ordering as a given, resulting from the bottom-up direction of composition. Proponents of the hierarchical approach in ecology
had already stated a similar concern: "Most biologists are already familiar
with the general idea of hierarchy through the concept of 'levels of organization' (cell, organism, population, community). However, even our infor-
mal discussion should make it clear that the term hierarchy is not restricted
to this simple sense. . . . It is clear that ecological organizations show hierarchical structure, but it should also be clear that the simple series is unlikely

to be useful across the range of observation sets and spatiotemporal scales involved in ecosystem analysis" (O'Neill et al. 1986, 60–61).

The questioning of interlevel relations stems also from the apparent disregard or lack of awareness, in mainstream biology and philosophy, about the contributions made by scholars working in the explicitly hierarchical framework. Evidence for this may be found by observing the bibliographical references mentioned in the literature, and the repeated complaint about the alleged lack of a detailed, precise, and articulated conceptual framework regarding the concepts of hierarchy or level of organization (see section 2).

The purpose of this chapter is not to respond to these particular accounts (mechanistic or other). Rather, they are mentioned as a symptom or a testimony of an odd imbalance between those areas of science where explicitly hierarchical approaches have been developed making use of interlevel causal relations and those philosophical perspectives so distrustful of the conceptual soundness of such kinds of relations. It is my claim that an oversight of the very relevant scientific approaches and a restricted rendition of the concept of levels of organization are two motives that help in our understanding the underlying rationale of that symptom.

This chapter is instead an attempt to clarify the concept of level of organization, to make it applicable across different conceptual schemes, and to establish a basis for coherent discussion and commensurability among varied and even alternative views on biological systems. In this sense, the goal is not to provide a technical definition or a precise and definitive taxonomy but, first, to assume a general and intuitive notion of levels and then furnish it with more specific content, attempting an articulation among different kinds and degrees of (interlevel) relation. According to the view advocated here, hierarchy is primarily understood as meaning "interlevel relation," and the concept of level of organization makes sense within such an explicit hierarchical framework.

2. Definitions and Classifications. Charge of Ambiguity and Plurality of Meanings

It is often stated that the concept of level is insufficiently developed and has acquired a plurality of meanings and uses in scientific literature, making it difficult to apply in practice. This is one of the reasons why many authors propose and develop their own concepts for explanatory purposes.

The recently emerged neomechanistic perspective on the levels of organization raised similar concerns: "The notion of *levels* is ubiquitous in discussions of science. Yet . . . it is unclear" (Bechtel 2008, 143); "Despite the ubiquity of levels talk in contemporary science and philosophy, very little has been done to clarify the notion" (Craver 2015, 23); "The term level is notoriously ambiguous" (Eronen 2015, 39).

This complaint about the ambiguity, lack of precision, and alleged resulting confusion of the term *level* seems to always accompany any critical discussion about the concept that such term denotes. Should we then admit that the current state of affairs is the same as more than fifty-five years ago, when the philosopher Mario Bunge voiced a similar concern: "As used in contemporary science and ontology, the term *level* is highly ambiguous. . . . The aim of the present paper is to list the usual and some possible meanings . . . of the word 'level,' to specify them briefly, to illustrate them and to propose some problems in which those concepts are involved" (Bunge 1960, 396)? Or that we are in the same situation as more than thirty-five years ago, when, despite his previous efforts, Bunge continued to lament the lack of "consensus on the significance of the terms 'level' and 'hierarchy,' which are used in a variety of ways and seldom if ever defined" and, again, set out to "remedy this situation" (Bunge 1979, 13)?

But the claims of vagueness of the notion of levels of organization appear to be unjustified and, in fact, quite far from the truth if we consider the rich literature devoted to the subject. Undoubtedly, there are many different versions of the term "level of organization" according to different theoretical perspectives (see note 4). This does invite some degree of ambiguity, especially if one disregards the specific theoretical contexts in which the concept of level plays an explanatory or modeling role. Further, there are indeed ambiguous formulations of the concept that steer away from epistemological precision. Nevertheless, a declaration of a general lack of clarity and precision leading to overall confusion as the trigger for generating additional definitions is certainly an overstatement. Moreover, such criticisms are hardly necessary, because having precise and theoretically sound concepts should not preclude the development of new or elaborated versions to answer specific theoretical demands in a particular field or area of research (or their more recent empirical findings or theoretical advances). This necessity is precisely what prompted the development of several different hierarchical theories in evolutionary biology, ecology, complexity and systems sciences, and some areas of cognitive science and will most likely pave the way for future advances.

We should recall—at least as a brief testimony—some antecedents. Even leaving aside the work in the late nineteenth century and early twentieth century on emergent evolution (George H. Lewes, John S. Mill, C. Lloyd Morgan, Charles D. Broad), there are at least two other earlier broad periods or clusters of theories that dealt with and contributed to the development of concepts akin to that of level of organization.

In the 1930s and 1940s, at least two separate yet partially connected theoretical departures, organicism and systems science, addressed fundamental aspects of biological organization in explicitly hierarchical terms (see Haraway 1976). On the one hand, researchers Joseph Needham, Alex B. Novikoff, James K. Feibleman, and others attempted to develop a scientifically sound organicist view that gave rise to the *theory of levels of integration* (Needham 1937; Redfield 1942; Feibleman 1954). On the other hand, von Bertalanffy and other biologists—for instance, Paul Weiss—had already begun developing a systems view, initially based on biological systems (later generalized as a *General System Theory*), in which a multilevel hierarchical approach was central. The latter view, unlike that of the purely organicist perspective, was able to connect with the ensuing period of further theoretical exploration of biological hierarchies (see Umerez 1994 and Etxeberria and Umerez 2006 for a more detailed treatment and further references).

The work on hierarchy theory flourished throughout the late 1960s and into the 1980s. In parallel with more general and epistemological discussions related to reductionism (i.e., Ayala and Dobzhansky 1974), detailed hierarchical theories were formulated, both as general approaches to biological organization (see, for instance, Whyte et al. 1969; Weiss 1971; Pattee 1973) and as more specific accounts of relevant phenomena in certain areas of biology—mainly evolutionary biology (Eldredge and Salthe 1984; Eldredge 1985; Salthe 1985; Grene 1987) and ecology (Allen and Starr 1982; O'Neill et al. 1986). All these contributions offered detailed and precise accounts of interlevel relations (Umerez 1994; Umerez and Moreno 1995).

In short, with the exception of the older work of the 1930s and 1940s, and the development of systems science, a hierarchical perspective had seen major advances in the organizational and evolutionary perspectives in the period between the late 1960s and the mid-1980s. Some of the proponents of these approaches have continued to defend hierarchical and multilevel views that are quite specific in their characterization of the sense in which they use "level of organization" and how they relate those levels among them in a hierarchical ordering.

In the philosophical domain, William Wimsatt had developed a distinctive approach in the 1970s (i.e., 1974, 1976; see Wimsatt 2007), with its roots in the work of Simon (1962). Wimsatt's view was further developed and transformed by later scholars, contributing to current neomechanistic perspectives.

As a result of this body of work from previous decades, there is a rich choice of definitions of *level* available to the contemporary scholar. For the purpose of this analysis, two representative examples are selected that emphasize different aspects regarding interlevel relation.

The first approach takes a more static compositional view, exemplified by Wimsatt's definition: "By level of organization, I will mean here compositional levels—hierarchical divisions of stuff (paradigmatically but not necessarily material stuff) organized by part-whole relations, in which wholes at one level function as parts at the next (and at all higher) levels." (Wimsatt 1994, 222). The second approach offers a more dynamic relational sort of definition, which encompasses composition, as in the following definition by Salthe: "Level: a representation of scale in a functional hierarchy such that higher levels regulate lower ones and lower ones give rise to higher ones. In the present application, the higher ones also include the lower ones as subsystems." (Salthe 1985, 295).

Significantly, a more recent definition—stemming from the first approach—connects both perspectives by subsuming the relational aspect under the idea of organization: "levels that are related as parts to wholes with the additional restrictions that the parts are components. The relata are mechanisms and components and the relationship is organization: lower level components are organized to make up a higher-level mechanism." (Craver 2009, 396).

The following section attempts to disentangle what is generally implicit in this idea of organization as used in the context of biological hierarchies.

The plethora of meanings of the concept of level enables some authors to develop taxonomies, classifying the diversity of the concept (see early examples in, e.g., Bunge 1960 or Whyte, Wilson, and Wilson 1969). See the meticulous analyses by Carl Craver (2007, 171ff; 2009, 389ff; 2015, 3ff) for more recent attempts to systematize the various versions of this fundamental concept.

Most—if not all—of these classifications of different meanings of *level* clarify how the term is used in different contexts, but they are presented as a juxtaposition of different meanings that are diligently distinguished, though typically they are not interrelated. As useful as these taxonomies are, it would also be valuable to provide ways of relating those different meanings in order to be able to establish a common ground for direct comparison among them. The view presented in the next section is an attempt to offer such a comprehensive, integrative approach.

3. Articulated View

The goal of the articulated view presented here is not to substitute other general approaches to definitions or taxonomies but to supplement them by trying to explicitly account for the trait organization, understood as a very specific kind of relationship among levels. Thus, the articulation is an attempt to disentangle the notion of "level of organization" by means of an epistemological reconstruction of the content of the concept, as a result of determining the kinds of relations among elements and their processes that are implicit when it is used in hierarchical approaches.

As a general starting point, levels are the various groups of elements connected with other such groups at a different scale[5] by virtue of a given relation and within a particular system. The only limitation implied by this intuitive concept is that the constitution of systems must be based on some type of *specific relation* among groups of elements (i.e., levels). The peculiarities of the relation in each particular case or the rendition of the concept of level is the object of a detailed epistemological reconstruction. This intuitive notion is taken as a heuristic starting point to further inquire into the nature of those specific relations and it is not, therefore, offering a particular definition or a criterion for a definite taxonomy. It should instead be complementary and compatible with several definitions and partitions into distinct levels of organization, since the present analysis should be applicable to any of them.

The analysis is clustered around five fundamental features that cover any possible meaning of the term and are ordered according to their degree of generality. These features are *composition, integration, emergence, control*, and *organization*. They represent, precisely, different kinds of relations among groups of elements or processes, which become articulated into an overall picture (Umerez 1994).

The reconstruction proceeds from the more general to the more specific. It starts from a potential common minimum for any concept of level of organization, related to the idea of composition, and gradually discloses further conditions, restrictive of that generality toward more specific and operative concepts. The idea is that in order to better understand biological *organization* we may attempt to consider organization as a cluster of different, but related, kinds of levels, each one with different types of traits and characteristics common to other systems (or ways of ordering things or phenomena) but which, when integrated, deploy the features of fully fledged biological organization, as in "levels of organization."

In terms of Salthe's *specification hierarchy* (1993), we could present those different kinds of levels according to the progressive specification of their relation:

{composition {integration {emergence {regulation {organization}}}}}

This articulation does not by itself support any particular definition, taxonomy, or ordering against any other. On the contrary, it should serve as an epistemological tool relevant to the analysis and contrast of all of them. The intention is to make explicit some characteristics about the relations among elements that give rise to levels, such as when they are involved in what is referred to as biological organization, with the ambition to clarify what any version of levels of organization in biological systems should (and, explicitly or not, typically does) fulfill.

The organization of biological systems—from the most simple to the most complex—entails instances of the aforementioned kinds of relations among the elements and processes involved. Those relations might be involved to varying degrees in each different system, and might give rise to a distinct ordering of elements in each case. It is important to note that this ordering is not universal or given, but always depends on the particular system, process, or operation that the explanation in terms of levels of organization is seeking to illuminate.

The articulation suggested herein serves as an additional tool for adequate comparisons among alternative hierarchical schemes and across different fields of inquiry.

3.1. Levels of Composition

First, we must analyze what "levels of composition" refers to. In this regard, at least four characteristics must be considered: (1) nestedness, (2) relation of partial ordering, (3) homogeneity and heterogeneity of component parts, and (4) the discreteness or continuity of the arrangement.

The relation among levels based on composition is the most general and basic, with the exception of *aggregation*, which is taken as the antithesis of organization (as in the aggregative case,[6] sensu Wimsatt 2000, 2006). In its generality, level of composition may be conceptually close to the notion of subsystem, but some further specifications are needed to clarify its meaning.

3.1.1. NESTEDNESS   Nestedness is understood as "the requirement that upper levels contain lower levels," or the property that "entities of smaller scale are enclosed within those of larger scale" (Allen and Starr 1982; Salthe 1985). It is unclear, however, whether the notion of composition should be restricted to physical inclusion or containments or may encompass wider compositional meanings, such as those derived from functional and dynamical properties. If the former is correct, then difficulties arise with how to describe the relation among organisms and demes or populations and ecosystems, or the relationship between membranes and cells. Levels of different composition are superposed, but by virtue of some specific relation (which must be explicit) that allows the proper use of level. For instance, in cases where the relation is functional (e.g., informational), not spatial, and the overall (organized) system can nevertheless be referred to as *composed of* those functionally related levels.

Since the concept of level of composition has to be applicable in these more complex cases, the minimal characterization of the relation of composition must be sufficiently broad in its definition to avoid mere nestedness as a mandatory condition.

3.1.2. RELATION OF PARTIAL ORDERING   In general, levels should maintain a partial ordering relation among them. In mathematical and formal terms, *order* is defined as the relation among members of a set according to which some members precede or follow others. A partial ordering on such a set is a relation $\leq$ that is "transitive, reflexive, and antisymmetric" (Blackburn 1994, 272). This order is set in each case with respect to a specific criterion or measure, such as containment, size, rate, enabling or constraining action, command, etc. Levels of composition show a partial ordering relation. This condition might seem too restrictive as a minimal condition because it imposes a very specific relation modality, but a careful analysis reveals that this condition is indeed necessary. Without a partial order relation, the very notion of level may become meaningless: levels that are reversible according to the same relation would be possible as well as different levels at the same very level, undermining the prospects of a cogent and theoretically useful sense for the concept.

This clearly does not imply that the particular order between two levels cannot be reversed according to different criteria (for instance, size or containment, as in the case of the genes with respect to the cell, against function, as in the case of gene expression with respect to cellular metabolism).

3.1.3. HOMOGENEITY OR HETEROGENEITY OF COMPONENTS   It is also necessary to address the issue of whether we are dealing with, at each level, just homogeneous (undifferentiated) or heterogeneous (differentiated) components. In this regard, for organizational perspectives (as well as mechanistic ones), it is important to note that at each level we are primarily dealing with differentiated parts, not just equivalent component parts. Therefore, in general, the relation of composition should be able to accommodate both cases, which would be distinguished later on according to further specifications of relations.

At this point, it is worth emphasizing one aspect that will be discussed in subsequent sections. Being able to accommodate both homogeneous and heterogeneous entities means that, as far as the characterization of composition is concerned, the two are indistinguishable. In other words, the *relation of composition* by itself does not distinguish between being composed of homogeneous or interchangeable components and being composed of heterogeneous and noninterchangeable components. If this distinction is relevant, as it happens to be in biological systems, some further specification will have to be added to account for it. This is precisely what explanations in the biological sciences consistently seek. An important part of biological knowledge consists of specifying the particular ordering and different roles of different kinds of components, and the way in which these heterogeneous components are integrated within their level *and* relate to those at contiguous levels.

3.1.4. DISCRETE LEVELS   Finally, and briefly, since it amounts to a question of convention in the meaningful use of words, it is important to emphasize that levels must be discrete, at least heuristically (even if they are artificially discretized from a continuum), in order to be a useful conceptual tool.[7]

3.2. Levels of Integration

This relation addresses the question of how *parts* are constituted into *wholes*, instead of just components into aggregates. Thus, it is a nuance that specifies a particular kind of relation of composition. All levels of integration are levels of composition in the broad sense established earlier (and they would consequently fulfill the simpler conditions to be aggregates), but not all levels of composition are levels of integration.

The idea of integration, in its most traditional sense, refers to the "making up or composition of a whole by adding together or combining the separate parts or elements" (OED). In the context of complex and biological systems, it means, first, that the relation by virtue of which a set of equivalent component parts produce a common dynamical pattern or a coordination of processes that holds them together or maintains a particular steady state. Second, it means that the relation by which the combination of various distinct component parts gives rise to a complex whole or a complex state with the same properties of relative stability and unified performance.

The relation of integration is the first step to exclude what, in terms of Bechtel (following Wimsatt's characterization of aggregativity), might be called the "null case" with which to confront other cases: "the null case in which organization is absent . . . components are put together but no order is imposed" (2008, 150n7). The relation of integration produces a coherent higher level (*cohesive* sensu Collier 1986) and may give rise to a form of self-contained unit, characterized by its own properties and an extreme integration of its parts (in complex systems). This feature is present in systems ranging from basic self-organized systems, such as far from equilibrium dissipative systems, where an ordered pattern at the global macro-level arises from dynamics of entities at the microlevel, to more complex systems, such as lipid compartments or protocells.

3.3. Levels of Emergence

If integration specifies the relation of wholes to its parts, the relation of emergence further specifies a particular kind of process of formation of (at least) some of these wholes. It classically implies the two features of nonpredictability and qualitative novelty. Though often unnoticed, it also classically implies the grounding of the higher levels on the lower levels, such that the existence of the higher levels is dependent on the existence of lower-level entities.

3.4. Levels of Regulation (Control)

Regulation is the relation by which higher levels, derived in some sense (through integration and emergence) from the lower ones, in turn exert some sort of influence on those very lower levels. This relation among levels should encompass both the disputed forms of downward causation (Campbell 1974), as well as those apparently more simple cases of cybernetic-like regulation (or control).[8]

3.5. *Levels of Organization*

The relation of organization among levels entails the combination of the other four, more encompassing, levels of specification, all of which are simultaneously present: "The organization of an entity refers to the arrangement of its component parts and their operations (functions) and to how they result in the capacities of the whole or the phenomena in which it appears. Often, organized entities are complex and hierarchical: their parts are themselves organized entities" (Etxeberria and Umerez 2013).

Therefore, when considering organization, we are referring to complex systems implying some form of interrelation among elements that goes well beyond mere composition: such systems manifest integrated global emergent properties, capable of regulating the behavior (dynamics) of their constituents.

4. Causal Plurality: Constraint

In addition to the two reasons that have been presented in the previous two sections (a disregard of developments in scientific hierarchical approaches and a restricted view of organizational relations), there is another and more general reason underlying the philosophical suspicion about interlevel relations that is related to the fundamental issue of causality. In fact, most criticisms that downplay the role of interlevel relations (especially top-down relations) in hierarchies are concerned with the problems of overdetermination or postulating more than one cause for the same effect.[9]

To admit that events or processes at higher levels have some decisive influence on the events or processes at lower levels may conflict with the alleged causal sufficiency of lower levels. Typically, this problem is framed in the context of the principle of the *causal closure of the physical domain*, which roughly states that "any physical event that has a cause at time t has a physical cause at t" (Kim 1993, 280). Such a view is grounded in physicalist ontology, asserting that all that exists is the physical world (rejection of nonphysical entities) (Blackburn 1994, 287). This principle is typically supplemented with the so-called problem of *causal explanatory exclusion*, that "seems to arise from the fact that a cause, or causal explanation, of an event, when it is regarded as a full, sufficient cause or explanation, appears to *exclude* other *independent* purported causes or causal explanations of it" (Kim 1993, 281).

A thorough examination of this issue is well beyond the scope of this chapter. Nevertheless, here again an imbalance may be detected between these philosophical reservations and the determination to overcome epistemological difficulties that characterize the explicitly hierarchical approaches adopted in evolutionary biology, ecology, or complexity and systems sciences.

Aside from the specific theoretical construction of each particular area and approach, the way to overcome these philosophical problems is sought, in most cases, through a common strategy that involves (1) the deployment and vindication of a more plural and complex understanding of causal relationships, made specific through (2) the intentional adoption of several additional concepts, such as the idea of constraint.

This effort is patently clear, for instance, in the development of the hierarchical theory of evolution, where the awareness of the need to expand the scope of causal interactions is present from the very beginning: "A hierarchical approach includes wider possibilities of causation within additional levels, as well as upward and downward causation between levels" (Vrba and Eldredge 1984, 169). As the following quote shows, the hierarchical approach has since been developing conceptual tools for a very detailed individuation and analysis of the causal factors involved in the appearance of evolutionary patterns: "Complex evolutionary patterns integrate variational dynamics of sorting processes that occur at different levels, with their effects propagated indirectly to other levels within the genealogical hierarchy via downward and upward causation. Hierarchy theory provides a theoretically and operationally unified framework for unraveling causal processes responsible for generating evolutionary patterns by identifying the involved individuals and their properties, hierarchical levels where these individuals reside, and their interactions within and across levels as well as between the two hierarchies" (Tëmkin and Eldredge 2015, 204).

There are also views that take a more general perspective, claiming that any appeal to selection process still requires an expansion of what counts as causal explanation: "My claim, thus, is that in selection processes these functional or relational properties can be causally efficacious, which means that properties other than physical properties can have causal powers. To put it in the form of a slogan: selection processes make the causal world exuberant" (Vicente 2013, 139–140).

Significant cases are also found even in some areas of molecular biology, though in the absence of scientific, explicit hierarchical approaches. For instance, the analysis of the folding of proteins aided by chaperones is couched in such terms that philosopher Alan Love interpreted it as an instance of interlevel, top-down causal explanation. In a review article, Christopher M. Dobson summarized the point: "It is apparent that biological systems have become robust not just by careful manipulation of the sequences of proteins but also by controlling, by means of molecular chaperones and degradation mechanisms, the particular state adopted by a given polypeptide chain at a given time and under given conditions. This process can be thought of as being analogous to the way in which biology regulates and controls the various chemical transformations that take place in the cell by means of enzymes" (Dobson 2003, 888).

Reflecting on this case, Love offers a straightforward reading in causal terms: "Explanations of protein folding that rely on chaperones are a form of top-down causal explanation. . . . The top-down causal explanation of protein folding is in terms of macromolecules and their components, and the hierarchical relations that apply to protein structure are delineated precisely" (Love 2012, 119).

In all of these examples, there is not only the recognition that in some cases an explanation based on physical properties as such is not appropriate but also an explicit acknowledgment that complex events or processes, such as many biological phenomena, often require or allow for complementary or alternative causal accounts that are not reducible. In sum, this amounts to questioning both the causal closure of the physical world and the principle of causal explanatory exclusion. To effectively address these issues, it is necessary to provide scientifically and empirically sound accounts and models capable of integrating interactions among levels of this sort.

The second aspect common to hierarchical approaches is the incorporation of special conceptual tools that supply the specific operational materialization of the relations among levels of organization. One such set of concepts may be summarized under an inclusive interpretation of the idea of constraint.

Constraint is a term that has slightly different meanings in different scientific disciplines. In several fields of biology and complexity sciences, *constraint* (among other terms with approximately the same meaning as *boundary condition*) is used to account for phenomena whose explanation is not exhausted by reference to the general and standard causal factors at one level, and requires taking into consideration factors at other levels. The development of more specific characterizations of constraint are attempts to spell out the interaction among levels of organization by deriving them from epistemologically legitimate concepts grounded in the physical sciences or in their explanatory stock.

In general, *constraint* "refers to a reduction on the degrees of freedom of the elements of a system exerted by some collection of elements, or to a limitation or bias on the variability or possibilities of change in the kind of such elements" (Umerez and Mossio 2013, 490). The action of constraint is inherently hierarchical because it does not interact *dynamically* (because it operates at a different rate) with the elements it is going to influence but interacts globally as a boundary condition or an initiating condition (Salthe 1985).

The concept in its usual meaning of "a limiting factor" has been routinely used to describe the operation of several natural and artificial devices or processes. Examples of constraints range from surfaces (e.g., table tops) to switches; from eddies or convection phenomena (such as Bénard cells) to other pattern formation cases or flocking behaviors, from membranes or enzymes to canalization or other developmental determinants, and from genetic or epigenetic instructions to sorting or other selective processes.

The unveiling of the implicit implications of the concept of constraint against the monist causal assumptions (those introduced at the beginning of this section) may be attributed to Michael Polanyi. He was the first author to discuss this notion as the distinguishing feature of biological organization (Polanyi 1968), even though he did not use the term *constraint*. Indeed, he used the more generic concept of *boundary condition* in order to introduce the idea of a dual control over the chemical processes occurring within an organism, where a higher level would additionally harness the dynamics of the lower level. He did, though, distinguish between two kinds of boundaries, test-tube type and machine type (Polanyi 1968, 1208), which indicates the intended generality of his attempt, since the latter expands significantly the scope of phenomena covered by the former.

This meaning of interlevel constraint, as an additional but necessary (complementary) ingredient to explain the workings of living systems, was further developed by Howard Pattee. By then he was investigating (in the context of the problem of the origin of life) how to account for the reliability of hereditary mechanisms, departing from a point of view grounded in physics (1966, 1967, 1968, 1969a, 1972). In these works, Pattee had already begun to use the concept of constraint as an appropriate explanatory tool, derived from physics (mechanics) to describe the functioning of hereditary systems as a combination of two separated dynamical realms at different levels. Quite naturally, he endorsed Polanyi's dual control perspective for his hierarchical theory, distinguishing structural from functional hierarchies (Pattee 1969b, 1973) and developing a broader theory of biological organization (see Pattee 2012; Umerez 2001, 2009). Additionally, the more technical distinction between holonomic and nonholonomic constraints (formulated according to their definition in physics) became fundamental for his formulation of those kinds of hierarchical relations (as well as for his treatment of the problems of the origin of life and heredity).[10]

The use of the concept of constraint and other related concepts as a limiting factor has become customary in many areas of the biological sciences and usually does not pose explicit epistemological issues. The early formulations of Polanyi and Pattee, though, help us to notice the underlying significance of such a concept to challenge the conventional view of causality: it makes room for a more pluralistic perspective, accepting as genuine input diverse forms of interaction among levels of organization.

5. Conclusions

In summary, it has been observed that some current philosophical approaches that are not rigidly reductionist tend nevertheless to downplay the significance of interlevel relations in biological and complex systems. In each case, the particulars of the rationale for such position may differ, but a conjunction of three related reasons constitute a common ground underlying most of them. These three reasons are (1) a partial disregard of the contribution of hierarchical approaches that (2) facilitates the perseverance of a compositional view that takes organization for granted and that (3) does not compel us to challenge the intimidation of potential causal overdetermination against top-down causal accounts.

To better frame the challenge, let us recall that Joseph Needham had already reminded us that understanding organization is the key concern of biological inquiry:

> Recognition of the objectivity and importance of organizing relations had always been an empirical necessity, forced upon biologists by the very subject matter of their science, but the issue was always confused by their inability to distinguish between the organization of the living system and its supposed anima. With the abolition of souls and vital forces the genuine organizing relations in the organism could become the object of scientific

study. . . . Today we are perfectly clear . . . that the organization of living systems is the problem, not the axiomatic starting point, of biological research. Organizing relations exist, but they are not immune from scientific grasp and understanding. On the other hand, their laws are not likely to be reducible to the laws governing the behavior of molecules at lower levels of complexity." (Needham 1937, 15–16 [1943, 242–43])

The articulated view presented in this chapter is an attempt to unfold the implicit content of the notion of level of organization as it is currently used in various areas of the biological and complexity sciences. Thus it is claimed that, in hierarchical approaches, *level of organization* designates a very definite form of compositional level that requires further conditions that specify the kind of composition involved. Those additional specifications of the peculiarities of the relation of composition include the more restricted relations of integration, emergence, and regulation among components that, collectively, give rise to the inclusive relation of organization as such.

The articulation through those five kinds of progressively restrictive relations among elements (composition, integration, emergence, regulation, organization) that give rise to kinds of levels allows us to distinguish between different perspectives and definitions of level without confronting them. Depending on the subject of inquiry, an intermediate kind of relation among those might be sufficient. Only the specification order needs to be preserved: each kind implies the less specific relations but not necessarily those that are more restrictive. Thus all of them imply relations of composition of some sort and, for instance, emergence or regulation must also entail integration. It is only when biological organization is involved that all kinds of relations are required. Conversely, the relation of composition does not imply any others and, hence, it is too weak, by itself, to account for properties or behaviors that depend on organization.

This is evident even in those approaches that are reluctant to admit stronger kinds of interlevel relations because their proponents are also compelled to specify that their explanations and models deal with "organized collections of components" (Craver and Bechtel 2007) and not with just "haphazard parts" (Craver 2008). In other words, the organized nature of those collections of components in biological systems does not derive from their sheer condition of being a component part but from something else that is more specific. According to the articulated view, this "something" is the relations of integration, emergence, and regulation: a

group of components acquires the condition of organized if its constitution (composition) involves some instance of those additional kinds of
relations.

The actual material implementation of this organizational disposition
is due to the action of some form of constraint or boundary condition over
the potential components or collections of them. As pointed out in the
previous section, reliance on constraints entails a thorough consideration
of the interaction among levels and addressing the potentially causal nature of such interactions, both bottom-up and top-down.

Acknowledgments

This chapter is based on earlier versions presented as oral communications
in the following conferences: ISPHSSB '11, CLMPS '11, SLMFCE '12.
The author wishes to thank the editors and an anonymous reviewer for
their help in improving this text. Funding for this work was provided by
grants GV/EJ IT590–13 from the Government of the Basque Country,
and FFI2011–25665 from the Ministerio de Economía y Competitivad
(MEC) and FEDER funds from the EC.

Notes

1. It was originally stated by Donald T. Campbell in 1974 in reference to the
action of natural selection: "(Downward causation) Where natural selection operates through life and death at a higher level of organisation, the laws of the higher
level selective system determine in part the distribution of lower level events and
substances. Description of an intermediate-level phenomenon is not completed
by describing its possibility and implementation in lower level terms. Its presence,
prevalence, or distribution (all needed for the complete explanation of biological
phenomena) will often require reference to laws at a higher level of organization
as well. Paraphrasing Point 1, for biology, all processes at the lower levels of a hierarchy are restrained by, and act in conformity to, the laws of the higher levels"
(Campbell 1974, 180).

2. Deflationary: Depriving the concept of substantive epistemological content,
claiming that it does not add anything in explanatory terms.

3. Eliminativism: A position that defends the convenience to eliminate from
our discourse what is considered to be a superfluous (or erroneous) entity or concept and to replace it with a more basic and fundamental one.

4. To mention a few, see Allen and Starr 1982, Eldredge 1985, O'Neill et al. 1986, Pattee 1973, Salthe 1985, Weiss 1971, Whyte et al. 1969.

5. Scale as a generic graded measure of some basic variable (size, time, rate, etc.).

6. "To be aggregative, the system property would have to depend upon the parts' properties in a very strongly atomistic manner, under all physically possible decompositions. . . . Aggregativity is the complete antithesis of functional organization" (Wimsatt 2000, 272; Wimsatt 2006, 675).

7. A very different issue, which is not going to be addressed here, is the general philosophical discussion regarding the ontological or merely epistemological status of levels of organization together with the more technical one about how accurately we can succeed in demarcating them (in "carving nature at its joints"). For instance, Allen and Starr (1982) take an explicitly epistemological stance whereas Salthe (1985) tends toward a more ontological one.

8. The issues involved in the understanding of these two last kinds of relations among levels, *emergence* and *regulation* (or *control*), are so wide, deep, and controversial that they deserve and require a full treatment that doesn't fit within the limits of this chapter. For now, the point is to stress the necessity to include them when dealing with any version of levels of organization, although here the analysis is limited to elaborate the more approachable but currently highly relevant relation of *composition*.

9. Overdetermination: An event is overdetermined if there exists more than one antecedent event, any of which would be a sufficient condition for the event occurring (Blackburn 1994, 274).

10. Holonomic constraints are auxiliary conditions that limit permanently the number of degrees of freedom of a system and are, therefore, the basis for structural hierarchies, while nonholonomic constraints are variable auxiliary conditions that limit in time the number of degrees of freedom of the system, being the basis for the functional hierarchies typical of living systems. The latter are dynamical structures that establish time-dependent relations among degrees of freedom but introduce a different temporal scale.

References

Ahl, Valerie, and Timothy F. H. Allen. 1996. *Hierarchy Theory. A Vision, Vocabulary, and Epistemology.* New York: Columbia University Press.

Allen, Timothy F. H., and Thomas B. Starr. 1982. *Hierarchy. Perspectives for Ecological Complexity.* Chicago: University of Chicago Press.

Ayala, Francisco J., and Theodosius G. Dobzhansky, eds. 1974. *Studies in the Philosophy of Biology. Reduction and Related Problems.* Berkeley: University of California Press.

Bechtel, William. 2008. *Mental Mechanisms.* London: Routledge.

Bechtel, William, and Robert C. Richardson. 1993. *Discovering Complexity. Decomposition and Localization as Strategies in Scientific Research*. Princeton, NJ: Princeton University Press.

Blackburn, Simon. 1994. *The Oxford Dictionary of Philosophy*. Oxford: Oxford University Press.

Brandon, Robert N., and Richard Burian, eds. 1984. *Genes, Organisms, Populations. Controversies over the Units of Selection*. Cambridge, MA: MIT Press.

Bunge, Mario. 1960. "Levels: A Semantical Preliminary." *Review of Metaphysics* 8, 3(51):396–406.

———. 1979. *Treatise on Basic Philosophy. Vol. 4: Ontology II: A World of Systems*. Dordrecht: Reidel.

Campbell, Donald T. 1974. " 'Downward Causation' in Hierarchically Organized Biological Systems." In *Studies in the Philosophy of Biology*, edited by Francisco J. Ayala and Theodosius G. Dobzhansky, 179–186. Berkeley: University of California Press.

Collier, John. 1989. "Supervenience and Reduction in Biological Hierarchies." *Canadian Journal of Philosophy* 14:209–34.

Craver, Carl F. 2007. *Explaining the Brain*. Oxford: Clarendon.

———. 2009. "Levels of Mechanisms: A Field Guide to the Hierarchical Structure of the World." In *The Routledge Companion to Philosophy of Psychology*, edited by John Symons and Paco Calvo, 387–99. London: Routledge.

———. 2015. "Levels." In *Open MIND: 8(T)*, edited by Thomas Metzinger and Jennifer M. Windt, 1–26. Frankfurt: Mind Group.

Craver, Carl F., and William Bechtel. 2007. "Top-Down Causation without Top-Down Causes." *Biology and Philosophy* 22:547–63.

Craver, Carl F., and Lindley Darden. 2013. *In Search of Mechanisms. Discoveries across the Life Sciences*. Chicago: University of Chicago Press.

Dobson, Christopher M. 2003. "Protein Folding and Misfolding." *Nature* 426:884–90.

Dupré, John. 2012. *Processes of Life: Essays in the Philosophy of Biology*. Oxford: Oxford University Press.

Eldredge, Niles. 1985. *Unfinished Synthesis. Biological Hierarchies and Modern Evolutionary Thought*. New York: Oxford University Press.

Eldredge, Niles, and Stanley Salthe. 1984. "Hierarchy and Evolution." In *Oxford Surveys in Evolutionary Biology*, edited by Richard Dawkins and M. Ridley, 1:182–206. Oxford: Oxford University Press

Eronen, Markus I. 2013. "No Levels, No Problems: Downward Causation in Neuroscience." *Philosophy of Science* 80(5):1042–52.

———. 2015. Levels of Organization: A Deflationary Account." *Biology and Philosophy* 30(1):39–58.

Etxeberria, Arantza, and Juan Umerez. 2006. "Organismo y organización en la Biología Teórica: ¿Vuelta al organicismo?" *Ludus Vitalis* XIV (26):3–38.

———. 2013. "Organization." In *Encyclopedia of Systems Biology*, edited by Werner Dubitzky, Olaf Wolkenhauer, Hiroki Yokota, and Kwang-Hyun Cho, 1,612–15. New York: Springer.

Feibleman, James K. 1954. "Theory of Integrative Levels." *British Journal for the Philosophy of Science* 5(17):59–66.

Grene, Marjorie. 1987. "Hierarchies in Biology." *American Scientist* 75:504–10.

Haraway, Donna J. 1976. *Crystals, Fabrics and Fields: Metaphors of Organicism in Twentieth Century Developmental Biology*. New Haven, CT: Yale University Press.

Kim, Jaegwon. 1993. *Supervenience and Mind: Selected Philosophical Essays*. Cambridge: Cambridge University Press.

Love, Alan. 2012. "Hierarchy, Causation and Explanation: Ubiquity, Locality and Pluralism." *Interface Focus* 2:115–25.

Machamer, Peter, Lindley Darden, and Carl F. Craver. 2000. "Thinking about Mechanisms." *Philosophy of Science* 67:1–25.

Mitchell, Sandra D. 2009. *Unsimple Truths. Science, Complexity, and Policy*. Chicago: University of Chicago Press.

Okasha, Samir. 2006. *Evolution and the Levels of Selection*. Oxford: Oxford University Press.

O'Neill, Robert V., Donald L. De Angelis, J. B. Waide, Garland E. Allen. 1986. *A Hierarchical Concept of Ecosystems*. Princeton, NJ: Princeton University Press.

Oppenheim, Paul, and Hilary Putnam. 1958. "Unity of Science as a Working Hypothesis." In *Concepts, Theories and the Mind-Body Problem. Minnesota Studies in the Philosophy of Science*, vol. 2, edited by Herbert Feigl, Michael Scriven, and Grover Maxwell, 3–36. Minneapolis, MN: University of Minnesota Press.

Needham, Joseph. 1937. *Integrative Levels: A Revaluation of the Idea of Progress*. Oxford: Clarendon Press.

Pattee, Howard H. 1966. "Physical Theories, Automata and the Origin of Life." In *Natural Automata and Useful Simulations*, edited by Howard Pattee, E. Edelsack, L. Fein, A. Callahan, 73–104. Washington, DC: Spartan.

———. 1967. "Quantum Mechanics, Heredity and the Origin of Life." *Journal of Theoretical Biology* 17:410–20.

———. 1968. "The Physical Basis of Coding and Reliability in Biological Evolution." In *Towards a Theoretical Biology 1, Prolegomena*, edited by Conrad H. Waddington, 67–93. Edinburgh: Edinburgh University Press.

———. 1969a. "Physical Problems of Heredity and Evolution." In *Towards a Theoretical Biology 2, Sketches*, edited by Conrad H. Waddington, 268–84. Edinburgh: Edinburgh University Press.

———. 1969b. "Physical Conditions for Primitive Functional Hierarchies." In *Hierarchical Structures*, edited by Lancelot L. Whyte, Albert G. Wilson, and Donna Wilson, 161–77. New York: Elsevier.

———. 1972. "Laws and Constraints, Symbols and Languages." In *Towards a Theoretical Biology 4, Essays*, edited by Conrad H. Waddington, 248–58. Edinburgh: Edinburgh University Press.

———. 1973. "The Physical Basis and Origin of Hierarchical Control." In *Hierarchy Theory: The Challenge of Complex Systems*, edited by Howard H. Pattee, 71–108. New York: G. Braziller.

———. 2012. *Laws, Language and Life*, edited by Joanna Raczaszek-Leonardi. Dordrecht: Springer.

Polanyi, Michael 1968. "Life's Irreducible Structure." *Science* 160: 1,308–12.

Redfield, Robert, ed. 1942. *Levels of Integration in Biological and Social Systems*. Biological Symposia, vol. 8. Lancaster, PA: Jaques Catell.

Salthe, Stanley N. 1985. *Evolving Hierarchical Systems: Their Structure and Representation*. New York: Columbia University Press.

———. 1993. *Development and Evolution: Complexity and Change in Biology*. Cambridge, MA: MIT Press.

Simon, Herbert A. 1962. "The Architecture of Complexity." *Proceedings of the American Philosophical Society* 106:467–82.

Tëmkin, Ilya, and Niles Eldredge. 2015. "Networks and Hierarchies: Approaching Complexity in Evolutionary Theory." In *Macroevolution. Explanation, Interpretation and Evidence*, edited by Emanuele Serrelli and Nathalie Gontier, 183–226. Dordrecht: Springer.

Umerez, Jon. 1994. *Jerarquías Autónomas. Un estudio sobre el origen y la naturaleza de los procesos de control y de formación de niveles en sistemas naturales complejos*. PhD diss., University of the Basque Country (UPV/EHU).

———. 2001. H. "Pattee's Theoretical Biology. A Radical Epistemological Stance to Approach Life, Evolution and Complexity." *BioSystems* 60(1/3):159–77.

———. 2009. "Where Does Pattee's 'How Does a Molecule Become a Message?' Belong in the History of Biosemiotics?" *Biosemiotics* 2(3):269–90.

Umerez, Jon, and Alvaro Moreno. 1995. "Origin of Life as the First MST. Control Hierarchies and Interlevel Relation." *World Futures* 45(2):139–54.

Umerez, Jon, and Matteo Mossio. 2013. "Constraint." In *Encyclopedia of Systems Biology*, edited by Werner Dubitzky, Olaf Wolkenhauer, Hiroki Yokota, Kwang-Hyun Cho, 490–93. New York: Springer.

Vicente, Agustin. 2013. "Where to Look for Emergent Properties." *International Studies in the Philosophy of Science* 27(2):137–56.

Vrba, Elisabeth S., and Niles Eldredge. 1984. "Individuals, Hierarchies and Processes: Towards a More Complete Evolutionary Theory." *Paleobiology* 10(2): 146–71.

Weiss, Paul A., ed. 1971. *Hierarchically Organized Systems in Theory and Practice*. New York: Hafner.

Whyte, Lancelot L., Albert G. Wilson, and Donna Wilson, eds. 1969. *Hierarchical Structures*. New York: Elsevier.

Wimsatt, William. 1974. "Complexity and Organization." In *PSA 1972*, edited by Kenneth F. Schaffner and Robert S. Cohen, 67–86. Dordrecht: Reidel.

———. 1976. "Reductionism, Levels of Organization, and the Mind-Body Problem." In *Conciousness and the Brain*, edited by Gordon G. Globus, Grover Maxwell and Irwin Savodnik, 199–267. New York: Plenum.

———. 1994. "The Ontology of Complex Systems: Levels, Perspectives, and Causal Tickets." In *Biology and Society: Reflections on Methodology. Canadian Journal of Philosophy*, supp., edited by Mohan Matthen and Robert Ware, 20: 207–74.

———. 2000. "Emergence as Non-Aggregativity and the Biases of Reductionisms." *Foundations of Science* 5:269–97.

———. 2006. Aggregate, Composed, and Evolved Systems: Reductionistic Heuristics as Means to More Holistic Theories." *Biology and Philosophy* 21(5): 667–702.

———. 2007. *Re-Engineering Philosophy for Limited Beings*. Cambridge, MA: Harvard University Press.

CHAPTER 4

Hierarchy: The Source of Teleology in Evolution

Daniel W. McShea

"Zebras evolved stripes in order to discourage biting flies." Evolutionists everywhere cringe. Language like that is forbidden in evolutionary discourse. We do not allow teleological talk, the "in order to" in that first sentence. Such language suggests that the evolution of zebra stripes was goal-directed, purposeful. Teleological language seems to imply that zebras in their evolution somehow sought out an end-state—deterring biting flies—or worse, that the end-state of deterring biting flies was the cause of the changes tending that way. Both are impossible. In our present worldview, species do not seek. Their present behavior is not guided by some vision of the future, for they have none. Nor does the future itself guide them. End-states do not reach back in time and cause evolutionary change. The future does not cause the past.

Still, there is a reason that we are tempted this way, a reason why we slip so easily into thinking about adaptive evolution in teleological terms. It is that the close fit of organism to environment seems to demand purpose. After Darwin, we think we know better, of course. We know that blind variation and selective retention creates the appearance of purpose. What we might say about the zebra is that in some ancestral population of unstriped or less-striped zebra ancestors, the variants with more striping were better able to deter flies (Caro et al. 2014), suffered fewer bites, perhaps contracted fewer diseases, and therefore left more surviving offspring, a process that was repeated over many generations leading to a descendant species with greater striping. Or something like that. Apparent teleology is explained by blind mechanism.

Oddly, and a little ironically, however, it turns out that our original pre-Darwinian intuitions were not far off. There is, I argue, a structural similarity between the process of adaptive evolution and all the other processes we commonly think of as teleological. Simple organismal tropisms, physiological homeostasis, the movement and patterning of cells in development, and even the behavior of human-made, goal-directed devices—all of these seemingly teleological systems are structured hierarchically. They are nested physically. They consist of a small thing that moves or changes inside a stable big thing, inside a field of some kind. In all of these systems, the apparent seeking behavior of the small thing is guided hierarchically, from the top down. The upper-level field directs the lower-level small thing moving within it. Likewise for evolving species. The species lineage leading to modern zebras is a small thing relative to its environment—in other words, relative to the environment within which it changes. That environment, that context, is a big thing. It is a kind of ecological field. And to the extent that a species evolves adaptively, it is guided from above by that ecological field.

In what follows, I explain why seemingly teleological systems are structured hierarchically in this way. I also argue that the teleological view of adaptive evolution has certain payoffs, and that it unsettles—in intriguing ways—our thinking about a process we imagine to be already well understood. For one thing, it unifies a wide range of biological processes, bringing together seemingly disparate aspects of behavior, physiology, development, and adaptation into a single explanatory scheme. For another, it provides a novel and expansive view of the process of adaptation, one that relegates natural selection to the role of mechanism, at least in principle replaceable by other mechanisms. And third, it highlights certain requirements for adaptive evolution, especially variation and environmental constancy, requirements that have been long recognized but have been underappreciated. In the teleological view, variation and environmental constancy emerge as special cases of requirements for teleological systems generally—namely, lower-level freedom and upper-level stability.

I am using the word *hierarchy* here in its structural sense—that is, to refer to nested physical objects, things within things. In my usage, hierarchy overlaps strongly with terms like integrative level (Feibleman 1954; Haber 1994), ecological hierarchy (Eldredge and Salthe 1984), scalar hierarchy (Salthe 1985, 2009), constitutive hierarchy (Valentine and May 1996), and compositional hierarchy (Wimsatt 1994). Importantly, however, hierarchy in my sense does not imply any functional relationship between parts and the whole, nor does it imply that wholes are emergent in any sense. A

helium-filled balloon, consisting of gas molecules plus the plastic skin that
encloses them, is a hierarchical system. So is a pile of rocks, although not
a very interesting one (see below). Also, hierarchy here does not include
hierarchies in time, in which one entity gives rise to several, which in turn
give rise to more, and so on. In other words, genealogical hierarchies are
excluded. Also excluded are specification hierarchies, such as postal addresses, and command or control hierarchies, such as military chains of command. In none of these is there necessarily any physical nesting of objects
within objects. (See Zylstra [1992] for a useful discussion of the distinctions
among the various aspects of hierarchy.) This is not to say that structural hierarchies cannot also have control-hierarchy features and vice versa—just
that control is not built into the concept of hierarchy that I'm using here.

Teleology and Hierarchy

One signature of all seemingly teleological systems is their persistence
(Sommerhoff 1950; Nagel 1979). A homing torpedo launched at a target ship shows persistence. If it deviates from the path toward the target
ship—say, if it momentarily detects a passing pod of whales—it returns to
a trajectory toward the ship. If a bacterium climbing a food-concentration
gradient deviates, say when local currents drag it to lower concentration,
it returns to an up-gradient trajectory. In human physiology, when blood
pressure falls too low, the secretion of renin induces angiotensin, which
constricts blood vessels, restoring the pressure. In a developing sea urchin embryo, primary mesenchyme cells migrate from the south pole of
the blastula to the subequatorial region, where they merge and secrete
the larval skeleton. If displaced from their trajectory toward the equator,
say by an experimenter, the cells return to that trajectory. Deviation and
return, error and correction, create the impression of an entity or a variable that is directed, headed somewhere, purposeful. It persists.

A related signature property of teleology is plasticity, the ability to find
a given trajectory from a wide range of alternative starting points (Sommerhoff 1950; Nagel 1979). The torpedo can find a trajectory toward the
target ship from any starting point within the sound field emanating from
the ship. The primary mesenchyme cells of the sea urchin embryo can find
their trajectory toward the equator from virtually any starting point within
the embryo.

Any theory of teleological behavior must account for these two properties (Nagel 1979). In an earlier paper (McShea 2012), I developed a

hierarchical view of teleology, arguing that persistence and plasticity in a biological entity are the result of what I call upper direction—that is, direction by a higher-level field within which the entity is immersed. The bacterium is upper directed in that its movements are directed by the food gradient within which it is immersed. The bacterium's persistence and plasticity are explained by the fact that the field is present over a large area, so that wherever the bacterium wanders, or wherever it starts, the field is there, directing the bacterium to an up-gradient trajectory. Direction, in other words, comes from something that is larger than and envelops the teleological entity (see also Feibleman 1954). In the human example, the persistence and plasticity of blood pressure is directed by a larger structure, an organ system that includes the kidneys and the blood vessels of the circulatory system. In the sea urchin, the persistence and plasticity of the primary mesenchyme cells is directed by a large-scale developmental structure, probably a gene activation field or a morphogen field (Ettensohn 1990; Ettensohn and McClay 1988; see Weiss 1971 and Korn 2002 on fields in development generally). The homing torpedo is upper directed in that it is guided by the sound field emanating from the target ship, a field that is larger than the torpedo and within which the torpedo is immersed.

I intend nothing mysterious by this notion of a field. Fields are physical structures. Their action on the entities they contain is direct and local, but fields can be present over a wide area. In some cases, the size of the field is evident only when the pattern of local effects is observed at larger spatial or temporal scales. For a single bacterium, the field would be evident if the bacterium deviated frequently from an up-gradient trajectory, so that its persistent behavior could be observed over a large spatial range. Or we could release a thousand bacteria into the gradient all at once and see the field instantaneously in the plastic and persistent behavior of the entire population. Again, fields as I understand them are physical—not idealizations, not abstractions, and not in any way transcendent. Or at least, they are no more transcendent than an ordinary gravitational field, or for that matter, a field of corn.

What about causation? I chose the phrase upper direction in order to distance this discussion from (what I see as) the unproductive debate in the philosophy of science over "downward causation." The debate is between reductionists who claim that all causation is necessarily lower level and antireductionists who leave some scope for higher-level causation. The tone of the argument here is antireductionist, but reductionists can rest assured that nothing mysteriously emergentist is being invoked. The causal processes at work in upper direction are no different from the ones routinely invoked in everyday explanation. Consider these examples of upper direction. (None are framed as teleological, although they could easily be modified to make them so.) The gas molecules in a round balloon are directed from above by the plastic skin of the balloon to remain within a fixed radius of the center. The movements of rats in a maze are directed by the walls of the maze, forcing the animals to stay inside the maze and restricting their movements to the walled paths. Cars are directed by the spatial pattern of the roads. National interest rates on home mortgages direct the decisions of individuals thinking about purchasing or selling homes, inducing or discouraging buying or selling. The balloon skin, the walls of the maze, the pattern of the roads, the national interest rate—all of these are fields, spatially distributed physical structures that direct the entities contained within them (gas molecules, rats, cars, and potential home buyers/sellers).

This notion of a field as directing lower-level behavior is a fairly commonplace one, although the language used varies enormously. In some cases, such as the bacterium pursuing food, we ordinarily call the field providing upper-level direction a gradient. In other cases, such as the rat in a maze, the field is a boundary, establishing limits on change or movement. In both cases, they might be described as upper-level "constraints" on the entities contained within them (Eldredge and Salthe 1984; Salthe 1985). Fields can also be thought of in some cases as contexts, which guide or channel their contained entities. In still other cases, such as the effect of interest rates, fields might be biases on the direction of change occurring within them. Here I use all these words as near synonyms. Fields can be gradients, biases, boundaries, constraints, or contexts, depending on the situation. I realize that I am violating some norms of usage here in using the words "field" and "upper direction" to cover concepts with very different connotations, some implying an active tendency and others a passive one, some implying limitation and others a driving force. For present purposes, however, the perhaps jarring effect of this violation is useful if it draws attention to the two features that all these different systems share: containment and the causal power that containment makes possible. In other words, it draws attention to hierarchy.

Notice that in many of the examples given, the fields are simple. The food field is a simple gradient. But an organ system governing blood pressure is a more complex structure. As will be seen, the ecological fields that direct evolutionary change may be more complex yet. The point is that the appearance of teleological behavior is not a function of the complexity of the field. Rather, it flows from the hierarchical relationship between the field and the contained entity.

So far, the analysis has considered only two levels, the field and the contained entity. But as Salthe (1985, 2009) and others have pointed out, complete causal stories for hierarchical systems typically require at least three levels: the focal entity, the next higher level, and the next lower level. Here the teleological entity lives at the focal level, the upper-level structure is the field, and the lower level is the domain of what we call mechanism (Feibleman 1954). Mechanisms play a critical causal role in teleological behavior. The behavior of a bacterium in a food field, or an individual in a housing market, is partly the result of its internal mechanisms. For the bacterium, these are its signal-transduction mechanisms. For the human individual, these are the motivations and other mental mechanisms underlying home-buying decisions. Lower-level mechanisms explain why and how the entity is affected by the field. I downplay these mechanisms here because in conventional explanations in biology they are commonly the sole focus, with the existence and causal necessity of upper-level structure taken for granted. I shift the focus in part to redress the imbalance.

Two Perhaps Obvious Further Requirements for Teleological Behavior

The first is stability. The upper-level field must be stable, or more precisely, it must be roughly constant on a timescale that is long relative to change or movement in the teleological entity. And it must be constant on a spatial scale that is large relative to the movements of the entity. If the food gradient varies greatly in space, the bacterium will become "confused," so to speak—unable to discern which way the gradient increases—and its trajectory will not be persistent. If a morphogen gradient is changing quickly, cells that are guided by it will not find their target locations.

The second requirement is freedom. The teleological entity must be at least somewhat free to change or move independently of the upper-level field. Persistence is error and correction, and an entity that has no freedom will make no errors and therefore will not appear to persist. If I throw a rock, a molecule within the rock rigidly follows the trajectory of the rock. The molecule is directed from above, by the rock of which it is a part, but in this case—as in all solid objects—the direction is too complete, too perfect, to support the appearance of teleology. Lower-level freedom is essential for teleological behavior.

A Near Requirement

In all the examples of teleological systems I discuss, including the evolutionary ones in the next section, the field is spatially continuous. The guiding field in the embryo occupies a discrete portion of space, and the primary mesenchyme cells behave teleologically within that space. As Campbell (1958), Simon (1962), and other systems thinkers have recognized, spatial continuity is what makes possible some of the most interesting properties of hierarchical systems. And it would seem to be important for teleology. At least, it is easy to see how a spatially continuous field can arise naturally and how it can provide direction wherever (within the continuous space) the teleological entity happens to wander. Still, spatial continuity is not an absolute in-principle requirement. One can imagine a field that is spatially dispersed, such as the web, which exerts its influence on teleological entities (people) through widely scattered web-connected devices.

Three Perhaps Necessary "Course Corrections"

First, there is no such thing, in our current worldview, as a system that is literally "goal-directed." No behavior is literally directed by something in the future, and from the perspective of a teleological entity, goals exist only in the future. If they were present for the entity now, they would not be goals. Concretely, the homing torpedo is not directed by the actual target ship. It cannot be, because the target ship is not present for it. The ship lies only potentially in its future. Rather, the torpedo is directed by a sound field emanating from the target ship, a field that is present for the torpedo right now and at every moment. Even human goal directedness is entirely governed by present motivations and ideas. (I have proposed elsewhere that the motivations driving seeming goal-directed behavior in us are brain states that are structured as fields of some kind [McShea 2012, 2013].) When we act purposively, we are governed by imagined or anticipated futures that are present in our minds right now. The actual future never causes anything in the actual present. It is for this reason that I modify the word *teleology* with *apparently* and *seemingly*. These words are not intended to imply that teleological behavior is somehow unreal. Persistence and plasticity are real and require explanation. *Apparently* and *seemingly* are simply an acknowledgment of the impossibility of future causation.

Second, it should be obvious, but perhaps needs saying anyway, that not all upper-level structures are capable of upper direction, and therefore not all hierarchically embedded entities behave teleologically. A pile of rocks is a higher-level structure, one that consists of the rocks that constitute it, but the pile cannot provide much direction to the contained rocks. One reason is that the pile does not have much integrity. The forces that hold the pile together are far too weak to provide upper direction, as would be obvious if, for example, one tried to move the pile as a whole. The whole thing comes apart. The parts, the rocks, would show no persistence. Their trajectories would not track the pile very well. The point is simply that many (most?) hierarchical systems are not especially teleological.

Finally, one might object to the whole line of argument here, pointing out that among the hierarchically structured systems that do show persistence and plasticity, some are not what we would call teleological. For example, a ball released from the lip of a bowl will find a trajectory toward the bottom from any point on the inside surface of the bowl. Further, once released, if it is displaced—by, say, a nudge from a finger—it returns to a trajectory toward the bottom. The ball is persistent and plastic, and it is directed by a larger field, gravity, within which it is immersed. But few would call its behavior teleological. One way to respond to this objection would be to tweak the conditions, requiring that the system in question attain a certain level of complexity—that they be difficult to understand—to qualify as teleological (McShea 2012), a threshold that the ball in the bowl does not reach. Along the same lines, I could argue that teleology is a matter of degree, that the bacterium in the food gradient is highly teleological while the ball in the bowl is just barely teleological. Or I could simply point out that providing necessary and sufficient conditions for teleology is not the mission here. This is not a project in analytical philosophy. It is not an explication of our usage of a term. Rather, it is an engineering analysis, in the style of Wimsatt (2007), intended to reveal how seemingly teleological systems actually work. And if—in addition to explaining the systems we are inclined to call teleological—that analysis also happens to cover some systems that do not move us to use that word, so be it. Perhaps we need to revise our language.

Adaptation and Hierarchy

For evolution, the hierarchical view invites us to take a top-down perspective—to look down, so to speak, from high above the species, with geography and time collapsed onto a scale suited to our imaginative capacities. From this perspective, looking down on a mid-Cenozoic horse lineage as it traverses the last forty million years, we see body size increasing and number of toes decreasing, molded by a chilling climate and the transformation of the animal's habitat from forest to open plains. This view invites us to see the ecological field, the combined biotic and abiotic context within which the lineage evolves, as a set of forces, or pressures, that act on the phenotype. Permitting myself considerable license with language, the ecological field is a pair of hands that envelopes each species, delicately shaping the phenotype—to the limited extent allowed by internal constraints and stochasticity—in a way that supports survival and reproduction. Nothing in this view contradicts the standard organism-centered or population-centered view of natural selection. It is simply a higher-level view of the same process.

I will try to make the contrast in perspectives more concrete. Consider the evolution of neck folding in turtles. In modern cryptodires and pleurodires, the neck has two hinged connections, one between each of two adjacent pairs of neck vertebrae, enabling them to fold the neck and pull the head under the protective umbrella of the shell. Mesozoic amphichelydean turtles did not have these hinges or this capacity (Rosenzweig and McCord 1991). From the standard perspective, we would say that in some ancient population of non–neck folders, random variation produced some individuals with proto-neck-folding capability, and because this capability provided some advantage, say some protection from predators, the variant individuals were more likely to survive and reproduce. Notice that even from this lower-level viewpoint, the persistence of the trajectory of the population would be evident. From generation to generation, random variation constantly threatens to carry the population off track, away from proto–neck folding. But the differential survival of individuals with proto-neck-folding capability brings the population back to an adaptive trajectory. (Notice that my use of the word *persistence* is different from the usual one in evolutionary discourse. Here it means error and correction [in the sense of Sommerhoff and Nagel], the signature of teleology, not endurance or mere survivorship.)

From an upper-level perspective, persistence is even clearer. Let us represent an evolving species as a time-series of points in a phenotype space. Also within the space, vectors representing the selective forces—for example, predation—imposed on the species by the environment point toward a neck-folding phenotype. With the passage of millions of years, the points become a line, roughly following the vectors and moving toward a local adaptive peak, neck folding. I say roughly because the intensity of

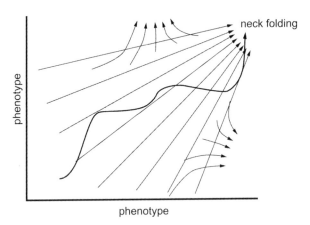

neck folding

phenotype

phenotype

FIGURE 4.1 A turtle species evolving under the direction of an ecological field that drives it generally in the direction of neck folding. See text for discussion.

predation doubtless varies a bit in time and space, and the phenotype is in any case prone to drift. So what we actually see is not a straight line from non–neck folding to neck folding but a trajectory with series of a deviations and corrections. We see persistence. And the cause, the source of overall direction, is the predator-containing ecological context within which the species is moving (figure 4.1). The size and stability of the upper-level ecological field are both critical for persistence. The field must extend over a large portion of the accessible phenotype space. In other words, neck folding must be advantageous over a large-range of variants. (Notice that while the representation here of the phenotype space is abstract, the interaction between phenotype and field—between a real organism and its ecology—is real and physical, not at all abstract.) In addition, neck folding must be advantageous, on average, over a sizable chunk of the physical space over which the population might wander, or over which it might propagate. Finally, the ecological field must be stable in time. Neck folding needs to be stably advantageous, on average, over the entire course of the adaptive process. If the source of advantage is predation, then predation pressure needs to be present most of the time. Of course, no ecological field is perfectly constant. More properly, the requirement is for stochastic constancy, leaving considerable latitude for random variation.

The nice thing about the turtle example is that turtles as a group also show plasticity. (Notice that *plasticity* is being used here in the Sommerhoff-Nagel sense—entities adopting similar trajectories from multiple alternative starting points—not in the usual biological sense, to refer to ecophenotypy.) The paleontological evidence indicates that neck folding evolved at least two times independently, in the cryptodires and in the pleurodires, taking somewhat different paths in each group (Rosenzweig and McCord 1991). The pleurodires fold their necks in a horizontal plane, while cryptodires fold them in a vertical plane, with the hinges located at different places along the neck, between different adjacent vertebral pairs, in the two groups. Still, the adaptations are convergent in the sense that they confer the same neck-folding capability. In evolution, teleological plasticity is convergence.

The upper-level view reveals a species directed, persistently and plastically, by its ecological context, by the ecological field within which it moves. The view emphasizes the causal role of the field. In contrast, the conventional lower-level view emphasizes the causal properties of the evolving organism or population—gene frequencies, developmental constraints, and the mechanistic details of the organism-environment interaction. Obviously those properties are important. Many terrestrial vertebrates with tough integuments evolved in an ecological field similar to the one in which turtles evolved, but they did not evolve neck folding. (Sadly, I should say, because it is hard to imagine a more endearing phenotype than a neck-folding baby hedgehog or pangolin.) Something in the genetic structure, the anatomy, the life history, and the developmental constraints and potentialities of turtles leant itself to variation of the appropriate sort to produce neck folding. And these factors are just as important in explaining the species' trajectory as the bacterial signal transduction pathways are in explaining the bacterium's trajectory. What the top-down view does is reduce all these factors to the level of mechanism. And as we zoom out, the mechanistic details disappear from view, leaving us with biological entities, species, with certain variational properties, moving in space and time and guided by the ecological fields within which they move. The details of the underlying mechanism are not ignored exactly, just black-boxed (Odum 1971; Valentine and May 1996).

The argument has so far been focused at the level of an evolving species and the evolution of adaptations at the level of phenotype. But the question naturally arises whether the same reasoning applies at larger scales, to macroevolutionary entities like clades and to properties such as speciation and extinction rates. At present, I don't see why not. The argument should be completely general, applying to evolving entities at all scales. Long-timescale entities also move and change within ecologies, and these ecologies have certain features that are stochastically constant, stable, on those timescales, and therefore ought to be able to provide consistent upper direction. The Paleozoic Era had an average climate, one that was warmer—for example—than the average climate of the Cenozoic. Ecological contexts become more general, with parameters that are defined more broadly, as the timescale increases. And they become perhaps harder to imagine, to think about in concrete terms. But there is no reason to think they become less real, less able to act causally on the entities they contain.

Importantly, nothing in this view requires that upper-level control be absolute. It is true that the walls of the maze absolutely prevent the rats from taking shortcuts or from wandering out of the maze altogether. But the direction that ecological fields provide to species is probabilistic. The field probabilistically biases the direction of species movement without precisely determining it, just as the gradient probabilistically biases the movement of the bacterium or the interest rate probabilistically biases the decision of the potential home buyer. In figure 4.1, the two smaller fields on the side are a recognition of the probabilistic nature of adaptation, of the fact that other adaptive pathways were undoubtedly available and neck folding was not inevitable.

Finally, nothing in the upper-level view implies that selection pressures are somehow organism independent. The ecological field that molds a species is a function of the properties of the species being molded, just as the fields that affect a moving particle are a function of the properties of the particle. An uncharged particle is not moved by an electric field. A turtle with no shell evolving in a predation-intensive environment would be under no pressure to evolve neck folding. A shell-less turtle is a different kind of turtle, with different ecologically relevant properties, and it therefore experiences a different ecological field, which presumably would take it in a different direction, perhaps toward greater speed afoot. And of course, as the phenotype changes, so do the relevant dimensions of the ecological field. As very small organisms become large, for example, gravity joins the ecological field, becoming relevant in ways that it is not for very small organisms. Thus as a species changes, the ecological field it experiences changes.

Adaptation and Teleology

There is a deep connection, I argue, between adaptive evolution and other teleological processes, including goal-directed behavior, physiology, and development in organisms. And the connection is not merely linguistic or metaphorical; it is structural. All these teleological processes have the same general physical structure—a smaller lower-level entity directed by a larger enveloping field. And this is the main virtue of the hierarchical view, the conceptual unification it achieves.

In addition to revealing a structural commonality among these disparate systems, this view also makes sense of apparent teleology. It pulls aside the wizard's curtain, revealing how the magic of seeming future causation works. It shows how hierarchically structured systems create the appearance of being directed by their outcomes—how the bacterium swimming up a food gradient can seem to be directed by a food source that it has not yet found (and may never find), how an acorn can appear to be guided in its development by the oak tree that it will become (but is not yet), how the striping of zebras can appear to evolve for the future purpose of deterring flies. The secret, the hierarchical view reveals, is that the end-state does not direct the teleological entity at all. What directs the entity is the field within which the entity moves, a field that is large and present for the entity at each moment, right now, as the entity is moving or changing.

The viewpoint also presents certain long-known necessary features of the adaptive process in a new light. Just as the food field must be large and stable in order to direct the bacterium, ecological fields must be large and reasonably stable in order to direct adaptation. In evolution, this is true in two senses. Ecological fields must be large and stable in space—that is, over at least the physical range of movement of the population. And finally, they must be "large" in time—in other words, roughly constant over the duration of the adaptive process. In sum, the field must be large enough to be present wherever the entity's wanderings take it and stable enough to be reliably present whenever the entity arrives there.

The complement of upper-level field stability is lower-level freedom. Just as the seemingly teleological behavior of the homing torpedo requires that it be able to move to some degree independently of the sound field of the target ship, so species must be able to move to some degree independently of the ecological field in which they are evolving. In evolution, lower-level freedom is guaranteed by the omnipresent tendency in organisms to vary. At the population level, this variation takes the form of mutation and drift. At a larger scale, it can be either drift or change under the influence of selective forces independent of the adaptation in question (McShea and Brandon 2010). Turtle necks were doubtless under selection for capacities other than folding, selection that must have deflected the evolution of folding at various times. From the perspective of the neck-folding field, these deflections also count as "errors." In any case, whatever the source of variation, a certain amount of it is necessary for persistence. To follow an error-and-correction trajectory, it must be possible to make errors. That said, it is worth noting that lower-level freedom must not be tuned too high. Too much freedom, too much variation and drift, and the ecological field would be unable to restore the species to its adaptive trajectory. Persistence would fail.

Finally, a point with uncertain consequences. It has been claimed (e.g., Rosenberg and McShea 2007) that natural selection is the only known route to adaptive evolution. Factors such as constraints and phenotypic plasticity have long been recognized as important influences on variation, and new routes to inheritance, such as epigenetic inheritance, have recently entered the evolutionary discourse. But none of these on its own is sufficient for adaptation. And so the original claim remains true: no process—other than blind variation and selective retention—is known that can produce a fit between organism and environment. The hierarchical view of adaptation does not challenge this. But it does open the door to other as-yet-unidentified possibilities. Natural selection is a lower-level mechanism. It is a mechanism by which organisms can be remodeled by the environment on evolutionary timescales. Any lower-level mechanism that allows for environmentally driven remodeling on long timescales would do. Perhaps there are self-organizing mechanisms that can produce the same sort of remodeling. In any case, what the hierarchical viewpoint reveals is that the heart of the process that creates the organism-environment fit, the factor that gives the process its teleological "feel," is the hierarchical relationship between the ecological field and the evolving species and not the particular mechanism by which the species is made malleable. In other words, adaptation in evolution does not, in principle, require natural selection. Of course, again, at present, it does, but simply because no other mechanism of malleability is known.

Conclusion

Kant famously said there would never be a Newton for a blade of grass—in other words, no one would ever provide a thoroughly mechanistic account of biological teleology (Rosenberg and McShea 2007). But, one might ask, doesn't natural selection provide just such an account, showing how all biological goal-directed systems could have arisen? Wasn't Darwin that Newton? The answer is both yes and no. Natural selection does explain the *origin* of organismal tropisms, physiological homeostasis, and the seeming goal-directedness of development. But it does not explain how these systems work, how they are able to create the appearance of goal directedness. It does not tell us how persistence and plasticity are achieved. The answer—the knockout punch following Darwin's jab at Kant—is hierarchy: large fields giving direction to smaller entities nested within them. Further, the process of natural selection is itself structured in this way. Selection is species changing within and directed by ecological fields. And it is that hierarchical structure that gives selection its teleological flavor.

This may seem like a strange inversion. We think ordinarily of selection as the source of hierarchy in biology rather than a consequence of it. For example, consider the trend in organismal hierarchy, in levels of organization, from bacterium to eukaryotic cell to multicellular individual to colony and society (McShea 2001). Conventionally, we imagine that selection is what drives these so-called major transitions (Maynard Smith and Szathmáry 1995). That might or might not be true. There are other ways that hierarchical structure can increase besides natural selection. For example, the natural aggregative tendencies of chemical systems, the slight bias toward bonding over dissociation, might be enough to power such a trend (Wicken 1987; see also Simon 1962 and Fleming 2012). But whether or not selection is the driving force, hierarchy must have come first. For one thing, higher-level units must arise before they can be selected (Fleming and Brandon 2015; Simpson 2012). For another—and what the argument here makes plain, I hope—selection itself requires hierarchy. In order to even think about selection in biological entities, we must first suppose them to be nested within ecologies and those ecologies to have causal powers. In other words, hierarchy is conceptually and ontologically prior to selection. It is more fundamental.

References

Campbell, Donald T. 1958. "Common Fate, Similarity, and Other Indices of the Status of Aggregates of Persons as Social Entities." *Behavioral Science* 3: 14–25.

Caro, Tim, Amanda Izzo, Robert C. Reiner Jr, Hannah Walker, and Theodore Stankowich. 2014. "The Function of Zebra Stripes." *Nature Communications* 5:

3,535. http://www.nature.com/ncomms/2014/140401/ncomms4535/abs/ncomms 4535.html.

Eldredge, Niles, and Stanley N. Salthe. 1984. "Hierarchy and Evolution." *Oxford Surveys in Evolutionary Biology* 1: 184–208.

Ettensohn, Charles A. 1990. "The Regulation of Primary Mesenchyme Cell Patterning." *Developmental Biology* 140: 261–71.

Ettensohn, Charles A., and David R. McClay. 1988. "The Regulation of Primary Mesenchyme Cell Migration in the Sea Urchin Embryo: Transplantations of Cells and Latex Beads." *Developmental Biology* 117: 380–91.

Feibleman, James K. 1954. "Theory of Integrative Levels." *British Journal for the Philosophy of Science* 5: 59–66.

Fleming, Leonore. 2012. "Network Theory and the Formation of Groups without Evolutionary Forces." *Evolutionary Biology* 39: 94–105.

Fleming, Leonore, and Robert Brandon. 2015. "Why Flying Dogs Are Rare: A General Theory of Luck in the Evolutionary Transitions." *Studies in History and Philosophy of Science Part C: Studies in History and Philosophy of Biological and Biomedical Sciences* 49: 24–31.

Haber, Wolfgang. 1994. "System Ecological Concepts for Environmental Planning." In *Ecosystem Classification for Environmental Management*, edited by Frans Klijn, 49–68. Dordrecht: Kluwer Academic Publishers.

Korn, Robert W. 2002. "Biological Hierarchies, Their Birth, Death and Evolution by Natural Selection." *Biology and Philosophy* 17: 199–221.

Maynard Smith, John, and Eörs Szathmáry. 1995. *The Major Transitions in Evolution*. Oxford: Oxford University Press.

McShea, Daniel W. 2001. "The Hierarchical Structure of Organisms: A Scale and Documentation of a Trend in the Maximum." *Paleobiology* 27: 405–23.

———. 2012. "Upper-Directed Systems: A New Approach to Teleology in Biology." *Biology and Philosophy* 27: 663–84.

———. 2013. "Machine Wanting." *Studies in History and Philosophy of Biological and Biomedical Sciences* 44: 679–87.

McShea, Daniel W., and Robert N. Brandon. 2010. *Biology's First Law: The Tendency for Diversity and Complexity to Increase in Evolutionary Systems*. Chicago: University of Chicago Press, 2010.

Nagel, Ernest. 1979. *Teleology Revisited and Other Essays in the Philosophy and History of Science*. New York: Columbia University Press.

Odum, Eugene P. 1971. *Fundamentals of Ecology*. Philadelphia: W. B. Saunders.

Rosenberg, Alex, and Daniel W. McShea. 2007. *The Philosophy of Biology: A Contemporary Introduction*. London: Routledge.

Rosenzweig, Michael L., and Robert D. McCord. 1991. "Incumbent Replacement: Evidence for Long-Term Evolutionary Progress." *Paleobiology* 17: 202–13.

Salthe, Stanley N. 1985. *Evolving Hierarchical Systems*. New York: Columbia University Press.

———. 2009. "A Hierarchical Framework for Levels of Reality: Understanding through Representation." *Axiomathes* 19: 87–99.

Simon, Herbert. 1962. "The Architecture of Complexity." *Proceedings of the American Philosophical Society* 106: 462–82.

Simpson, Carl. 2012. "The Evolutionary History of Division of Labour." *Proceedings of the Royal Society B: Biological Sciences* 279(1726): 116–21.

Sommerhoff, G. 1950. *Analytical Biology*. London: Oxford University Press.

Valentine, James W., and Cathleen L. May. 1996. "Hierarchies in Biology and Paleontology." *Paleobiology* 22: 23–33.

Weiss, Paul A. 1971. "The Basic Concept of Hierarchic Systems." In *Hierarchically Organized Systems in Theory and Practice*, edited by Paul Weiss, 1–43. New York: Hafner.

Wicken, Jeffrey S. 1987. *Evolution, Thermodynamics, and Information*. Oxford: Oxford University Press.

Wimsatt, William C. 1994. "The Ontology of Complex Systems: Levels of Organization, Perspectives, and Causal Thickets." *Canadian Journal of Philosophy* 20 (supplementary): 207–74.

———. 2007. *Re-Engineering Philosophy for Limited Beings: Piecewise Approximations to Reality*. Cambridge, MA: Harvard University Press.

Zylstra, Uko. 1992. "Living Things as Hierarchically Organized Structures." *Synthese* 91: 111–33.

CHAPTER 5

Three Approaches to the Teleological and Normative Aspects of Ecological Functions

Gregory J. Cooper, Charbel N. El-Hani, and Nei F. Nunes-Neto

1. Introduction

The biological realm is divided into two interacting hierarchies: the economic/ecological hierarchy and the genealogical/reproductive hierarchy. The former involves a kind of Darwinian economy of nature, where the currency is flow of matter and energy. The latter is about the transmission of heritable information (see Tëmkin and Eldredge 2015 for a contemporary exposition and references). Briefly, the economic/ecological hierarchy involves the following nested sequence: proteome—cell—organism (soma)—population (avatar)—community (biocenosis)—biosphere. The genealogical/reproductive hierarchy involves this nested sequence: genome—cell—organism (germline)—deme—species—monophyletic taxon—tree of life.

As one would expect, the science of ecology is devoted primarily to investigating the economic/ecological hierarchy (hereafter we will refer to it simply as the economic hierarchy). As a discipline, ecology is composed primarily of two subfields. Population/community ecology studies patterns of distribution and abundance in populations of organisms and how these populations come together to form ecological communities (i.e., populations of distinct species occurring in the same habitat). Ecosystem ecology studies the flow of energy and matter within communities situated in abiotic environments (i.e., ecosystems). Another name for this branch of ecology is biogeochemistry (see Schlesinger and Bernhardt 2013 for a comprehensive view of the field). There are also a number of bodies of inquiry devoted to bridging these two fields. Two prominent examples are ecological stoichiometry (Sterner and Elser 2002) and the biodiversity and ecosystem function research program (Tilman et al. 2014). Ecological stoichiometry studies the balance of chemical elements in ecological interactions. Research on biodiversity and ecosystem function studies the relationship between biological diversity (especially species diversity) and ecosystem processes such as productivity. Finally, there is a branch of ecology—evolutionary ecology—that studies the interaction between the two hierarchies, primarily at the organismic level (see Cooper 2003 for further details on the disciplinary structure of ecology).

The concept of function is deployed in ecology in a variety of ways. The next section explores these various uses, arguing that the most prominent and ecologically distinctive usage involves the concept of ecosystem function. The remainder of the chapter explores the extent to which three established philosophical analyses of function can illuminate the concept of ecosystem function. The three philosophical approaches examined are the causal role theory, the etiological theory, and the organizational theory.

2. Function in Ecology

Ecologists use the term *function* in a variety of ways. As Jax (2005) points out, sometimes it is just a way to refer to causal processes or activities. In this usage, it is a part of the familiar form–function dyad (Wouters 2003). The fox, for instance, is composed of various morphological attributes—that is, its form. It also engages in various activities, some voluntary and some not, such as eating mice or heart contraction. These causal activities are sometimes called functions, but there is nothing distinctively ecological about this use of function. It simply describes what happens at a particular level, in this example primarily the organismic level, in the economic hierarchy.

A second usage of the term *function* concerns the classification of various species into functional types or functional groups. In this case, different species are categorized collectively on the basis of shared features. The features may be morphological, physiological, or behavioral. For example, plants in arid environments have certain shared morphological features, such as thick and waxy leaves for avoiding desiccation. Certain Amazon fishes have become air breathers in response to the prevalence of low-oxygen habitats. The shared features reflect a response to a common type of selective environment, invoking questions about the nature of the selection pressures involved and the adaptive responses of the affected species. Thus this use of function is grounded in evolutionary biology rather than in ecology and is directed toward the genealogical/reproductive hierarchy (hereafter we will refer to it simply as the genealogical hierarchy).

Functional groups are sometimes viewed in a different way, however. First, they can be viewed in terms of the causal role they play in shaping the ecological communities where they reside. The shared features that underlie the common categorization are themselves functional rather than morphological, physiological, or behavioral. For example, pollinators are grouped in terms of their shared capacity to facilitate sexual reproduction of plants, despite the fact that a variety of morphological and behavioral mechanisms are deployed to facilitate this result. The predatory activities of the fox mentioned above places that animal in the category of a mesopredator, a group that can have important impacts on the structure of ecological communities (e.g., mesopredator release; see Ritchie and Johnson 2009). The focus here is once again on the economic hierarchy, in this case primarily on the impact of populations on the structure of ecological communities.

A second approach to functional groups appeals to different types of biochemical activities, as in the case of nitrogen-fixing bacteria. Often this type of categorization is done on the basis of very general trophic relationships, resulting in a familiar classification of primary producers, herbivores, secondary consumers, tertiary consumers, and decomposers. As with the previous usage, this use of function implies a contributory role to a larger system. In this case, the larger system is an ecosystem and the causal role will take the form of an impact on ecosystem processes, such as productivity or the cycling of material (e.g., nitrogen and carbon). As mentioned above, there is, for example, a large body of literature seeking to understand the functional role of species diversity in various ecosystem processes, such as productivity. While there remains much controversy, certain general conclusions do seem to be well supported—for instance, there is ample evidence that increased species diversity at the level of primary producers enhances ecosystem productivity (Cardinale et al. 2011).

The introduction of ecosystem processes invokes what is the most widespread use of the concept of function in ecology—ecosystem function. Ecosystems are characterized by fluxes of material and energy. From a biogeochemical perspective, the interplay of these fluxes is the basis of all life on the planet. Consider the most basic of ecosystem processes: primary production. It involves the capacity of certain organisms, given the appropriate enabling conditions, to transform solar radiation into biomass. If either the organisms or the enabling conditions are not in place, then this capacity will not be exercised. That is the sense in which both the primary producers and the enabling conditions (e.g., appropriate levels of nutrients) play a functional role in the ecosystem function of primary production. Thus there are really two concepts of function at work here. First, there is the functional role that various ecosystem components (both biotic and abiotic) play in enabling the ecosystem to exercise its capacities (i.e., to manifest the various processes of which it is capable) and, second, there is ecosystem function understood as the actual manifestation of these capacities by ecosystems. There is clearly some type of hierarchical structure involved here, even though ecosystems are not part of either the economic or genealogical hierarchies (Tëmkin and Eldredge 2015), because they also contain abiotic components. In keeping with the idea that ecosystem ecology can also be described as biogeochemistry, the relevant hierarchy involves the interaction between the biological and physicochemical realms. We will have more to say about this hierarchy in section 4.

There is one final sense of ecosystem function that must be clarified. Ecosystems deliver valued services (e.g., crop pollination, water filtration) that have an important bearing on human well-being. Sometimes ecosystem function is associated with the delivery of these valued services and functional breakdown is associated with the interruption of these services. Since, however, this use of function is embedded in a social-ecological context, we will not discuss it here. For more details, we refer to readers to other works on the topic (Kareiva et al. 2011; de Groot et al. 2002; Jax 2005; Nunes-Neto et al. 2014).

Ecologists rarely talk about function in the sense of activity—the first usage of function we distinguished. Reference to functional groups in all three of the senses distinguished above is more common. The first of these senses, functional groups defined in terms of shared adaptive strategies, is essentially an extension of the study of adaptation at the organismic level. There is nothing distinctively ecological about this use of function and we will not discuss it further. The final two usages of function concern functional groups as playing a causal role, and they are distinctively ecological. Sorting populations into functional groups in terms of the causal role they play in structuring ecological communities (e.g., pollinator, apex predator, mesopredator) is certainly something ecologists commonly do. However, the use of functional language to describe these roles is much less common. By far, the most prevalent use of function in ecology is in reference to ecosystem function, and we will confine our attention to this notion in the discussion that follows. Before examining the application of specific theories of function to ecosystem function, we provide a brief overview of the relevant philosophical landscape.

3. Philosophical Theories of Function: The Problems of Teleology and Normativity

The concept of function has been a source of heartburn for philosophers of science going way back to the logical empiricists (e.g., Hempel 1965; Nagel 1961). The basic problem is that functional ascription seems to get the causal order backward. Causes typically explain their effects, but the case of functions is apparently just the reverse. Why are baby copperhead snakes born with forked tongues? Because, in the future, these snakes will rely on those forked tongues to find the edges of scent trails, which will be important for locating both food and mates. The future utility appears to be explaining the contemporary presence of a trait. Call this *the problem of teleology*. A second mystifying factor is that baby copperheads are, in some sense, supposed to be born with forked tongues. In other words, to ascribe a function to a trait is more than saying what it does; it is saying what it is supposed to do, what it, in some sense, should do. When that does not happen, what follows is a functional breakdown—a malfunction. Call this *the problem of normativity*. Though distinct, it is clear that the two problems are closely related. Furthermore, both have the potential to render functional discourse scientifically problematic. Much of the motivation behind philosophical analyses of function is the desire to "naturalize" function—that is, to show that it does not invoke ideas that go beyond a widely shared scientific worldview.

Over the past forty years, an extensive philosophical literature has developed dealing with these two problems and with the notion of functional explanation more generally. Most philosophers now agree that there are two major candidates for a satisfactory account of function (see Neander 2015; though she disagrees with the consensus, she does make a convincing case for its existence). As Neander points out, the two candidates can be explained in terms of Mayr's (1961) distinction between proximate and ultimate questions, or how-questions and why-questions. Briefly, how-questions seek to understand the causal mechanism at work in a phenomenon, whereas why-questions ask about the historical processes that brought it into existence. One approach to function, the causal role strategy, focuses on the former. On this view, functional accounts describe the causal role that component elements play in the larger systems in which they are embedded. For many who espouse this approach, the teleology and normativity of function are legacies of a bygone era that we must move leave behind. The other candidate, the etiological approach, views functional accounts as answers to why-questions. On this view, the associated teleology and normativity is what makes functional discourse both distinctive and explanatory. In terms of the two hierarchies, the causal role approach is most naturally associated with the economic hierarchy, whereas the etiological approach is focused on the genealogical hierarchy. Finally, there is a newer approach, which we will discuss below: the organizational account of function, which seeks to go beyond the abovementioned consensus by seeking something of a synthesis between the established contenders. We turn now to an examination of the prospects for each of these views to elucidate ecological function.

4. A Causal Role Approach to Ecosystem Function

This approach to function was introduced by Robert Cummins (1975), and it has generated a vast literature. In this section, we will rely only on an adapted version of the causal role (CR) approach developed by Craver (2001). Craver locates CR functions in a hierarchical framework. Specifically, he identifies three levels: the contextual, the isolated, and the constitutive. The *isolated level* lies in the middle of this hierarchy; it simply describes what an entity is independent of the higher and lower levels. The *constitutive level* is at the bottom level of the hierarchy and gives a mechanistic explanation of the entity's behavior in terms of the activities of its component parts. The *contextual level* is at the top of the hierarchy and identifies the functional role of the entity in the context of the activities of a broader containing system. There appear to be strong parallels between Craver's hierarchical structure and a hierarchical framework developed by Salthe (1985; see Tëmkin and Eldredge [2015] for a contemporary version). Craver describes the levels in his hierarchical framework as "perspectives"

rather than "levels of nature," taking a more epistemological than ontological view, but one is tempted to conclude that the underlying reality that
these perspectives represent is structured in much the same way as that
described by Salthe. In the analysis that follows, we will identify Craver's
isolated, contextual, and constitutive levels with the focal, upper, and lower
levels (respectively) as described in Tëmkin and Eldredge (2015).

As mentioned earlier, ecosystems are not part of either the economic
or the genealogical hierarchies. Why they are absent from the genealogical
hierarchy is an issue we return to below. Their absence from the economic
hierarchy is due to the fact that they contain abiotic elements (Tëmkin and
Eldredge 2015). The relevant hierarchy for ecosystems involves the interaction between the biota and physicochemical characteristics of the planet.
A detailed discussion of this hierarchy is beyond the scope of the chapter,
but we will describe some aspects of it.

The isolated level simply describes what ecosystems do (i.e., the capacities they have). Ecosystems are biogeochemical systems. As such, they are
characterized by the biogeochemical process of productivity, which includes
nutrient cycling and energy flow. Furthermore, this process can be broken
down into component processes. The cycling of nitrogen, for example, includes decomposition, denitrification, and nitrogen fixation. The constitutive level provides a causal explanation of these ecosystem capacities in
terms of the activities of their constituent components. For example, cycling of matter would be explained through reference to the capacities of
the functional groups that contribute to this system capacity. Consider the
carbon cycle: the producers (e.g., terrestrial plants) fix the carbon in their
biomass, contributing to the first step in the cycle; the consumers (e.g., herbivores) utilize the plants, ingesting their carbon biomass, contributing to
the second step of the cycle; and, finally, the decomposers (e.g., soil organisms, such as arthropods and bacteria) contribute to the degradation
of carbon biomass, resulting from the wastes or dead bodies of plants and
herbivores, contributing, in turn, to the third step in the carbon cycle. The
decomposers create conditions in the soil for the regeneration of plants
and the incorporation of carbon. Finally, the contextual level identifies the
functional role of these ecosystem capacities in the manifestation of capacities of a larger containing system.

Craver's hierarchical approach applies straightforwardly to ecosystems
when ecosystems are the isolated level and ecosystem components are the
constitutive level. When we move up the hierarchy, making ecosystems
the constitutive level, problems emerge. What are the larger containing

systems to which ecosystems contribute? Is it the landscape? The regional ecosystem? The biome? The entire planet? Consider the following example: The mangrove forests along the coast of Florida appear, at least phenomenologically, to embody a relatively distinct type of ecosystem. It is a highly productive system that serves as a breeding and nursery habitat for many species that carry out most of their life history out in the open ocean. It also appears to play an important functional role for the adjacent terrestrial landscape, buffering it from extreme ocean events and providing essential habitat for a variety of birds, reptiles, mammals, and other organisms. Focusing on the aquatic biota, one might say that the functional role of mangrove forests within the broader oceanic ecosystem is to deliver high productivity inputs (and, perhaps, biodiversity maintenance) for that broader system. From this perspective, the mangrove forest is the constitutive level and the broader oceanic ecosystem is the isolated level. But what constitutes the broader ocean ecosystem? The various species for which mangrove forests are important will vary dramatically with regard to the spatial scale associated with their life history and the life histories of species with whom they interact. If we ascend another level up the hierarchy and adopt the broader oceanic ecosystem as the constitutive level, then the problem is even more intractable. What is the contextual level for that ecosystem? We are confronting here manifestations of the vexing problem of ecosystem individuation, to which we will return in the final section.

What does the hierarchical approach look like if we move down rather than up in the hierarchy, taking some particular ecosystem component as the isolated level and the ecosystem as the contextual level? Consider, for example, the primary producers. As mentioned above, there is evidence that increased species diversity at the level of primary producers enhances ecosystem primary productivity. What determines, at the constitutive level, the diversity of primary producers in an ecosystem? Here we make contact with one of the venerable questions of ecology. Ecologists have been struggling to understand the causes of species diversity at least since Hutchinson (1959) famously asked the question "Why are there so many kinds of animals?" In general, when we make the biotic components of ecosystems the isolated level, investigations of the constitutive level will involve an entire range of familiar ecological inquiries, from ecological stoichiometry to metacommunity ecology.

Causal role theorists tend to ignore the issues of teleology and normativity. For some, that will render this approach to ecosystem function untenable. We will return to the costs associated with giving up these ideas in the concluding section. These issues aside, however, we can inquire whether the CR approach illuminates ecosystem function. As pointed out above, the CR analyses of function are most naturally seen as addressing the how-questions of biological organization. If we are concerned with how the various ecosystem components contribute to the manifestation of ecosystem level capacities, the CR approach appears to give the appropriate answers (Odenbaugh 2010). A similar conclusion is warranted if we move one level down and ask how more fundamental ecological entities and processes contribute to the manifestation of the capacities of ecosystem components. If, however, we move a level up and attempt to see ecosystems as themselves constitutive elements of broader systems, then, as we have seen, the results are not as compelling.

As mentioned earlier, defenders of the etiological approach regard the teleology and normativity of functional discourse as essential, and they would not be happy to leave them behind. We now turn to that approach to see whether it can underwrite a more robust notion of ecosystem function.

5. The Etiological Approach to Ecosystem Function

One way to solve the problems of both teleology and normativity is to turn to history. In the past, there was variation among copperheads in the degree to which their tongues were forked. Those with a more pronounced forking of the tongue were more effective at locating the edges of scent trails and following those trails. This gave them a fitness advantage over their less-forked brethren and, because tongue morphology was a heritable trait, forked tongues became fixed in the population. The baby copperheads are born with forked tongues not because that trait will be useful in the future but because it was useful in the past. Furthermore, the fitness-enhancing effects of forked tongues provide a normative standard for evaluating the tongue morphology. Tongues that are not sufficiently forked are detrimental to survival and reproductive success.

This approach to analyzing function has come to be known as the etiological view. It comes in many forms, which we will not explore here (see McLaughlin 2001 for an overview). The important point here is the way in which the etiological view uses a history of selection to naturalize both the teleological and the normative components of functional ascription. It does so by providing a unified answer: the fitness-enhancing effects of past tokens explain the presence of the contemporary type, and the success with which the contemporary tokens embody the type provides the normative standard for evaluation. Put another way, the etiological approach solves both problems by answering the why-question for the functional trait in question.

Can the etiological approach solve the problems of teleology and normativity for ecosystem function in a similar fashion? Has natural selection shaped the functional capacities of ecosystems in the way that it has shaped the functional capacities of snakes? Although some biologists have argued that natural selection does operate at the level of ecosystems (Dunbar 1960), this does appear, at present, to be the minority view. The problem is that the conditions for natural selection—differential reproduction based on heritable variation in fitness-relevant traits—do not appear to hold for ecosystems. That is presumably why ecosystems do not appear in the genealogical hierarchy. Since we cannot hope to resolve this issue here, we settle for a conditional conclusion: the etiological approach can use a history of selection to naturalize the teleology and normativity of ecosystem function only if ecosystem-level selection can be demonstrated. If, on the other hand, ecosystem-level selection cannot be substantiated, then this would also cause problems for the attribution of a (normative) functional role to ecosystem components. Garson (2013) claims that system components can only be viewed as having a mechanistic function if the overall system of which they are a part can be seen as functional in a normative sense as well (see Kitcher 1993 for a similar view). If this is correct, then the failure to underwrite normative function at the ecosystem level will infect the ecosystem components as well.

The conditional conclusion just reached might appear too strong. Even if natural selection cannot be demonstrated at the ecosystem level, it might be possible to argue that natural selection operating at the organismic level creates a kind of functional organization at the ecosystem level that can be seen as both teleological and normative. Consider, for example, ecosystem resistance, understood as the capacity to maintain structural organization in the face of perturbation or disturbance. If an ecosystem is regularly subjected to a certain perturbation regime, then one might expect the organisms in the ecosystem to become adapted to that regime. However, if the organisms that comprise the ecosystem persist, then the ecosystem will persist as well. As Golley (1993, 196) puts it, "Eventually, the biota evolves to fit a certain sequence of environmental disturbances thereby enhancing the system's capacity to resist." Our mangroves can be used to illustrate. Lugo (1990) indicates that particular mangrove forests along the coast of Florida are impacted by hurricanes about once every twenty years on average. The life history characteristics of the individual mangrove trees (i.e., their patterns of growth and reproduction) have adapted to this disturbance regime, enabling them to persist. Since the mangrove trees are the structurally dominant feature of a mangrove forest, this enables the mangrove forest ecosystem to persist. While this argument may have some force for the particular capacity of ecosystem resistance, it is not clear how it could be extended to the broader suite of ecosystem capacities, such as productivity and nutrient cycling.

A more promising approach for using organism-level selection to ground functional organization at the ecosystem level is the appeal to niche construction. Niche construction refers to the capacity of organisms to shape the selective environments in which they live. The mangroves just mentioned are an example. Mangroves are ecosystem engineers, displaying niche constructing activities that shape the selective pressures faced both by themselves and by many other species in the system. Another prominent example of a niche constructing organism is the beaver, with its propensity to construct ponds and dams. Niche construction (and its ecological counterpart, ecosystem engineering) might supply the missing link between ecosystem function and evolution—powerful niche constructing activities of organisms may shape the biotic channels through which biogeochemical processes flow and, since these activities are shaped by natural selection, the biogeochemical processes are as well. But while there are optimistic views on this possible integration of ecosystem function and evolution (e.g., Laland et al. 1999), there are also authors who recognize the importance of niche construction but are skeptical about the capacity of the niche constructing activities of individual organisms to influence large-scale ecosystem variables (e.g., Tyrrell 2013). Our conditional conclusion needs to be modified: the etiological approach can use a history of selection to naturalize the teleology and normativity of ecosystem function only if the ecosystem level selection can be demonstrated or if it can be demonstrated that the niche constructing activities of organisms shape ecosystem-level capacities, such as productivity and nutrient cycling.

There is a final possibility that warrants mention. We have been discussing ways to apply natural selection at the ecosystem level to ground a functional account. A quite recent attempt in this direction is found in Dussault and Bouchard (2016), who propose that role functions in ecology should be interpreted based on a modified version of Bigelow and Pargetter's (1987) evolutionarily forward-looking account, which defines functions as effects that enhance an organism's propensity to survive and reproduce. Following up on Bouchard's persistence-based account of ecosystem evolution (2013, 2014), Dussault and Bouchard interpret persistence in terms of stability and resilience and, then, propensity as propensity to persist, in order to accommodate the obvious fact that ecosystems do not reproduce themselves. This leads them to what they call the persistence enhancing propensity (PEP) account of role functions in ecology, according to which the function of a given item x in an ecosystem E is to F if, and only if, x is capable of doing F and x's capacity to F contributes to E's propensity to persist. But notice that, despite being based on natural selection, their approach is not really an etiological theory of function, since it is forward- rather than backward-looking. Therefore, even if one accepts the PEP account, this would not be a vindication of the etiological approach to the teleology and normativity of ecological functions.

There is, finally, a key conceptual hurdle that stands in the way of the attempts to reconcile ecosystem function and evolution just discussed. It is based on the familiar distinction between selection *of* and selection *for* (Sober 1984). The evolutionary process needed to underwrite the teleological and normative aspects of function is selection *for* (Neander 1991), whereas the selection-shaping ecosystem capacities described above, assuming that the case can even be made, is selection *of*. Thus the case for using natural selection to underwrite a normatively rich conception of ecosystem function has not been substantiated. Let us now turn to a final option, the organizational approach to ecosystem function.

6. The Organizational Approach to Ecosystem Function

The organizational accounts of function address the explananda of both etiological and causal-role approaches and, thus, propose a synthesis of the two previous attempts to deal with functional explanation (Schlosser 1998; Collier 2000; Mossio et al. 2009; Saborido et al. 2011). On the one hand, the organizational approaches explain the existence (in the sense of persistence) of the items of functional ascription. On the other hand, they explain a systemic capacity to which the function of this same item (as a part of a containing system) contributes. Here we will focus only on the organizational approach proposed by Moreno, Mossio, and Saborido (see Mossio et al. 2009; Mossio and Moreno 2010; Saborido et al. 2011; Moreno and Mossio 2015).

For Mossio and colleagues, a trait (T) has a function in the organization (O) of a system (S) if and only if the following conditions (Cn) are satisfied:

C1: T contributes to the maintenance of the organization O of S

C2: T is produced and maintained under some constraints exerted by O

C3: S is organizationally differentiated (Mossio et al. 2009, 828)

Let us explain this definition through an example. First, pumping blood is the function of the heart (T), because it is the heart's contribution to the maintenance of the organization (O) of the system (S), say, a heart-bearing organism, where the heart's capacity for pumping blood is related to the distribution of blood and nutrients. This is a bottom-up contribution or influence (from T to O of S). Second, in a complementary way, the heart is produced and maintained by the organization of the body (which includes all other organs). In order to exist, the heart depends, directly or indirectly, on the channeling of blood by the veins, on the exchange of gases by the lungs, and on the detoxification function of the liver, among other activities. Thus we are pointing to the conditions of the existence of the heart, which are found in the organization of the body as a whole. This is a top-down contribution or influence (from O of S to T). Third, the heart's contribution to the organization (O) of the body (S) is a differentiated one, since it does not overlap with that of the liver or lungs.

A comparison to the etiological and causal role perspectives will help illuminate the organizational account of functions. While the causal role approaches typically try to eliminate teleology and normativity from functional ascriptions, the etiological approaches conflate these two problems by trying to solve them in a unified way. One of the important contributions of the organizational approach proposed by Mossio, Moreno, and Saborido is that it teases apart these two issues. According to this approach, the organizational closure is the criterion that grounds both the teleological and the normative dimensions of the functional ascriptions in biology. However, those two problems are solved by means of two different naturalizing formulations. Teleology is naturalized by assuming that in the current circular organization of the biological system, the functional effect of the item is, at least in part, one cause of its own existence. Thus the functional effect explains the existence of the entity, avoiding problems related to the appeal to explanatory capacities of functional effects in history (as it happens with the etiological approaches). In sum, one effect (function) is one of the causes and, consequently, the opposition between causality and teleology is resolved. While from an etiological point of view, function explains the existence (Wright 1973) or the current evolutionary maintenance (Godfrey-Smith 1994) of a functional item based on natural selection, in the organizational approach, the existence of the item is explained through the current organization in which it is embedded. As Mossio and colleagues argue, "organizational closure justifies explaining the existence of a process by referring to its effects: a process is subject to closure in a self-maintaining system when it contributes to the maintenance of some of the conditions required for its own existence. In this sense, organizational closure provides a naturalized grounding for a teleological dimension: to the question 'Why does X exist in that class of systems?', it is legitimate to answer 'Because it does Y'" (Mossio et al. 2009, 825).

Normativity, in turn, is naturalized by assuming that what a given entity is supposed to do (its function) is an effect that contributes, even if only indirectly, to its own existence, because that contribution supports the existence of other entities within the organizationally closed system on which the very entity to which function is being ascribed depends. Thus it is the activity of the functional entity itself, as well as of the whole system, that offers a criterion to distinguish malfunctions from functions. The norm is intrinsic to the closed organization of the system:

> Similarly, organizational closure grounds normativity. Because of the organizational closure, the activity of the system has an intrinsic relevance for the system itself, to the extent that its very existence depends on the effects of its own activity. Such intrinsic relevance . . . generates a naturalized criterion to determine what the system is 'supposed' to do. In fact, the whole system (and its constitutive processes) "must" behave in a specific way, otherwise it would cease to exist. Accordingly, the activity of the system becomes its own norm or, more precisely, the conditions of the existence of its constitutive processes and organization are the norms of its own activity. (Mossio et al. 2009, 825)

Let us come back to the example of the heart. Why does the heart exist? Because it pumps blood. Among the many effects of the heart, this is what the heart should do in order to exist and, thus, this effect is taken to be the heart's function. The pumping of blood by the heart (an effect of its existence) explains its own existence, because the pumping of blood allows the distribution of blood and nutrients through the body, contributing to the maintenance of other organs, to the conditions of existence of the body, and, ultimately, to the very conditions of existence of the heart. In sum, since this philosophical approach is no longer committed to a historical selective narrative for the origins of the functional items,[1] the existence or persistence of the functional item can be explained by its effect in the context of the current organization. Ontogenetically, the function (i.e., one among the effects of the item) is the cause.

Here the idea that the effect is the cause[2] is crucial, but it should be formulated more precisely. In general, a closure means that a sequence of processes forms a causal loop. When these processes are purely physical or chemical—as, for instance, in the cycle of water in the prebiotic Earth—we are talking about the closure of processes (see Nunes-Neto et al. 2014). In turn, a closure of constraints is present when complex biological structures constrain the flow of matter and energy, reducing its degree of freedom in a way that contributes to the maintenance of the organization of the whole system.

Following Moreno and Mossio (2015), we can say that closure of constraints is a specific mode of dependence between at least two constraints (in a set of constraints), in which closure is realized by a system that produces some of the constraints, harnessing its underlying dynamics. This means not only that the constrained processes—that is, the physical-chemical flow—form a causal loop but that this loop is achieved because the constraints, generated in the system, influence each other, so as to achieve closure. In formal terms, according to Moreno and Mossio (2015), a set of constraints (C) realizes closure if, for each constraint (C_i) belonging to C, (1) C_i depends directly on at least one other constraint of C (C_i is dependent), and (2) there is at least one other constraint C_j belonging to C, which depends on C_i (C_i is an enabling condition). For instance, the heart is maintained because of the constraining action it performs on the flow of matter, channeling the blood's way in the body so as to contribute to the maintenance of other constraints (exerted by the lungs, the liver, etc.). These other constraints are also, in turn, enabling conditions to the heart in such a way that the constraints close the organization of the body. If one constraint is eliminated or is malfunctioning, the organizational closure of constraints of the whole system may be broken and the system may collapse.

The organizational perspective was applied to ecological systems by Nunes-Neto et al. (2014, 131), who defined an ecological function as "a precise (differentiated) effect of a given constraining action on the flow of matter and energy . . . performed by a given item of biodiversity, in an ecosystem closure of constraints." The main element here is the idea of an ecosystem closure of constraints, which is the basis to ground the teleology and normativity of functions, from a naturalized organizational perspective.

To expand on this point, let us take the example of a minimal ecosystem with three functional groups. First, consider an ecosystem with two hierarchical levels (i.e., in a hierarchy of control; cf. Ahl and Allen 1996), the level of the items of biodiversity[3] (the constraints) and the level of the flow of carbon atoms (the processes). In our minimal ecosystem, the items of biodiversity are three functional groups: producers, consumers, and decomposers of organic matter. The producers of organic matter (plants) constrain the flow of carbon atoms, reducing its degree of freedom, through the absorption of carbon dioxide and by carbon fixation in photosynthesis. The flow of carbon atoms becomes more determinate, more harnessed, as those atoms become part of the plant biomass (sometimes for centuries or millennia, depending on the producer species). Part of this biomass (leaves, fruits, sprouts, etc.) is eaten by the consumers (herbivorous animals), which further channels the flow of carbon when carbon atoms become part of their biomass. When the consumers and producers die, the animal carcasses and plant leaves, fruits, and twigs constitute the organic matter that is further processed by decomposers (soil organisms, such as arthropods and bacteria), transforming it into available nutrients for plants, closing the cycle.

In more formal terms, there is a mutual dependence between these constraints. For instance, let us take the example of the consumers. They constrain the flow of matter in a way that reduces its degree of freedom, and with this effect in the ecosystem as a whole, they create enabling conditions to the existence of the decomposers. So, on the one hand, they are *enabling conditions* to the decomposers. On the other hand, the consumers are *dependent* on the producers of organic matter and on the very decomposers, since they mobilize nutrients to the producers.

The organizational account of ecosystem function is a promising approach, but it faces challenges as well. The power of the CR approach is most evident when dealing with ecosystem components. When applied to ecosystem function at the level of biogeochemical processes, it also runs squarely into the difficult problem of ecosystem individuation.

The organizational account combines the strength of the CR approach with a natural account of the normativity of function. It also attempts to account for the teleology of function, and here there is a potential problem. The problem concerns history.

Etiological accounts are historical in nature. The teleological problem (the future causing the past) is solved by putting the causation in the past, but this has led to some problems in attributing functions to contemporary traits. By focusing on contemporary organizational structure, the organizational approach avoids these problems. But there may be a trade-off. The causal loops of the organizational approach avoid the problem of the future causing the past, but they may face the problem of the past causing the present. In other words, there are questions about how these self-maintaining systems come into existence—questions for which the etiological approach has an answer. In this debate, supporters of the organizational approach could make some points. First, the etiological approach doesn't really account for the origin of functional traits either, following also a criticism made by Cummins (2002) and Cummins and Roth (2010). And second, the organizational approach is sensitive to the role of natural selection in shaping the spread of traits, which would be possible because of the assumption that function is an a priori condition for the very Darwinian competition of variant complex organizations (where the functional items are embedded). This, however, does not mean that the organizational approach resolves all the problems concerning ecosystem function.

The situation around the issue in current debates in philosophy of biology and ecology is quite complex, and we would certainly demand more elaboration and discussion and, probably, a pluralistic perspective on accounts of ecological function.

7. The Upshot

In this chapter, we discussed three approaches to the teleological and normative aspects of ecological function, with a focus on ecosystem functions. According to our analysis, the etiological approach to functions might ground the teleology and normativity of ecosystem functions if natural selection could be applied to ecosystems. It is clear that this application of natural selection is controversial at best, but the situation would look different if one succeeded in building a convincing approach to function, as discussed above. The causal role approach works nicely as an account of how various ecosystem components generate ecosystem processes, but it runs into trouble with ecosystem function at the level of whole ecosystems. It also abandons teleology and normativity. The former may be a disappointment for the growing body of ecologists who view ecosystems as, in some sense, self-organizing systems (Desjardins et al. 2015) and thus as systems that tend to develop toward and maintain some preferential states. Often these ideas are developed in the context of complex adaptive systems theory (Levin 1998) and the worry can be put as follows: If there is no normativity to ecosystem function, then how are we to interpret the "adaptive" in complex adaptive systems? The lack of normativity will also cause problems for those environmental ethicists (e.g., Rolston 1987) who view ecosystems as the proper objects of moral considerability. More generally, the lack of teleology and normativity will not satisfy those who feel that ecosystems are goal-directed systems capable of impairment.

The organizational approach grounds the teleology and normativity of ecosystem functions in the ecosystem closure of constraints, based on the mutual dependences between items of biodiversity. This approach also runs afoul of the problem of ecosystem individuation. It is a newer approach that is still experiencing growing pains and, as such, still has some internal issues to work out.

We will conclude with two final points. First, the fact that ecosystem individuation is a problem across the board might be taken to suggest that ecosystem function, at the level of whole ecosystems, is not really full-blown function after all, but more on the order of ecosystem activity—something akin to the first sense of function discussed in this chapter. Second, the fact that there are strengths and weaknesses associated with each of the options examined might be taken to support pluralism with respect to ecosystem function in much the same way that pluralism has become an attractive option in the function literature overall.

Particularly, concerning pluralism, we think that definitional issues, like the one on what is an ecosystem function, are not relevant as definitional issues per se—only if they point to different explanatory agendas (see, e.g., Brigandt and Love 2012). So the important issue is how the way we define concepts lead to explanatory expectations and, thus, to research programs. From this standpoint, a pluralistic view about functions and their connection to evolutionary issues is interesting, given that different ways of defining function can lead to different but interesting (even if sometimes incommensurable) research agendas and explanatory expectations. Thus, to point to what might be the roles of etiological, causal role and organizational accounts with regard to how we deal with functional explanation in ecology seems interesting. The tricky methodological issue is how to combine a proper philosophical articulation of the different positions. Maybe to recognize the diversity of approaches and how each of them helps illuminate one aspect or dimension of the issue (such as ecosystem function) is the best we can reach. Maybe this is an advantage rather than an obstacle—the richness of ideas leaves us with more possibilities for thinking about complex ecological systems.

Notes

1. In some sense, this means that for the organizational approach, there is a separation between functions and remote causes, just as it is proposed in the causal role approach. From the point of view of the organizational approach, functions are understood as proximate causes, but this does not mean that the evolutionary history has no importance. Although we point here to this difference (as it could be put by the organizational perspective), the identification of functions as proximate or remote causes certainly deserves further consideration.

2. This is the idea of closure—that is, of a loop in the sense that X causes/has an effect on Y, which then causes/has an effect on X ($X \to Y \to X$).

3. For Nunes-Neto et al. (2014, 132), the expression item of biodiversity refers to "a biological entity or activity directly relevant for the maintenance of an ecosystem, by actively participating in, at least, one constraining action within this same ecosystem." The reference is not only to species composition but to entities or activities that are considered biodiverse in contemporary ecology. So, this concept—which can be understood as intrinsically hierarchical and scale-free (just as the ecosystem concept)—"includes in its domain . . . the following items: morphological or physiological traits, organisms, populations, species and functional groups" (132).

References

Ahl, Valerie, and T. F. H. Allen. 1996. *Hierarchy Theory: A Vision, Vocabulary, and Epistemology*. New York: Columbia University Press.

Bigelow, John, and Robert Pargetter. 1987. "Functions." *Journal of Philosophy* 84(4): 181–96.

Bouchard, Frédéric. 2013. "How Ecosystem Evolution Strengthens the Case for Functional Pluralism." In *Functions: Selection and Mechanisms*, edited by Philippe Huneman, 83–95. Dordrecht: Springer.

————. 2014. "Ecosystem Evolution Is about Variation and Persistence, Not Populations and Reproduction." *Biological Theory* 9(4): 382–91.

Brigandt, Ingo, and Alan C. Love. 2012. "Conceptualizing Evolutionary Novelty: Moving beyond Definitional Debates." *Journal of Experimental Zoology (Molecular and Developmental Evolution)* 318B: 417–27.

Cardinale, Bradley J., Kristin L. Matulich, David U. Hooper, Jarrett E. Byrnes, Emmett Duffy, Lars Gamfeldt, Patricia Balvanera, Mary I. O'Connor, and Andrew Gonzalez. 2011. "The Functional Role of Producer Diversity in Ecosystems." *American Journal of Botany* 98(3): 572–92.

Collier, John. 2000. "Autonomy and Process Closure as the Basis for Functionality." *Annals of the New York Academy of Science* 901: 280–90.

Cooper, Gregory J. 2003. *The Science of the Struggle for Existence: On the Foundations of Ecology.* Cambridge: Cambridge University Press.

Craver, Carl F. 2001. "Role Functions, Mechanisms, and Hierarchy." *Philosophy of Science* 68(1): 53–74.

Cummins, Robert. 1975. "Functional Analysis." *Journal of Philosophy* 72(20): 741–65.

————. 2002. "Neo-Teleology." In *Functions: New Essays in Philosophy of Psychology and Biology*, edited by Andre Ariew, Robert Cummins, and Mark Perlman, 157–72. Oxford: Oxford University Press.

Cummins, Robert, and Martin Roth. 2010. "Traits Have Not Evolved to Function the Way They Do Because of a past Advantage." In *Contemporary Debates in Philosophy of Biology*, edited by Francisco J. Ayala and Robert Arp, 72–85. Chichester: John Wiley & Sons.

de Groot, Rudolf S., Matthew A. Wilson, and Roelof M. J. Boumans. 2002. "A Typology for the Classification, Description and Valuation of Ecosystem Functions, Goods, and Services." *Ecological Economics* 41(3): 393–408.

Desjardins, Eric, Gillian Barker, Zoë Lindo, Catherine Dieleman, and Antoine C. Dussault. 2015. "Promoting Resilience." *The Quarterly Review of Biology* 90(2): 147–65.

Dunbar, M. J. 1960. "The Evolution and Stability in Marine Environments Natural Selection at the Level of the Ecosystem." *American Naturalist* 94(875): 129–36.

Dussault, Antoine C., and Frédéric Bouchard (2016). "A Persistence Enhancing Propensity Account of Ecological Function to Explain Ecosystem Evolution." *Synthese*. doi: 10.1007/s11229-016-1065-5.

Garson, Justin. 2013. "The Functional Sense of Mechanism." *Philosophy of Science* 80(3): 317–33.

Godfrey-Smith, Peter. 1994. "A Modern History Theory of Functions." *Noûs* 28(3): 344–62.

Golley, Frank B. 1993. *A History of the Ecosystem Concept in Ecology: More than the Sum of Its Parts.* New Haven: Yale University Press.

Hempel, Carl G. 1965. *Aspects of Scientific Explanation and Other Essays in the Philosophy of Science.* New York: Free Press.

Hutchinson, G. Evelyn. 1959. "Homage to Santa Rosalia, or Why Are There So Many Kinds of Animals?" *American Naturalist* 93(870): 154–59.

Jax, Kurt. 2005. "Function and 'Functioning' in Ecology: What Does It Mean?" *Oikos* 111(3): 641–48.

Kareiva, Peter, Heather Tallis, Taylor H. Ricketts, Gretchen C. Daily, and Stephen Polasky. 2011. *Natural Capital: Theory and Practice of Mapping Ecosystem Services*. Oxford: Oxford University Press.

Kitcher, Philip. 1993. "Function and Design." In *Midwest Studies in Philosophy Volume XVIII*, edited by Peter A. French, T. E. Uehling Jr., and Howard K. Wettstein, 379–97. South Bend: University of Notre Dame Press.

Laland, Kevin N., F. John Odling-Smee, and Marc W. Feldman. 1999. "Evolutionary Consequences of Niche Construction and Their Implications for Ecology." *Proceedings of the National Academy of Science*, USA 96(18): 10242–47.

Levin, Simon A. 1998. "Ecosystems as Complex Adaptive Systems." *Ecosystems*. 1: 431–36.

Lugo, Ariel E. 1990. "Fringe Wetlands." In *Forested Wetlands: Ecosystems of the World*, edited Ariel E. Lugo, Sandra Brown, and Mark M. Brinson 15, 143–69. Amsterdam: Elsevier.

Mayr, Ernst. 1961. "Cause and Effect in Biology." *Science* 134(3489): 150–6.

McLaughlin, Peter. 2001. *What Functions Explain: Functional Explanation and Self-Reproducing Systems*. Cambridge: Cambridge University Press.

Moreno, Alvaro, and Matteo Mossio. 2015. *Biological Autonomy: A Philosophical and Theoretical Enquiry*. Dordrecht: Springer.

Mossio, Matteo, and Alvaro Moreno. 2010. "Organizational Closure in Biological Organisms." *History and Philosophy of the Life Sciences* 32(2–3): 269–88.

Mossio, Matteo, Cristian Saborido, and Alvaro Moreno. 2009. "An Organizational Account of Biological Functions." *British Journal for the Philosophy of Science* 60(4): 813–41.

Nagel, Ernest. 1961. *The Structure of Science: Problems in the Logic of Scientific Explanation*. New York: Harcourt, Brace & World.

Neander, Karen. 1991. "Function as Selected Effects: The Conceptual Analyst's Defense." *Philosophy of Science* 58(2): 168–84.

———. 2015. "Functional Analysis and the Species Design." *Synthese*. doi:10.1007 /s11229-015-0940-9.

Nunes-Neto, Nei F., Alvaro Moreno, and Charbel N. El-Hani. 2014. "Function in Ecology: An Organizational Approach." *Biology and Philosophy* 29(1): 123–41. doi:10.1007/s10539-013-9398-7.

Odenbaugh, Jay. 2010. "On the Very Idea of an Ecosystem." In *New Waves in Metaphysics*, edited by Allan Hazlett, 240–58. New York: Palgrave Macmillan.

Ritchie, Euan G., and Cristopher N. Johnson. 2009. "Predator Interactions, Mesopredator Release and Biodiversity Conservation." *Ecology Letters* 12: 982–98.

Rolston, H. III. 1987. "Duties to Ecosystems." In *A Companion to A Sand County Almanac: Interpretive and Critical Essays*, edited by J. Baird Callicott, 246–74. Madison: University of Wisconsin Press.

Saborido, Cristian, Matteo Mossio, and Alvaro Moreno. 2011. "Biological Organization and Cross-Generation Functions." *British Journal for the Philosophy of Science* 62(3): 583–606.

Salthe, Stanley N. 1985. *Evolving Hierarchical Systems: Their Structure and Representation*. New York: Columbia University Press.

Schlesinger, William H., and Emily S. Bernhardt. 2013. *Biogeochemistry: An Analysis of Global Change*. Waltham, MA: Academic Press.

Schlosser, Gerhard. 1998. "Self-Re-Production and Functionality: A Systems-Theoretical Approach to Teleological Explanation." *Synthese* 116(3): 303–54.

Sober, Elliott. 1984. *The Nature of Selection: Evolutionary Theory in Philosophical Focus*. Cambridge, MA: MIT Press.

Sterner, Robert W., and James J. Elser. 2002. *Ecological Stoichiometry: The Biology of Elements from Molecules to the Biosphere*. Princeton, NJ: Princeton University Press.

Tëmkin, Ilya, and Niles Eldredge. 2015. "Networks and Hierarchies: Approaching Complexity in Evolutionary Theory." In *Macroevolution: Explanation, Interpretation, Evidence,* edited by E. Serrelli and N. Gonthier, 183–286. Dordrecht: Springer.

Tilman, David, Forest Isbell, and Jane M. Cowles. 2014. "Biodiversity and Ecosystem Functioning." *Annual Review of Ecology, Evolution and Systematics* 45: 471–93.

Tyrrell, Toby. 2013. *On Gaia: A Critical Investigation of the Relationship between Life and Earth*. Princeton, NJ: Princeton University Press.

Wouters, Arno G. 2003. "Four Notions of Biological Function." *Studies in History and Philosophy of Biological and Biomedical Sciences* 34(4): 633–68.

Wright, Larry. 1973. "Functions." *Philosophical Review* 82(2): 139–68.

PART 2

Hierarchical Dynamics

Process Integration across Levels

Information and Energy in Biological Hierarchical Systems

Ilya Tëmkin and Emanuele Serrelli

As a process, biological evolution encompasses phenomena across levels of biological hierarchies that directly or indirectly channel, alter, or interrupt the flow of information as it is transmitted from ancestors to descendants. Such formulation necessitates a general concept of information that is physically interpretable and hierarchically structured. Due to a plethora of meanings that the term *information* has acquired in different contexts, establishing a single general concept of information proved to be elusive. Therefore, we propose a substantive, admittedly restricted, notion of information that pertains to explaining biological evolution by satisfying the following criteria: the information must (1) have material basis, so it can be stored; (2) be amenable to copying (replicating); (3) be interpretable in the economic context of energy and matter flow (i.e., capable of eliciting a discriminating outcome when used); and (4) be distributed across levels of the genealogical hierarchy.

Energy, or the capacity of a physical system to perform work, is a principal causal agent in evolution because the fluctuations in the flow of energy and matter through the entities in the economic hierarchy are causally responsible for the survival and differential propagation of the entities in the genealogical hierarchy of replicators, thus channeling or disrupting the flow of information that shapes the historical pattern of life. Because the spectrum of processes involving energy storage and transformation extends greatly beyond those of biological systems, it is necessary to circumscribe precisely the energy budgets pertinent to living entities and, more specifically, to evolutionary processes.

Evolution and Information

Information is a physical state of a biological system that has an effect upon the system's persistence in a given environment. A system is a network of functionally interdependent and structurally interconnected components, comprising an integrated structure, capable of enduring in a relative stable state over a prolonged period of time. Evolving biological systems are organized in the context of scale into a nested compositional hierarchy, a pattern of relationships, where entities at one level are composed of parts at lower levels and are themselves nested within more extensive entities. The macrostate, or the state of the system at the level at which a particular phenomenon is observed (the focal level), results from the configuration of interacting lower-level component parts ("microstates") and the constraints imposed by the higher levels. The physical implementation and construction of information is context specific, depending on the systemic properties and the forms of interaction, specific to a focal hierarchical level. Examples of information-bearing states of biological systems at different hierarchical levels include the sequence of nucleotides in a DNA molecule, a spatial configuration of interacting cells in a tissue, and social and spatial population structure.

Evolving biological entities are spatiotemporally constrained systems (i.e., *individuals* in a philosophical context). Hence, due to their finite duration, information persists only as long as the entity that stores this information exists and transmits it to a subsequent generation through some sort of replicating process (e.g., cell division, reproduction, speciation). For evolution to occur, such replication process must be proliferative and hereditary—that is, generating variation in transmitted information by producing multiple descendants that might differ from its ancestral template. Because in a hierarchically nested system entities replicate only as parts of the more inclusive entities, the historical fate of information at a given level depends not only on interactions at the focal level but also on information processing and replication of genealogical individuals at other levels.

Information Processing

Most generally, information processing in evolving systems takes the form of sorting, where variation in the physical states of genealogical individuals at a given level manifests in properties that are causally linked to their differential birth and death rates in a given economic context, ensuring that only some of the variants persist and transmit information to the next generation. Darwinian natural selection is but one special type of sorting, resulting from nonrandom differences in fitness, or replication success, that operates on phenotypic properties at the organismal level. Other examples of sorting include selection at supraorganismal levels (e.g., kin/group selection) and stochastic processes (e.g., genetic drift and species drift). Because biological systems are hierarchically nested, the sorting of individuals at higher levels recursively results in the sorting of individuals at lower levels.

Sorting works by interpreting the information expressed by genealogical entities in such a way that affects the survival of the interactors, or biological open systems involved in continuous processes of energy and matter exchange with their surroundings in order to maintain their steady state. The economic success of interactors is causally responsible for the persistence and the differential propagation of replicators. As replicators (genealogical entities), interactors (economic, or ecological entities) comprise a nested hierarchy, established by scalar differences in process rates. Thus evolution occurs at the nexus between the economic and genealogical hierarchies, where the dynamic cyclical interactions in the former are translated into an irreversible, historical pattern of the latter. The interaction of individuals in the economic hierarchy with their environment is causally related to differential replication of genealogical individuals.

Due to hierarchical structure, the information content of entities at a focal level is constrained by the boundary conditions set up by the dynamics of entities at the upper level, so that only some of the possible states derived from alternative configurations of lower-level components are actually realized. Therefore, information processing at the focal level is causally related to information processing at other levels.

Information Coding

Information coding defines the rules of correspondence between the dynamic and informational phenomena in complex evolving systems, where these classes of processes are decoupled. The relationship of entities in the genealogical hierarchy allows only unidirectional, time-irreversible control of information flow, which reflects the historical nature of biological systems and yields a static hierarchy of classification. The economic systems form a hierarchy of control, allowing for dynamic interactions of entities within and across levels through upward and downward causation. The economic and genealogical hierarchies represent, respectively, the spatial and temporal dimensions of the organic realm. This fundamental dissimilarity in the nature of the relationships among replicators and interactors prevents exact one-to-one correspondence of entities and levels between the genealogical and the ecological hierarchies. As a result, coding is required for establishing a causal link between the informational and functional domains of life. Despite the fact that the classic example of coding in biological systems is the translation of RNA, there is a growing empirical evidence of different coding systems at other levels of biological organization (e.g., splicing, signal transduction, animal behavior).

The ultimate meaning, or function, of coding in biological systems amounts to ensuring a faithful interpretation of information in the economic context in such a way that the physical state and modes of self-maintenance that proved successful (i.e., persistent and robust to environmental perturbations) for the past generations are reproduced in the present. Importantly, some degree of error and misinterpretation (as commonly found, for instance, in DNA replication) contributes to expanding the variation in the next generation. A failure or the lack of capacity to interpret the code, which might result from perturbation of the environment or the replicating system itself, leads to demise—death or extinction—of a genealogical system, resulting in a loss of information and, hence, the irreversibility of evolution.

Information and the Structure of the Organic World

Conceptualizing information as essential physical attributes of hierarchically nested genealogical individuals—which by means of coding and computation establish a causal link between the time-irreversible control of information flow and the dynamics of energy and matter exchange of the entities of an economic hierarchy—opens new frontiers in our understanding of biological evolution. It concretizes a clouded concept of information in a restricted substantive formulation as structured data, amenable to interpretation and capable of generating a meaningful outcome. Shifting away from a more traditional, information theory based concepts, the semiotics-inspired notion of information brings into focus the physical nature of the causal connection between the genealogical and economic hierarchies of life, the heart of evolutionary phenomenology.

Energy in Biological Hierarchy

Biological entities in the ecological, or economic, hierarchy are open, or nonequilibrium, systems that ultimately rely on abiotic environment for the continuous influx of energy—mainly in the form of reduced carbon and photons of light—required for their survival (maintenance of structural and functional integrity) and their increase in size (expansion). Through highly regulated exchange of energy and matter with their surroundings, they manage to maintain highly organized states that are far from thermodynamic equilibrium. The nature and complexity of organization of a particular biological system, and the associated energetic cost are determined by the rates and mechanisms of energy and matter flow specific to the scalar level of the system in question (such as molecular interactions in metabolic networks within individual organisms and trophic interactions—such as predator-prey relationship—in an ecological community food web). As the energy available to biological systems is in abundant supply, a trend of a continuously increasing biomass, accompanied by increases in metabolic activity, was reported for some groups of animals on a global scale and for the entire duration of the Phanerozoic eon.

The expansion of a living system is attained by proliferative replication of its parts, or constituting systems at a lower level of a nested compositional hierarchy. As far as the energy budgets are concerned, the lion's share of consumed energy is invested into the maintenance of organization and expansion, accompanied by waste dissipated into the surroundings, whereas comparatively small amount of energy is invested into information replication, transmission, and storage. Thus not only the control of energy and matter flow is causally linked to genealogical processes, but the direction of the causation is specified from higher-level economic entities to lower-level genealogic constituents. For instance, speciation (i.e., replication at the species level) is largely, albeit indirectly, triggered by the disruption or destabilization of ecological networks at the more inclusive biotic assemblage level (such as communities and more encompassing ecological entities), rather than by maximization of the control of trophic energy among individual organisms in a population due to natural selection at the lower level. Undoubtedly, processes at the molecular, organismal, and population levels are important evolutionary factors in generating evolutionary novelty, shaping anagenetic trends, and driving population-level dynamics—providing necessary initiation conditions, enabling sorting at higher levels—but they do not drive (or directly cause) speciation in the vast majority of cases.

Supported by ample empirical data and consistent with theoretical predictions, the architecture of energy transforming biological systems composed of hierarchically nested, complex networks is capable of buffering extreme extrinsic perturbations, producing a pervasive pattern of stability in living systems across scale (evidenced, for instance, in long-term persistence of biological communities and relative stasis of phenotypic attributes of a species across geological time scales).

Energy and Evolution

Evolution occurs at the nexus between the economic and genealogical hierarchies as a response to perturbations that are sufficient to disrupt the steady state of biological energy flow networks, so that the buffering mechanisms preventing the cascading effects across hierarchies fail. Altered or interrupted pathways of energy flow directly affect the integrity and survival of living systems and, as a consequence, their ability to expand through proliferative replication of their constituent subsystems, thereby interfering with or channeling the transmission of information. The principal causal factors, capable of perturbing biological networks of energy and matter flow and triggering an evolutionary response, are abiotic environmental factors of cosmic and planetary scale that translate into climatic, lithological, and geochemical processes acting across hierarchical levels.

In this light, evolution is directly explicable in terms of energy transformations that affect the origin, transmission, and sorting of information at any level of organization. For example, the notion of fitness, the principal criterion of natural selection—that is, the sorting process at the level of individual organisms—is conceptualized as the control of consumed ("trophic") energy, which ultimately translates into maximizing viability and fecundity and leading to an increase in the size of the population. Thus fitness can be equated with expansive energy. A failure to maximize the control of trophic energy will lead to a decrease (negative expansion) of the entity, resulting in organismal death or demise of a population.

At the level of a species, according to the Red Queen hypothesis, a change in the cumulative trophic energy control of all conspecific individuals

is balanced by equal and opposite net change in the realized absolute fitness of all interacting species within a biocenosis. Such presumed compensatory interavatar interactions reflect the empirical data on the resilience
of biocenoses to extrinsic substantial perturbations. However, sufficiently
strong perturbations, capable of disrupting the buffering capacity of energy flow networks, elicit differential responses related to avatar diversity
and connectivity, resulting in secondary extinctions and food-web fragmentation, with a large detrimental effect on an overall community stability.

References

Bambach, Richard K. 1993. "Seafood through Time: Changes in Biomass, Ener-
getics, and Productivity in the Marine Ecosystem." *Paleobiology* 19: 372–97.
Barbieri, Marcello. 2002. *The Organic Codes: An Introduction to Semantic Biol-
ogy*, 1st ed. Cambridge: Cambridge University Press.
Brooks, Daniel, and E. O. Wiley. 1988. *Evolution as Entropy: Toward a Unified
Theory of Biology*. Chicago: University of Chicago Press.
Collier, John. 2008. "Information in Biological Systems." In *Philosophy of Infor-
mation*, edited by Pieter Adriaans and Johan van Benthem, 763–87. Dordrecht:
Springer.
Eldredge, Niles. 1986. "Information, Economics, and Evolution." *Annual Review
of Ecology and Systematics* 17: 351–69.
Eldredge, Niles, and Stanley Salthe. 1984. "Hierarchy and Evolution." In *Ox-
ford Surveys in Evolutionary Biology*, vol. 1, edited by Richard Dawkins and
M. Ridley, 182–206. Oxford: Oxford University Press.
Emmeche, Claus, and Kalevi Kull, eds. 2011. *Towards a Semiotic Biology: Life Is
the Action of Signs*. London: Imperial College Press.
Floridi, Luciano. 2011. *The Philosophy of Information*. Oxford: Oxford University
Press.
Oyama, Susan. 2000. *The Ontogeny of Information: Developmental Systems and
Evolution*, 2nd ed., rev. and expanded. Durham: Duke University Press.
Pattee, Howard H. 1970. "The Problem of Biological Hierarchy." In *Towards a
Theoretical Biology 3, Drafts*, edited by Conrad H. Waddington, 117–36. Edin-
burgh: Edinburgh University Press.
Tëmkin, Ilya, and Niles Eldredge. 2015. "Networks and Hierarchies: Approach-
ing Complexity in Evolutionary Theory." In *Macroevolution: Explanation, In-
terpretation and Evidence*, edited by Emanuele Serrelli and Nathalie Gontier,
183–226. Switzerland: Springer International Publishing.
Van Valen, Leigh. 1976. "Energy and Evolution." *Evolutionary Theory* 1: 179–229.

Summaries for Part 2

Contributions in this second part focus on the dynamic relationships among entities at different scales, with particular attention, on the one hand, to the identification of significant entities in evolution (e.g., gene, genome, developmental modules) and, on the other hand, to the emergent origin of entities and individuals in the history of life.

Ryan Gregory, Tyler A. Elliott, and ***Stefan Linquist*** use the example of Transposable Elements (TEs) in the human genome to demonstrate the necessity of a multilevel approach to explain genome features and their variation all across the living world. They propose an element-level perspective (similar to the classic gene's-eye view) on TEs, understanding them as biological entities that reproduce, compete, and evolve. Such perspective is able to take TEs' variability into account and discern their possible "functions" at different levels of biological organization among the much wider set of their "effects." Organism and population perspectives on TEs, without an integration with the element-level perspective, produce explanations that are blind to variation, very coarse-grained in attributing functions, ultimately incomplete, and sometimes even logically problematic (like, for example, the explanation of TEs as functional for future evolvability). A multilevel approach can tease apart four separate domains of scientific questioning about TEs—that is, their origin, persistence, abundance, and phylogenetic distribution—and lead to the discovery of parameters, mechanisms, relationships, and functions to produce a much more powerful evolutionary explanation.

Silvia Caianiello focuses on organismal-suborganismal levels (referred to as the "phenotypic hierarchy" by the author) of biological hierarchies and explores the nontrivial relationship between genetic underpinnings and epigenetic phenomena that ultimately results in generation of an organism-level phenotype. This theme is of particular importance to the development of the hierarchy theory of evolution because the theory was largely developed in attempt to account for macroevolutionary phenomena and generally paid limited attention to suborganismal processes. The chapter provides a sound summary of theoretical developments placed squarely into a hierarchical perspective and forms a sound basis for future development of the hierarchy theory at the organismal and suborganismal level, elucidating the causality of phenotypic determination.

Telmo Pievani and *Andrea Parravicini* focus on the complex relationships between hierarchy theory of evolution as advocated in this volume and a long-standing set of diverse and contentious issues described under the rubric of multilevel selection. The authors provide a concise historical outline of the multilevel selection debates—sometimes very conflictual—with the rise, fall, and rehabilitation of the idea that units other than organisms can undergo processes of selection, or at least sorting, up to the current undisputed pluralistic approach. Pievani and Parravicini also connect multilevel selection with the framework of the "major transitions" of evolution, showing that this area of study, on one hand, can be seen as another multilevel view of evolution (with a strong diachronic dimension) and, on the other hand, concerns the origin of new systems, individuals, and levels. Selectionist multilevel approaches are then put into relation with the ontological stance of the hierarchy theory of evolution, showing the ecological extension and the much broader generality of the hierarchy theory of evolution in disentangling the processes and mechanisms of evolution.

The contribution of *Pavličev et al.* focuses on emergence in biological systems and proposes some theses that are relevant to the hierarchy theory of evolution. The chapter attempts to associate one kind of emergence (third order emergence in a classification by T. Deacon) with the origin of novel individuals. The term *individual* is used by the authors in such a general way as to encompass "individual" body parts (characters), developing individual organisms, "individual" cultural patterns, and emergent entities characterizing evolutionary major transitions. Following a brief theoretical introduction, the chapter delves into examples from two domains: the evolutionary differentiation of suborganismal structures and human cultural diversification. Based on their analysis, the authors derive the conditions for the origin of an individual—the existence of historical lineages of structures, the presence of relations of feedback and cooperativity among parts, and the operation of self-maintaining robustness in relation to environmental influences. Additionally, the authors observe the frequency of switch-like thresholds in the context-sensitivity of complex systems. Context-sensitivity of living systems (e.g., robustness to only a limited range of environmental conditions) is what confers the possibility of further emergence of individuals and properties. Topological properties are important to explain the origin of individuals, although they are not sufficient, as other properties pertaining to the parts (e.g., reaction norms of species in an ecosystem) are determining factors.

CHAPTER 6

Why Genomics Needs Multilevel Evolutionary Theory

T. Ryan Gregory, Tyler A. Elliott, and Stefan Linquist

Introduction

The iconic image of a human karyotype may suggest otherwise, but in fact, chromosomes do not align themselves neatly alongside their matched partners and position themselves conveniently in rank order by size. Real genomes are far less cooperative than this, which explains why a property as fundamental as the number of human chromosomes was incorrectly accepted as 2n = 48 for more than thirty years before it was accurately assessed as 2n = 46 in the mid-1950s (Harper 2006). A few more decades passed before it could be determined that human chromosome 2 represents a fusion of two ancestral chromosomes that remain separate in the other apes (who do, indeed, have 2n = 48 chromosomes; Yunis and Prakash 1982; IJdo et al. 1991).

Similar difficulties have arisen in efforts to determine other basic features of the genome, including the number of protein-coding genes. In this case, some early calculations had suggested a maximum of a few tens of thousands of protein-coding genes (e.g., Spuhler 1948; King and Jukes 1969; Ohno 1972), but a finding that the human genome does not contain so many more genes than that of a fruit fly or roundworm was nonetheless considered rather surprising to many modern authors (e.g., Hahn and Wray 2002; Claverie 2003; Pennisi 2007). Prior to the availability of the draft human genome sequence, most genomicists were (literally) willing to bet that the number of human genes was much larger (Pennisi 2003). Current estimates place the number of protein-coding genes in the human genome at approximately 20,000, though this number continues to be revised (usually downward) in light of new analyses (e.g., Pertea and Salzberg 2010; Ezkurdia et al. 2014).

Perhaps the simplest characteristic of all, the quantity of DNA in the genome, has proven to be one of the most challenging to understand. This is because the human genome, at approximately 3.2 billion megabase pairs (Mbp) of DNA per copy, is simultaneously both confusingly small and perplexingly large—small as compared to genomes such as those of certain aquatic salamanders or lungfishes, which may be thirty to forty times larger than our own (even an onion's genome contains five times more DNA than a human's), but large in the sense that it contains vastly more DNA than is obviously required to specify the construction of even a complex mammal. Indeed, less than 2 percent of the human genome is made up of protein-coding genes. The rest is composed of a diverse array of noncoding DNA.

All told, eukaryotic genome sizes range about 60,000-fold. Among animals, the range is still a staggering 7,000-fold, and even among the vertebrates alone it is about 350-fold (Gregory 2016). In other words, organisms of similar complexity and, one might assume, genetic requirements, may differ by orders of magnitude in the quantity of DNA in their nuclei. Genome size bears no relation to any intuitive notions of organismal complexity, a fact that perplexed biologists for decades prior to the discovery of noncoding DNA. At that point, there was no longer any reason to expect large genome sizes to be associated with more complex developmental programs or organismal body plans. Yet the discovery of noncoding DNA of numerous types raised several new questions regarding its origin, mechanisms of its gain and loss, its impacts (or perhaps functions) in terms of organismal biology, and the reasons for its widely varying representation among eukaryotic genomes. Most of these questions remain unresolved despite decades of active research.

Most recently, there has been heated debate regarding the question of the proportion of the human genome that is "functional" (as opposed to representing "junk DNA"). Much of this has been spurred by claims—namely, that 80 percent of the human genome is "functional"—made in publications reporting the results of the Encyclopedia of DNA Elements (ENCODE) project and associated media blitz (e.g., Ecker 2012; ENCODE Project Consortium 2012; Pennisi 2012). Some of the strongest criticism has focused on the definition of *function* and in the (extremely lax) criteria used by ENCODE in ascribing functions to nongenic elements in the genome (e.g., Doolittle 2013; Graur et al. 2013). Additionally, there has often been a failure—by ENCODE and many others—to consider how the biological characteristics and evolutionary histories of genetic elements themselves may explain some of their observed properties, independent of (or, indeed, in contrast to) organism-level accounts.

In this chapter, we provide a brief overview of the rationale for adopting a multilevel evolutionary approach to understanding large-scale genome evolution. We also highlight some examples of how this perspective can result in the generation of alternative hypotheses and lead to differing explanatory outcomes compared to the traditional organism-centric view.[1]

The Jungle in the Genome

If protein-coding regions represent a small portion of even a modestly sized genome such as our own, then what makes up the difference—both within our genome and among genomes of widely varying sizes? Aside from protein-coding exons, eukaryotic genomes are home to a range of sequences of varying descriptions, including defunct gene copies (pseudogenes), noncoding regions interspersed among the exons (introns), highly repetitive strings of short sequences (satellite DNA), and other such elements. However, it is becoming increasingly apparent that the dominant component of most eukaryotic genomes, and the one that has the greatest influence on genome size, is the category of sequences known as transposable elements (TEs).

Transposable elements, as the name implies, are sequences that are capable of moving (transposing) from one location in the genome to another, often leading to an increase in copy number in the process. TEs are broadly categorized into two classes according to their basic mode of transposition. Class I TEs, also called retroelements or retrotransposons, are first transcribed into RNA, which is then reverse transcribed back into the genomic DNA at another location. In this sense, Class I elements utilize a "copy and paste" mechanism, with the original source element remaining in its previous location as a new copy is inserted elsewhere. Class II TEs, or DNA transposons, make use of a "cut and paste" mechanism in which the element itself is excised from the genome and inserted somewhere else, without passing through an RNA intermediate. Some TEs are fully autonomous in that they encode the enzymes necessary for their own transposition. Others require access to the transpositional machinery of unrelated elements. For example, short interspersed nuclear elements (SINEs) are dependent on long interspersed nuclear elements (LINEs) for their transposition ability.

Whereas the human genome contains perhaps 20,000 protein-coding genes, it is host to more than 3 million copies of TEs, including more than one million copies of the primate-specific SINE known as *Alu* and more than 500,000 copies of the LINE called *LINE-1*. Identifiable TEs make up about half of the human genome, although the majority of these are no longer active and instead are more akin to molecular fossils. TEs also seem to be the primary contributors to variability in genome size across taxa, although the specific composition and distribution of TEs in different genomes may vary considerably (Elliott and Gregory 2015a,b). Explaining differences in TE abundance and diversity therefore represents a primary requirement for achieving an understanding of genome form, function, and evolution.

TEs and Selection at Multiple Levels

Transposable elements are often characterized as "parasites" of the genomic "host," and indeed many are known to act as disease-causing mutagens and are actively suppressed by various genomic systems. In other cases, former TEs have been co-opted by the genome to serve important regulatory, developmental, or other roles at the cellular and organismal levels. Much of the debate over how much of the human genome is functional at the organism level is based on whether one views TEs as parasites (or, at best, as harmless commensals) or as integrated components of a complex regulatory system (or, at least, as mutualists). In turn, the tendency to fall into one camp or the other often reflects an organism-centric perspective on TEs—that is, they are evaluated in terms of being either detrimental, neutral, or beneficial from the perspective of the organism harboring them.

These organism-oriented options also drive the formulation of many single-parameter models of genome size evolution, such as the mutational hazard hypothesis (TE insertions are mildly deleterious and the factor determining genome size is host population size; Lynch and Conery 2003; Lynch 2007), the mutational equilibrium model (TEs are not under strong selection and the factor that shapes genome size is the ratio of DNA deletion over insertion; Petrov et al. 2000; Petrov 2002), and the nucleoskeletal theory (genome size is under positive selection for its impact on nucleus size, which coevolves to match adaptive changes in cell size; Cavalier-Smith 1978). Each of these perspectives has been criticized as insufficient, by itself, to account for observed differences in genome size (e.g., Gregory 2001, 2003, 2004b; Linquist et al. 2015).

The alternative to considering the properties and constituents of genomes only from an organismal point of view is to incorporate additional levels of organization into a multilevel evolutionary framework. Most obviously, this would involve the addition of TEs themselves as biological entities that reproduce, compete, and evolve within the genome and interact and coevolve with the host genome in which they reside. Doing so does not negate the importance of organism-level selection in shaping the genome, as some TEs are indeed selected for or against based on their effects on the organism.

Moreover, TEs contribute along with all other types of sequences the total DNA content of the nucleus, which itself is well known to correlate positively with cell size and negatively with cell division rate in many taxa. These cell-level relationships, in turn, can lead to correlations between genome size and features such as metabolic rate (e.g., due to surface area to volume effects on gas exchange rates), developmental rate (e.g., due to slower individual cell divisions when more DNA must be replicated), body size (e.g., when cell number is relatively constant and cell size affects overall body size), and by extension various aspects of developmental, physiological, and ecological lifestyle (e.g., flight, metamorphosis), as well as environmental tolerances and geographical distribution (for reviews, see Gregory 2005b; Bennett and Leitch 2005). This alone would be sufficient to justify a view involving more than one level, as there are likely to be many cases of intragenomic evolutionary processes driving changes in genome size in one direction (e.g., selection among TEs for increased transposition rate) that are counteracted by organism-level selection in another direction (e.g., selection against genome expansion in birds with high metabolic demands and a related requirement of small cells). This aspect of multilevel thinking in genome biology has been discussed in detail previously (Gregory 2004a, 2005a, 2013). In the remainder of this chapter, we explore some additional ways in which adding a TE-level perspective can have significant impacts on how genomic phenomena are investigated and understood.

Existence Is Not Evidence of Function

In the absence of a multilevel perspective, it has been common to argue that the sheer abundance of noncoding DNA suggests that it must have an as yet undescribed organism-level function. Were it not so, the reasoning goes, then natural selection (operating at the organism level) would have eliminated it long ago. The classic "selfish DNA" papers (Doolittle and Sapienza 1980; Orgel and Crick 1980) were authored in part as a response to this widespread notion. As the authors of the selfish DNA papers rightly noted, the fact that many genetic elements are capable of autonomous replication means that there is automatically another plausible explanation for their existence without them necessarily being functional for organisms—namely, that they are there because they are good at being there. The authors were careful to note that *some* of these elements were likely to have been co-opted into functional roles, but that this was not a requirement for understanding their existence, even in massive numbers.

By way of analogy, we might contemplate the fact that the human gut contains at least as many bacterial cells as there are cells comprising the remainder of the human body (Gill et al. 2006; Sender et al. 2016), and these contain one hundred times more genes than are found in the human genome (Gill et al. 2006). It may be tempting to assume that these bacteria provide a valuable digestive service to their human hosts, or else surely they would have been purged. Perhaps we should even consider humans as "superorganisms," most of whose genes happen to be packaged inside bacterial cells. To be sure, the makeup of the gut microbiome has important impacts on human health, and it is likely that removal of large quantities of bacteria from the digestive tract would have noticeable consequences. But we must also keep in mind that bacteria are organisms in their own right, and that from their perspective, the human gut represents prime habitat in which to live and reproduce. Some of the bacterial species in the gut can be highly detrimental if they acquire certain genes or become too abundant. A great many may simply be there because they are good at being there. A strictly human-centric view of this bacterial community would disregard this obvious possibility.

Origin, Persistence, Abundance, and Distribution

For any given trait, we might inquire as to when, how, and why it arose, but also why it continues to occur, why it is either common or rare, and why it is found in certain lineages and not others. The fact that the functions of traits may shift, or that nonfunctional features may acquire a functional role, means that the reasons for a trait's origin and for its continued persistence, its abundance, and its phylogenetic distribution may all be different (e.g., Gould and Lewontin 1979; Gould and Vrba 1982). Feathers provide a clear example in this regard, as they must have originally arisen for one function (e.g., thermoregulation) in avian dinosaurs before being co-opted and modified for a different function (e.g., sexual displays, gliding) prior to and during the evolution of flight. In modern birds, feathers play a variety of roles, including thermoregulation, flight, and sexual selection—studying just one of these would not provide an accurate understanding of when or why feathers first evolved, or how they and their functions have changed over evolutionary time.

The same principle applies to discussions of transposable elements, and taking an element-level perspective rather than focusing solely on the organism allows us to ask each of these evolutionary questions individually. Consider the aforementioned *Alu* element, the most numerous type of TE in the human genome (for a review of *Alu* biology, see Batzer and Deininger 2002). In terms of its origin, *Alu* appears to have arisen from a duplication of a 7SL RNA gene near the origin of the primates. However, this is distinct from the question relating to its abundance—for example, why is it present in such enormous numbers in the human genome? How did it become so abundant, and why do so many copies persist? Answers to these questions are likely to involve parameters such as transposition rates, mechanisms of preventing deletion by the host genome, relationships with other elements (e.g., LINEs, on which *Alu* depends for transposition), potential organism-level functions of individual elements, and impacts on cells and organisms in the aggregate of genome size.

Recognizing Element-Level Variability

Another important consequence of the element-level view is that it is allows an acknowledgment of variability among TEs. Failure to recognize differences among individual copies of the same TE and/or distinct families of TEs can affect various aspects of genomic analysis. For example, the widely cited mutational hazard hypothesis of genome evolution (Lynch and Conery 2003; Lynch 2007) was criticized early on for treating all elements in essentially the same way—namely, as being mildly deleterious. This is important because the core of the hypothesis is that a deciding factor of whether such a mildly deleterious insertion is removed by purifying selection is whether population size is large or small—that is, whether selection or drift holds sway. As Charlesworth and Barton (2004) pointed out, TEs differ markedly in their level of "virulence," and thus neither the propensity to spread within the genome nor the potential damage done upon insertion is equal across all elements. Even a significantly detrimental element can persist and increase in number if it highly active for the same reason that deadly viruses can still spread through a population if their rates of transmission to new hosts are sufficiently high.

A similar tendency to paint all genomic elements with the same broad brush was observed under the ENCODE project. In this case, criteria that might have been useful for identifying functional protein-coding regions (e.g., transcription, proximity to a regulatory region, chromatin modification) were applied across the genome. Any element that gave a positive result on any one of these assays in even one cell type at least one time was classified as having a "biochemical function" (and contributed to the much-criticized claim that 80 percent of the human genome has an identifiable function). As we have pointed out previously, this is akin to calibrating one's metal detector in a jewelry store and then assuming that every item detected on the beach must be gold (Linquist et al. 2014). The fact is that not all elements are the same, and that some—notably TEs—may exhibit features associated with function in protein-coding genes for other reasons. An obvious example would be the fact that retrotransposons are transcribed, which is a necessary step in their mode of transposition and need not reflect any organism-level function at all.

A recognition of element-level variability can provide some much-needed perspective on claims of TE function in another important sense. Far too many papers adopt an opening premise that TEs were "long dismissed as useless junk," but that there is now evidence that they serve important roles. Notwithstanding that the first part of the claim is patently untrue—there was no time in which potential functions for noncoding elements were widely dismissed out of hand—the second aspect of the claim is problematic in its adoption of an "all or nothing" stance. That is to say, finding evidence of function for a small number of particular TE copies does not license a conclusion that all copies of that element (let alone all TEs) are functional as well. To cite just one example among many, Wang et al. (2007) begin their paper on endogenous retroviruses (ERVs, a type of TE) with the customary claim that such elements were "once deemed 'junk DNA,'" and proceed to argue that they are not mere junk because they function as distributed binding sites for p53 proteins. Indeed many ERVs may serve this role, but in this paper only 1,509 of 319,000 ERV copies (0.5 percent) that were found include the relevant binding site. About 90 percent of the ERVs are represented only by "solo

LTRs," long terminal repeats at the end that remain after the rest of the element has been lost. Moreover, ERVs, like various other types of TEs, have been implicated as causes of disease. An element-level perspective would suggest that these TEs have accumulated within the genome for some other reason (perhaps, again, because they are good at accumulating), and then a small fraction of them have since been co-opted for use by the host genome.

Effects versus Functions

There is an important distinction to be made between *effects* and *functions* in genome biology (for reviews, see Doolittle et al. 2014; Elliott et al. 2014). Effects can be positive or negative and need not reflect the outcome of natural selection. Functions, under more robust definitions, are the subset of effects that were in fact shaped by natural selection. A classic example is provided by the heart—adding to total body mass is an effect of the organ, whereas pumping blood is its function. In like fashion, it is clear that transposable elements have various effects on organisms, from acting as mutagens when they insert into genes to influencing cell size via their combined effects as part of total bulk DNA content, but one cannot conclude that these elements are functional by merely observing that they have effects.

There can be a strong temptation to ascribe functions to TEs whenever they are seen to have some positive effect on the host genome. Without a multilevel perspective that includes TEs in their own right, it may be difficult to see any other possible explanations. Alternative interpretations present themselves when the multilevel view is adopted, however. This is well illustrated by the example of stress-induced TE activation. It has been known for some time that cellular stress, through wounding or thermal or chemical stressors, can lead to otherwise quiescent TEs being expressed (e.g., Arnault and Dufournel 1994; Wessler 1996; Grandbastien 1998; Capy et al. 2000). Under an organism-centric perspective, it would be easy to consider this as evidence that TEs play a role in mediating the genome's response to stress. However, when the element level is considered, additional explanatory hypotheses arise. For example, it is possible that the TEs are silenced by use of a coarse-grained mechanism such as methylating large swaths of DNA. If genes relevant to the stress response are also turned off during normal conditions in the same way in nearby locations, then accessing these genes during times of stress may inadvertently lead to the escape of previously suppressed TEs. It is also possible that a stressed host simply becomes incapable of controlling the activity of virulent TEs. In either case, it would be a mistake to attribute the activation of TEs during cellular stress to them having an adaptive role in host stress response (Elliott et al. 2014).

Evolvability and Future Benefit versus Interlineage Selection

Thus far, our discussion of the multilevel perspective has focused on TEs, genomes, and organisms as the interacting levels of organization. However, it is possible to extend this further to include higher levels as well. Doing so is important when considering a common (mis)interpretation of TE function as having to do with generating evolutionary novelty. As noted, TEs may have substantial mutagenic effects through their insertional activities. They may also represent sources of genetic material that can be co-opted by the host genome into various regulatory or structural roles. And their presence as highly repetitive sequences may instigate chromosome-level rearrangements that generate further variation. In this regard, TEs may play a very important role in providing the raw material upon which natural selection can act, potentially opening up novel evolutionary trajectories (e.g., Biémont and Vieira 2006). Yet it would be a fallacy to consider this potential evolutionary effect to be the function of TEs—natural selection possesses no foresight, and features cannot be maintained today because of the benefit that they may confer tomorrow (Gregory 2009). Unfortunately, more than a few genomicists have fallen into this trap. In order to make sense of these long-term evolutionary consequences without using flawed reasoning about future benefits, it is necessary to invoke a process of interlineage selection. This also requires a recognition of the distinctions between origin, persistence, abundance, and distribution described earlier.

A clear example of this is given by sexual reproduction. Sex involves significant costs, including the requirement to find a mate and the passing on of only half of one's DNA to each offspring. Yet it evolved, persisted, and has become very common across eukaryotes. Various benefits can be identified, such as diversifying the genetic makeup of one's offspring, allowing recombination that makes it possible to purge deleterious alleles, and enabling faster evolutionary change in response to challenges posed by pathogens. These benefits and others may help to explain why sexual reproduction persists, but they do not answer the question of why sex arose in the first place. Why sexual reproduction is so common may have yet another explanation—for example, sex increases the rate of speciation and/or reduces the risk of extinction such that sexual lineages outreproduce and/or outlast asexual lineages. This would be an example of interlineage selection and is distinct from the genome- and organism-level effects of sex. Such interlineage selection relates—retrospectively—to the reasons why certain lineages have branched or survived more than others. This is fundamentally different from assuming that sex exists because it leads to more diversification in the future.

Likewise, TEs may indeed contribute to evolvability by fostering mutation and recombination, but this cannot be why they arose or why they persist in the genome. Rather, the relevant question would be historical in nature: Are the lineages that diversified the most and/or survived the longest the ones that had more TE-generated variation? Similar lines of reasoning have been applied to other types of noncoding DNA (Doolittle 1987, 1989). The important issue from the standpoint of this chapter is that arguments based on evolvability are nonsensical when they invoke future benefits but make for useful hypotheses when they properly incorporate multilevel thinking.

Concluding Remarks

The study of genomes is beset by challenges both technical and conceptual. Even the most fundamental properties of the genome, such as chromosome number and gene count, took decades of work to elucidate. More complex aspects, such as the factors that have shaped genome size and sequence content, remain open questions after more than sixty-five years of study. There remains considerable disagreement in the literature regarding the proportion of the human genome that is functional at the organism level—a debate that has often been hampered by a narrow focus on the organism as the only relevant level of biological organization.

In this chapter, we have endeavored to show how a multilevel perspective is important in investigating the characteristics of the genome and their biological significance. This multilevel approach allows the generation of new hypotheses and helps us to avoid both false dichotomies (e.g., either all TEs are functional or all are "junk") and faulty reasoning (e.g., TEs are maintained in the genome because they may someday prove useful).

Eukaryotic genomes are complex, and much remains to be understood about their basic biology and evolution. Adopting a multilevel framework allows a fuller exploration of the critical issues involved in the ongoing effort to achieve this understanding.

Notes

1. For more detailed and technical discussions of issues surrounding function and levels of selection in genome evolution, see Gregory (2004, 2005, 2014); Linquist et al. (2013); Doolittle et al. (2014); and Elliott et al. (2014).

References

Arnault, C., and I. Dufournel. 1994. "Genome and Stresses: Reactions against Aggressions, Behavior of Transposable Elements." *Genetica* 93: 149–60.

Batzer, Mark A., and Prescott L. Deininger. 2002. "*Alu* Repeats and Human Genomic Diversity." *Nature Reviews Genetics* 3: 370–80.

Bennett, Michael D., and Ilia J. Leitch. 2005. "Genome Size Evolution in Plants." In *The Evolution of the Genome*, edited by T. Ryan Gregory, 89–162. San Diego: Elsevier.

Biémont, Christian, and Cristina Vieira. 2006. "Junk DNA as an Evolutionary Force." *Nature* 443: 521–24.

Capy, P., G. Gasperi, C. Biémont, and L. Bazin. 2000. Stress and Transposable Elements: Co-evolution or Useful Parasites? *Heredity* 85: 101–6.

Cavalier-Smith, T. 1978. "Nuclear Volume Control by Nucleoskeletal DNA, Selection for Cell Volume and Cell Growth Rate, and the Solution of the DNA C-Value Paradox." *Journal of Cell Science* 34: 247–78.

Charlesworth, Brian, and Nick Barton. 2004. Genome Size: Does Bigger Mean Worse? *Current Biology* 14: R233–R235.

Claverie, Jean-Michel. 2001. "What If There Are Only 30,000 Human Genes?" *Science* 291: 1255–57.

Doolittle, W. Ford. 1987. "What Introns Have to Tell Us: Hierarchy in Genome Evolution." *Cold Spring Harbor Symposia on Quantitative Biology* 52: 907–13.

———. 1989. "Hierarchical Approaches to Genome Evolution." *Canadian Journal of Philosophy* 14 (supplementary): 101–33.

———. 2013. "Is Junk DNA Bunk? A Critique of ENCODE." *Proceedings of the National Academy of Sciences of the United States of America* 110: 5294–5300.

Doolittle, W. Ford, Tyler D. P. Brunet, Stefan Linquist, and T. Ryan Gregory. 2014. "The Distinction between 'Function' and 'Effect' in Genome Biology." *Genome Biology and Evolution* 6: 1234–37.

Doolittle, W. Ford, and Carmen Sapienza. 1980. "Selfish Genes, the Phenotype Paradigm, and Genome Evolution." *Nature* 284: 601–3.

Ecker, Joseph R. 2012. "ENCODE Explained." *Nature* 489: 52–55.

Elliott, Tyler A., Stefan Linquist, and T. Ryan Gregory. 2014. "Conceptual and Empirical Challenges of Ascribing Functions to Transposable Elements." *American Naturalist* 184:14–24.

Elliott, Tyler A., and T. Ryan Gregory. 2015a. "Do Larger Genomes Contain More Diverse Transposable Elements?" *BMC Evolutionary Biology* 15: 69. doi:10.1186/s12862-015-0339-8.

Elliott, Tyler A., and T. Ryan Gregory. 2015b. "What's in a Genome? The C-Value Enigma and the Evolution of Eukaryotic Genome Content." *Philosophical Transactions of the Royal Society B*, in press.

ENCODE Project Consortium. 2012. "An Integrated Encyclopedia of DNA Elements in the Human Genome." *Nature* 489: 57–74.

Ezkurdia, Iakes, David Juan, Jose Manuel Rodriguez, Adam Frankish, Mark Diekhans, Jennifer Harrow, Jesus Vazquez, Alfonso Valencia, and Michael L. Tress. 2014. "Multiple Evidence Strands Suggest That There May Be as Few as 19,000 Human Protein-Coding Genes." *Human Molecular Genetics* 23: 5866–78.

Gill, Steven R., Mihai Pop, Robert T. DeBoy, Paul B. Eckburg, Peter J. Turnbaugh, Buck S. Samuel, Jeffrey I. Gordon, David A. Relman, Claire M. Fraser-Liggett, and Karen E. Nelson. 2006. "Metagenomic Analysis of the Human Distal Gut Microbiome." *Science* 312: 1355–59.

Gould, Stephen J., and Richard Lewontin. 1979. "The Spandrels of San Marco and the Panglossian Paradigm: A Critique of the Adaptationist Programme." *Proceedings of the Royal Society of London B* 205: 581–98.

Gould, Stephen J., and Elizabeth S. Vrba. 1982. "Exaptation—A Missing Term in the Science of Form." *Paleobiology* 8: 4–15.

Grandbastien, M.-A. 1998. "Activation of Plant Retrotransposons under Stress Conditions." *Trends in Plant Science* 3: 181–87.

Graur, Dan, Yichen Zheng, Nicholas Price, Ricardo B. R. Azevedo, Rebecca A. Zufall, and Eran Elhaik. 2013. "On the Immortality of Television Sets: 'Function' in the Human Genome According to the Evolution-Free Gospel of ENCODE." *Genome Biology and Evolution* 5: 578–90.

Gregory, T. Ryan. 2001. "Coincidence, Coevolution, or Causation? DNA Content, Cell Size, and the C-Value Enigma." *Biological Reviews* 76: 65–101.

———. 2003. "Is Small Indel Bias a Determinant of Genome Size?" *Trends in Genetics* 19: 485–88.

———. 2004a. "Macroevolution, Hierarchy Theory, and the C-Value Enigma." *Paleobiology* 30: 179–202.

———. 2004b. "Insertion-Deletion Biases and the Evolution of Genome Size." *Gene* 324: 15–34.

———. 2005a. "Macroevolution and the Genome." In *The Evolution of the Genome*, edited by T. Ryan Gregory, 679–729. San Diego: Elsevier.

———. 2005b. "Genome Size Evolution in Animals." In *The Evolution of the Genome*, edited by T. Ryan Gregory, 3–87. San Diego: Elsevier.

———. 2009. "Understanding Natural Selection: Essential Concepts and Common Misconceptions." *Evolution: Education and Outreach* 2: 156–75.

———. 2013. "Molecules and Macroevolution: A Gouldian View of the Genome." In *Stephen J. Gould: The Scientific Legacy*, edited by Gian Antonio Danieli, Alessandro Minelli, and Telmo Pievani, 53–71. Milan: Springer Italy.

———. 2016. Animal Genome Size Database. http://www.genomesize.com.

Hahn, Matthew W., and Gregory A. Wray. 2002. "The G-Value Paradox." *Evolution & Development* 4: 73–75.

Harper, Peter S. 2006. *First Years of Human Chromosomes: The Beginning of Human Cytogenetics*. Bloxham, UK: Scion.

Hasler, Julien, and Katharina Strub. 2006. "*Alu* Elements as Regulators of Gene Expression." *Nucleic Acids Research* 34: 5491–97.

Ijdo, J. W., A. Baldini, D. C. Ward, S. T. Reeders, and R. A. Wells. 1991. "Origin of Human Chromosome 2: An Ancestral Telomere-Telomere Fusion." *Proceedings of the National Academy of Sciences USA* 88: 9051–55.

King, Jack L., and Thomas H. Jukes. 1969. "Non-Darwinian Evolution." *Science* 164: 788–98.

Linquist, Stefan, Karl Cottenie, Tyler A. Elliott, Brent Saylor, Stefan C. Kremer, and T. Ryan Gregory. 2015. "Applying Ecological Models to Communities of Genetic Elements: The Case of Neutral Theory." *Molecular Ecology* 24: 3232–42.

Lynch, Michael. 2007. *The Origins of Genome Architecture*. Sunderland, MA: Sinauer.

Lynch, Michael, and John S. Conery. 2003. "The Origins of Genome Complexity." *Science* 302: 1401–4.

Ohno, Susumu, 1972. "So Much 'Junk' DNA in Our Genomes." In *Evolution of Genetic Systems*, Edited by H.H. Smith, 366–70. New York: Gordon and Breach.

Orgel, Leslie E., and Francis H. C. Crick. 1980. "Selfish DNA: The Ultimate Parasite." *Nature* 284: 604–7.

Pennisi, Elizabeth. 2003. "A Low Gene Number Wins the GeneSweep Pool." *Science* 300: 1484.

———. 2007. "Why Do Humans Have So Few Genes?" *Science* 309: 80.

———. 2012. "ENCODE Project Writes Eulogy for Junk DNA." *Science* 337: 1159–61.

Pertea, Mihaela, and Steven L. Salzberg. 2010. "Between a Chicken and a Grape: Estimating the Number of Human Genes." *Genome Biology* 11: 206.

Petrov, Dmitri A. 2002. "Mutational Equilibrium Model of Genome Size Evolution." *Theoretical Population Biology* 61: 533–46.

Petrov, Dmitri A., Todd A. Sangster, J. Spencer Johnston, Daniel L. Hartl, and Kerry L. Shaw. 2000. "Evidence for DNA Loss as a Determinant of Genome Size." *Science* 287: 1060–62.

Sender, Ron, Shai Fuchs, and Ron Milo. 2016. "Revised Estimates for the Number of Human and Bacteria Cells in the Body." *bioRχiv* http://dx.doi.org/10.1101/036103.

Spuhler, J. N. 1948. "On the Number of Genes in Man." *Science* 108: 273–80.

Wang, Ting, Jue Zeng, Craig B. Lowe, Robert G. Sellers, Sofie R. Salama, Min Yang, Shawn M. Burgess, Rainer K. Brachmann, and David Haussler. 2007. "Species-Specific Endogenous Retroviruses Shape the Transcriptional Network of the Human Tumor Suppressor Protein p53." *Proceeding of the National Academy of Sciences of the USA* 104: 18613–18.

Wessler, S. R. 1996. Plant Retrotransposons: Turned on by Stress. *Current Biology* 6: 959–61.

Yunis, Jorge J., and Om Prakash. 1982. "The Origin of Man: A Chromosomal Pictorial Legacy." *Science* 215: 1525–30.

CHAPTER 7

Revisiting the Phenotypic Hierarchy in Hierarchy Theory

Silvia Caianiello

1. The Tenets of the Hierarchy Theory and the Role of the Somatic Hierarchy

Since the 1980s, the hierarchy theory of evolution supports a multilevel approach to the complexity of the living world. Drawing from the epistemological premises of the theory of complex hierarchical systems, it emphasized that for developing a "general theory of biology," it is necessary to delve into a "theory of hierarchical levels—how they arise and interact" (Eldredge and Vrba 1984; cf. Vrba and Gould 1986; Allen and Starr 1982; Salthe 1985; Eldredge and Salthe 1984; Cordeschi 2010). This approach pointed out the failure of modern synthesis to "capture multiple levels of biological organization" (Love 2006), upholding that evolutionary processes operate over multiple "scalar" levels. In this view, levels are composed of a succession of nested entities characterized by different spatiotemporal ranges, and entities at different levels are distinguished by diverse "natural frequencies" in the rate of their constitutive processes, which grow increasingly slower moving upward (Simon 1962). Because of this gap in timescales, cross-level interactions—unlike the direct causal dynamics taking place at the same ontological level—are framed as constraints (from lower level upward) and boundary conditions (when higher level processes affect lower level ones), both acting by restricting the space of possibilities (Allen and Starr 1982; Salthe 1985; Eldredge and Salthe 1984; Auletta et al. 2008). Furthermore, entities at different levels could vindicate the status of individuals, or "spatiotemporally bounded historical entities," rather than classes. "Glued" together by the property of more-making, they exhibit the "cohesion, unity, and integrity characteristic of organisms" (Eldredge 1985, 10, 144).

The task of the hierarchy theory from the start was both analytic and synthetic:

1. To evaluate the individual contribution of processes occurring at disparate lev- els of biological organization to the origination and maintenance of life's diver- sity in evolution

2. To integrate them into an encompassing explanatory framework by disentan- gling a wide phenomenology of cross-level hierarchical interactions—both re- lating patterns to lower level processes and supporting the privileged status of downward with respect to upward causal relationships (Eldredge 1985, 178)

In contrast to the older system theory's approaches and their "tradi- tional emphasis on isomorphisms among levels" (Salthe 1985, 134; Sal- the 1988), the hierarchy theory recognized the existence of "emergent" properties and proper dynamics at each level. This crucial opening, which uncoupled "differential success" from any "causal claim" (Gould 2002), enabled the hierarchy theory to consider also nonselectionist dynamics, both upward and downward, as the origin of "sorting" (Vrba and Gould 1986; Lieberman and Vrba 1995; Gregory 2004). The emphasis on "sort- ing" offered a richer perspective than the one represented by the multi- level selection debate (cf. Okasha 2006 for a review); it substantiated the programmatic "plea for pluralism" by Gould and Eldredge (1977), a plea that has not been missed in the actual multilevel approach to biological complexity (cf. Mitchell 2003).

The timing for a "detailed" development (Eldredge and Salthe 1984, 184) of the hierarchy theory on the biological and evolutionary ground was not accidental, nor was the disciplinary field from which it was launched, paleontology. Since the late 1970s, Gould, Eldredge, Vrba, and others be- came the fiercest opponents of genetic reductionism as represented by the ultra-Darwinism (which focuses on genes as the only relevant level of evo- lution) as well as of the "extrapolationist" approach of the Modern Syn- thesis, which generalizes microevolutionary processes to macroevolution.

At the time, the primary commitment of the hierarchy theory was to vindicate the distinctness of macroevolutionary phenomena from micro- evolutionary processes, a thesis still widely represented in contemporary evolutionary developmental biology (evo-devo; Davidson 2011; Stern and Orgogozo 2009, 21; Stern 2011). Focused as it was on the extension of biological, "Darwinian" individualities to the higher levels of the scalar hierarchy, the hierarchy theory at its onset tended to downplay the evolutionary relevance of levels and entities below the organism and above the gene. In this way, the so-called somatic or phenotypic hierarchy—the "proteins, organelles, cells, tissues, organs, and organ systems that make up the bodies of individual organisms"—could be deprived of evolutionary significance (Eldredge 1985, 134) or even conflated with an "integrated genotype-phenotype" organismic entity (Salthe 1985, 237–40). Similar was the destiny of "hierarchy of homology," set aside as a merely descriptive record of evolutionary history (Eldredge 1985, 192). This framework, however, was partly shattered by the unexpected finding of a shared gene regulatory toolkit for animal development—comprising not only similar developmental genes but more integrated and functional aspects of ontogenies: conserved molecular, cellular, and developmental processes formerly hidden beyond the veil of incomparable anatomies (Carroll et al. 2001; Duboule 2010). Provided with a new mechanistic basis (Roth 1984; Wagner 1989), the notion of homology came to encompass a new causally active role in the making of metazoan diversity (Amundson 2005, 8).

However, as the modern synthesis came, since the debate on the levels of selection, to gradually embrace a multilevel, hierarchical approach to evolution (Okasha 2006; 2012), even the hierarchy theory has moved forward. It has happened throughout a dense and yet unrecorded history of dialogical contributions, which has taken increasingly into account the evolutionary role of entities and processes at organizational levels below that of the organism (cf., among others, Wake and Larson 1987; Doolittle 1988; Lieberman et al. 1993; Rollo 1994; Salthe 2012; Tëmkin and Eldredge 2014). As the debate for the extension of the modern synthesis has lately underscored the need for a theory of phenotypic evolution—and, above all, of phenotypic variation as nonrandom and only indirectly related to genetic variation (Pigliucci and Müller 2010; Kirschner and Gerhart 2010, 276)—the time is ripe for a refinement and revision of the so-called somatic or phenotypic hierarchy, its levels of variation and constraint, and their evolutionary bearing.

Such a refinement might reassess some of the major commitments of the original theory. First, the search for "stable levels of packaging of information" (Eldredge 1985, 190) might be fruitfully applied to the phenotypic hierarchy. Second, the original quest for the "general rules by which variation evolves at each level of the genealogical hierarchy" (Eldredge and Vrba 1984) might encompass also the "structured variability" (Toussaint and von Seelen 2007) and differential evolutionary stability displayed by discrete entities at different phenotypic levels (Rollo 1994, 114).

This chapter reviews some reasons and implications for the reassessment of the evolutionary bearing of the so-called somatic or phenotypic hierarchy and of the semantic functions of its layered architecture.

2. The Developmental Dynamics of Information

The increasing awareness of the asymmetry between genetic and phenotypic evolution, even beyond the onset of genome-wide association studies (Moczek 2012, 3; Harrison et al. 2012; O'Malley et al. 2015, 11; Neumann-Held and Rehmann-Sutter 2006; Weinberg 2014; Callebaut et al. 2008), has fostered an organizational approach to evolution (Rollo 1994) focused on the distinctive role played in the informational process by the "different levels of hierarchical organization that intervene between the genotype and the phenotype" (Savageau 2001, 143; cf. Wilkins 2007, 8592; Noble et al. 2014, 2242). Defining this role is the challenge that must be met for envisaging a theory of biological organization suited to explain how, in living systems, "not only the constituent parts are reproduced, but also the organizational constraints that produce the distinctive order of the whole" (Goodwin 2006, 341; cf. Rollo 1994, 109; Fontana and Buss 1996; Bechtel 2007).

A consistent part of contemporary evo-devo has pushed forward a "developmentalist challenge" (Schaffner in print), which gainsays the restriction of informational capacity to genes, contesting that DNA sequences per se have no fixed meaning but rather become truly informational only in context. Biologically meaningful information would be namely produced only in the context-dependent expression of the genome at different organizational levels. This conceptual shift has brought to the fore the role of developmental systems in structuring and "even establishing" the information content of genes (Neumann-Held and Rehmann-Sutter 2006, 5).

Getting rid of the early, too simplistic application of the information theory in biology, what has come to light is that a fuller account of the informational process must include the receiver and encompass the process of information selection (Auletta et al. 2008; see Oyama 1985; Jablonka 2002). A crucial misunderstanding was engendered by a loose application of Shannon's theory of communication in biology. This latter in fact presupposes a message already well defined at the input and focuses solely on the faithfulness of its transmission in the presence of disturbances. This model facilitated considering biological information as an "inherent property of inputs" (Oyama 1985, 76–77)—that is, genes as the repository of transmissible information. The result was an undue shift from the statistical notion of information, understood in terms of number of alternatives and reduction of uncertainty, to a simplified liner model of causal agency.

A more matching model for biological systems is rather represented by Wiener's information theory. Wiener distinguished carefully between "information taken brutally and bluntly" and "semantically significant information," the one "which gets through to an activating mechanism in the system that receives it." This mechanism calls "on the whole of past experience in its transformations, and these long-time carry overs are not a trivial part of its work" (Wiener 1989, 81, 93–94). As stressed by Auletta et al. (2008), in biological phenomena, the input represents only a source of variety, while the selection (the meaning as the effective use of the information, in short, its biological function) is at the end of the information processing, the only phase in which some kind of control and noise filtering occurs.

A different formulation, more fine-tuned to reflect a hierarchical approach, is the distinction between "molecular syntax" as an ideally "complete account of molecular interactions" (already structured by inherent rules) as opposed to "molecular semantics"—that is, the specific context in the developing system that determines the biological role of the recruited lower level elements (Laubichler and Wagner 2001).

For instance, homologous genes in *C. elegans* (*lov-1*) and humans (PKD-1 and PKD-2), the proteins of which code for the same kind of receptors, come to perform different functions as they become part of quite different higher-level mechanisms in the two species. The same lower level "informational input" will thus change its biological meaning—its function—when embedded into different higher-level contexts (Weber 2005, 185). But even more complex entities can be recruited by still higher organizational levels, so as to drastically change their resulting functional role. An example thereof is the high evolutionary conservation of the same basic mechanisms of ontogenetic patterning of the avian limb, which produce quite different final phenotypes in different species because of the expansion of processes placed on secondary and tertiary levels of organization (Müller and Wagner 1991, 247). The "receiving apparatus"—which, in Wiener's terms, is "superimposed on the transmission line" so that it can filter the quantity of information needed "to serve as the trigger for action" (Wiener 1989), or the resulting biological function—can thus be located at different organizational levels of the phenotype.

To summarize, the issue of contextuality imposes a fuller account of the informational process, emphasizing the relational ontology of levels, which selects biologically meaningful information only as the result of a dynamics of codetermination (Bertolaso 2015; Bertolaso and Caianiello 2016). This interaction between the syntactic and semantic levels is doubly constrained. First, by the specific properties of the lower level entity or collective recruited into a higher level functional context (Wilkins 2007, 8591)—what might be equated to the "lower level constraints" as understood in the hierarchy theory—and, second, by all "the systemic properties (spatial, regulatory, dynamical) of higher level (cellular, organismal) biological entities as well as their evolutionary history" (Laubichler and Wagner 2001, 55), which might be equated to the "boundary conditions" in hierarchy theory. These two forms of nondynamic causality represent the extent of Salthe's "triadic system" framework (Salthe 1985), which, scaled at the infraorganismal levels, requires a correct identification of the organizational level at which semantically meaningful information is veritably produced.

It must be stressed, however, that in this view, organizational layers do not feature only as external boundary conditions, "contexts of meaning" (Pattee 1970, 123), or as ecological microenvironments, with which tightly integrated entities in the phenotypic hierarchy horizontally interact at their own level (Korn 2002). Rather, the different levels of phenotypic organization at play in the semantic process ought to be envisaged as constitutively part of the informational process (Oyama 1985, 77).

To feature how biological information can be organized in a "complex hierarchy ranging from DNA sequences and chromatin regulation to cellular signaling and tissue/organ organization" (Walker et al. 2013, 2), it is also necessary to overcome the restriction of information solely to digital codes. Molecular semantics entails the deployment of a rich array of analog coding processes, even more so at the subcellular level (Hoffmeyer 2006). The spatial dimension of the organized structure of developing systems becomes thus an integrative part of the informational process.

3. Phenotypic Hierarchy and the Evolution of Development

Phenotype is a multileveled concept (Newman 2003, 171) whose layers are related in nonlinear fashion to their genetic underpinnings. The function of genes must be associated to precise organizational levels (Deutsch 2010, 157; Shapiro 2013): "Genes at one level make molecular features at various phenotypic levels" (Korn 2002, 219). As stressed in the recent manifesto of the evolutionary systems biology, phenotypic transitions require a mechanistic and evolutionary analysis of the "genotype-phenotype relationships at multiple levels" (Soyer and O'Malley 2013, 606, my italics; cf. Mykles et al. 2010).

This multilevel extension of the phenotype matches the multilevel deployment of developmental processes at different organizational layers (Love 2006), the many stages and levels of organization that comprise the full extent of morphogenesis, from the regulatory genome (Davidson 2006) to interacting cells, tissues, and organ systems.

The phenotypic hierarchy arises from the nonlinear mapping of genotype and phenotype at multiple levels. Therefore, it cannot be unified into a unique compositional framework with genes as more fundamental, Lego-like elements at the lowest level. In fact, only a small subset of the genome is relevant for biological organization (Rollo 1994, 8–9; Gregory 2004). According to the widely accepted cis-regulatory, or "expressionist," paradigm (Vinicius 2010, 86ff.), few clusters of regulatory genes are relevant for developmental and morphological evolution, and mutations in cis-regulatory elements are regarded as the predominant mechanism underlying the evolution of development and form (Carroll 2008; but see Lynch and Wagner 2008). From the subcellular to the organismal level, the phenotypic hierarchy appears, therefore, to be a "procedural" hierarchy, involving regulatory and organizational functions (Love 2006)—the kind of hierarchy that, as pinpointed by Grene (1987, 505–6) only entails a veritable dynamic perspective.

It is, however, mostly on the evolutionary scale that the levels of such procedural hierarchy display their autonomous significance. Ontogeny in itself cannot be adequately conceived as a hierarchic process in a scalar sense (Minelli and Peruffo 1991; Wagner 2014, 420), although differences in the timing of developmental processes at different scales are considerable, as well as mechanistically crucial in the temporal ordering of morphogenesis (Alon 2007; Alderson and Doyle 2010). Nonetheless, global homeostasis across the networked subsystems of development is ensured by recursive combinatorial regulation (Niklas 2015), and local deviations are corrected by means of genetic feedbacks (Balavoine 2014). The timescale at which organizational levels do matter is rather the one of evolution of development and of the independent evolutionary diversification of units at different layers of biological organization (genetic pathways, mechanisms of development, morphological structures; Minelli and Fusco 2013). It is only at this scale that dissociable "modules" at different levels display their potential for developing a "basic autonomy" (Bechtel 2007).

Erwin and Davidson have elaborated a powerful model for representing the mapping between genomic organization and the evolution of phenotypes. They relate the architecture of the developmental genome to the magnitude of phenotypic evolution, from the megaevolutionary level of body plans to the macroevolutionary level of speciation. They argue that developmental gene regulatory networks (dGRN) exhibit a sort of spatial hierarchical structure, with an invariant core (kernels) responsible for the stability of body plans—whereby changes in lower (more peripheral) regulatory circuits match evolution at lower taxonomic levels—with the mechanisms for speciation featuring at the lowest level of the differentiation gene batteries (Davidson and Erwin 2006; Erwin and Davidson 2009). Macroevolution and microevolution would thus no longer be divided only by the nature of the changing genetic elements, with cis-regulatory mutations responsible for morphological variation across species and mutations in coding genes for variation within species (Stern and Orgogozo 2009, 750). It is rather the hierarchical position of the mutating elements in the dGRN architecture that determines the nature of the novelty they can produce.

The hierarchical order that drives evolution at different taxonomic levels may not be collapsed to the level of the DNA sequence and cannot, therefore, support the self-sufficiency of a "detailed analysis of cis-regulatory changes in genes that are direct targets of sequence level selection" (Davidson 2011, 35).[1] Even if all evolutionary changes in morphology are cis-regulatory, and the logic of their evolution rests on detectable principles, the evolution of body plan cannot be reduced to a record of cis-regulatory changes. In fact, such an uniformitarian view does not take into account the properties of gene regulatory networks and the new mechanistic factors that they bring into play. The developmental logic of evolutionary change is generated rather "at the system level . . . in the architecture of developmental gene regulatory networks, far from micro-evolutionary mechanism" (Davidson 2011, 35).

It is questionable how far the evolutionary role of phenotypic organization can be fully reflected in the mirror of evolving developmental gene regulatory networks or exhausted in the parallel, but strictly one way, hierarchies of genomic and phenotypic organization. The doubt that something else is required to account for evolution (Morange 2015), and the notion that evo-devo is at risk of losing, in the predominance of the developmental genetic approach, the dimensionality of a veritable hierarchical perspective is widespread (Love 2015; Wake 2015).

A quite different approach to macroevolution, albeit sharing the same antiuniformitarian commitment (Erwin 2011), is offered by the theory of evolutionary transitions. I will refer in particular to the version of the theory elaborated by Szathmáry and Maynard Smith since the late 1990s, which, borrowing Griesemer's concept of reproducer, provides a powerful account of how the evolution of the "packaging of information" cannot be split from the evolution of phenotypic organization, and furthermore lays the ground for overcoming the distinction between replicators and interactors.

The theory holds that evolutionary transitions are marked by the emergence of new higher-level individuals. The first major transition represents the shift from the limited heredity of "replicators" (simple autocatalytic systems)—which replicate only a limited amount of types and do not yield enough variability to undergo microevolution—to the potentially unlimited heredity of "reproducers," starting from "template replicators" such as oligonucleotides, which, exploiting the new mechanism of "modular" replication, can make an indefinite number of different types. The hereditary differences due to their inexact heredity fit the requisite of variability and, thus, reproducers prove better candidates as units of evolution than sheer replicators (Jablonka and Szathmáry 1995; Szathmáry and Maynard Smith 1997).

All further evolutionary transitions in individuality after replicators concern therefore only reproducers, and each such transition represents a further step in the direction of unlimited heredity. The condition of possibility of such novel individuals is the emergence of a new mode of development, which changes the way information is stored and transmitted. "Reproduction requires both copying and development" (Szathmáry and Maynard Smith 1997, 556), and higher-level individuals in evolutionary transitions arise only by the establishment of a new developmental capacity that can be transmitted to their offspring.

The new mode of development entails a new packaging of information, which irreversibly changes the role of "old-fashioned replicators," although their primary system of inheritance is maintained in parallel with the newly evolved one. Insofar as they are "embedded in an evolving hierarchy of levels of reproductive organization," replicators at each transition are further "locked in the deepest recess of a hierarchy of prisons," from which they manage to escape from time to time (Griesemer 2000).

Reproducers at different levels act both as "ecosystems" of lower level units (Szathmáry and Maynard Smith 1995)—exerting downward control for subduing "genetic interests of component entities in the direction of cooperation" (Griesemer 2000)—and as individualized entities, themselves subject to selection. This view further elaborates Gould's claim that units of selection must be interactors, or "actors within the gut of the mechanism," as persistence and replication are necessary but not sufficient conditions for acquiring the causal status of evolutionary individuals (Gould 2002, 619). Reproducers merge, for Griesemer, the ecological properties of interactors to the multiplicative property of replicators, linking more closely evolution and development.

Consequently, in this view, the "units of evolutionary transition" are more than just units of selection; they do not explain merely "evolution at levels" but "the evolutionary origin of new levels" (Griesemer 2002, 212). As units of evolution, other properties come to the foreground: first, their organizational properties and, second, their variational properties, which are constantly enhanced since the earliest transition to the "unfaithfulness" of unlimited heredity.

4. Homology and Individualization

Homology is a hierarchical concept (Roth 1994, 331). To be biologically meaningful, it must be related to a specific level of the biological hierarchy (Bolker and Raff 1996). Molecular, developmental, histological, and even behavioral or functional traits (Müller 2003) comprise the strictly phenotypic hierarchy of homology—"strictly" because the task of tracing back homologies to a "continuity of information" (Van Valen 1982) at the genetic level has been increasingly challenged by the findings of developmental genetics since the 1990s (Wagner 2014, 88). Even across phenotypic levels, it appears that we are just "courting dilemma" when we try to define homologues moving from homologies at lower levels (Sander and Schmidt-Ott 2004, 74). Morphological homologues can differ in their mode of development, so that their "sameness" cannot always be ascribed to particular mechanisms of development (Müller 2003; Wagner 2014, 90). Therefore, a hierarchical approach to homology has to abandon any linear, compositionally nested representation of levels (Minelli 1998). The cases in which classical morphological homologies emerge from the cross-level integration of factorial homologies (Minelli and Fusco 2013) at all relevant levels appear more exceptions than the rule, operationally relevant mostly as null model to investigate the extant vagaries of parallel evolution.[2]

Such "individualized units of phenotypic evolution," and of evolutionary phenotypic change (Wagner 2014, 245), exhibit a remarkable autonomy from their mechanistic underpinning: "Genetic control, molecular makeup, cell populations, inductive interactions, and ontogenetic trajectories could all be modified by evolution, while the resulting homologues are maintained" (Müller 2003, 56).

Therefore, the hierarchical approach to homology is both epistemologically and ontologically relevant for evolutionary investigation.

Its epistemological relevance can be assessed in that the level-dependent description is highly informative and operationally necessary for setting the correct framework of comparison. "Homology can only be ascertained from phylogenetic comparison of characters within the same level of organization" (Müller 2003, 56).

Its ontological relevance is related to the evidence that the different levels of organization at which homologous characters appear can exhibit a remarkable independency in evolution as well as in evolutionary timescales. Homologues are "heritable building units of the organism" (Müller 2003, 63), endowed with individual evolutionary stability and, consistently with the mechanistic biological homology concept, exhibit a genuine identity rather than mere resemblance. Their identity is preserved at "the relevant layer of developmental individualization," while "the complex of interactions within which they play their individual role" is highly variable (Minelli 1998, 345).

As "units of phenotype organization for higher organisms with a high tendency to retain identity for long periods of time," many homologues (from cell types to tissues, organs and body parts) reach a macroevolutionary life span, characterizing major metazoan or plant clades (Wagner 2015, 334). Their macroevolutionary relevance is enhanced by the shared assumption that the individualization of homologues is but the very process of origination of evolutionary novelties (Müller and Wagner 1991).

Homologues, as "heritable building units of the phenotype" (Müller 2003, 63), raise the riddle of a heritability that is faithfully developmentally reconstructed at the same phenotypic level, without necessarily involving the same lower level processes.

Wagner has recently advanced a mechanistic theory for accounting for the complex relationship between identity and diversification, invariance and plasticity, at the basis of homologous characters. He argues that their identity is maintained by the invariance of a "core character identity network" (ChIN), which represents the most conserved part of the gene regulatory network underlying character development. The historical identity and evolutionary stability of the ChINs are due to the evolved coadaptation of transcription factors, which tightens their cooperative action (Wagner 2015, 340).

Drawing a formal analogy with the distinction between genes and alleles, Wagner explains the relationship between character identity and character states—that is, the accessible state that is assumed by an homologous character due to adaptive modification or other kinds of evolutionary changes in the genes that orchestrate development—by means of the position of the ChIN in the respective developmental gene regulatory network. The ChIN is namely found at the intermediate tier between a highly variable upper layer (genes involved in determining the positional information for a group of cells) and a lower tier of downregulated target genes (Wagner 2014, 96ff.).

Thus the diversification of character states—as well as functions—across different species is the outcome of a compromise between identity and variability; the character identity becomes evolutionarily subordinated to the new complex of interactions within which the homologue plays its individual role—that is, to the whole of the developmental regulatory network. At the same time, the ChINs, as repository of homological identity, do play a causal role as to the intrinsic evolvability of the character, delimiting the possible range of character states that is compatible with the integration of the coevolved elements comprising the ChINs.

However, the riddle of individualization is not entirely solved by this mechanistic framework, as it does not exhaust all possible cases, especially the cases when the material "discontinuity" across levels is such that "it is questionable whether there are dedicated gene regulatory networks" underlying the development of morphological homologs (Wagner 2014, 76). In such instances, the divide between molecular syntax and higher order semantics has increased in evolution at the point that their material continuity is cut off. The lately discovered ubiquity of developmental system drift has unveiled the extent at which, even in closely related taxa, homologous characters have diverged in their morphogenetic or gene regulatory underpinnings (True et al. 2001). Even mechanistic explanations of the neutral rewiring of regulatory interactions by means of compensatory substitutions (Wray 2013) cannot but reinforce the assumption of the relative autonomy evolved by discrete evolutionary units at a wide range of phenotypic levels.

An alternative, or complementary (Wagner 2014, 96), explanatory device for coping with this autonomy is the "organizational homology concept," which prospects a three-stage process in the evolution of homologies: generation—integration—autonomization (Müller and Newman 1999).

The organizational homology concept is based on an "epigenetic character initiation" thesis, according to which the triggering change for the establishment of new homologues occurs in the "epigenetic context in which genetic evolution takes place" (Müller and Newman 1999, 69). It assumes that even in the "genetically stabilized and heritable developmental systems" of the Mendelian world, new homologues arise as a byproduct of the evolutionary modifications of developmental systems as their parameters trespass critical thresholds. The integration process, which progressively locks the new element into the body plan, provides the maintenance of the epigenetic conditions responsible for its emergence, as experimentally corroborated by the reversibility of phenotypic characters as such conditions change (Müller 2003, 62). The epigenetic origination of homologues precedes their genetic fixation by stabilizing selection.

It is, however, the autonomization phase—the increasing independence of homologues from the "underlying developmental, molecular and genetic processes"—that meets best the challenge of accounting for the causal power of such "individualized constructional elements in the evolution of organismal forms" (Müller and Newman 1999, 57), adding to their individuality the status of a causal factor.

At the stage of autonomization, homologues become "attractors" of further morphological design, organizing and ordering factors of the evolution of form. As attractors of a dynamical morphogenetic system, they force the trajectories of the system even in face of changing adaptive conditions, thus self-preserving their evolutionary persistence. They not only become "more influential for the further path of morphological evolution than . . . the biochemical circuitry that controls their developmental formation," but they can also shape the very evolution of their own mechanistic basis. The organizational role they play "in heritable, genetic, developmental, and structural assemblies" becomes the very cause of their evolutionary robustness as autonomized elements (Müller and Newman 1999, 64–65).

Both the tripartite evolutionary model of the organizational homology concept and Wagner's ChIN perspective emphasize the individuality of homologues. It is Wagner, however, who makes the strongest philosophical commitment on the status of homologues as qualified individuals. As "phenotypic units of evolutionary change," homologues are "spatiotemporally bounded." They "are inherited, and thus, form lineages, as species do . . . and they evolve"; they have "a limited historical lifetime; they originate (i.e., are a novelty at some point during evolutionary history) and may disappear in some lineages"; they "change as a unit due to [their] causal cohesion" (Wagner 2014, 23, 253, 419).

Wagner is aware that to support the ontological notion of homologues as evolutionary individuals, a new metaphysics is needed. He is equally aware that this new metaphysics must cope with the resurgent dangers both of internalism, the pitfall of emphasizing a directionality originating from intrinsic features of the living system, and of essentialism. His proposal is to move from an extension of the notion of "natural kinds," relating individualization to "a cluster of properties that are correlated in their presence or absence because of some kind of homeostatic mechanism." Such a criterion for individuality would not be necessarily dependent on "intrinsic features and mechanisms"—in fact, homeostatic mechanism can be extrinsic to the natural kind itself, such as ecological factors affecting species cohesion, or developmental and functional integration of homologues due to their increasing interdependence in evolution (generative entrenchment; Wagner 2014, chap. 7; cf. Schank and Wimsatt 1986).

It appears, however, that a revision is needed not only to preserve the ontological status of homologues but because former notions of biological individuality—the same that were crucial in the elaboration of the original hierarchy theory of evolution, such as Ghiselin (1977)—have lost their empirical basis after the discovery of highly conserved developmental genetic mechanism shared among a wide range of species (Wagner 2014, 237).

The evolutionary mosaicism exhibited by complex organisms, as they are "composed of differentiated, historically continuous building blocks" (Wagner 2015, 334), makes it harder to derive their historical continuity as evolutionary individuals directly from the "tight internal cohesion and functional interaction among subparts" (Vrba and Eldredge 1984, 149, 155) when subparts—such as evolutionary modules at different levels— are credited with their own evolutionary individuality. Furthermore, the evolutionary mosaicism of modular complex organisms is itself a prerequisite for adaptive evolution (Lewontin 1978). Taking into account the structured nature of phenotypic variability, as in the account of the modular logic of the genotype-phenotype maps proposed by Wagner and Altenberg 1996, has provided a key to approach the long-standing riddle of complex adaptations.

The evolutionary relevance of the organismal level, defended by Eldredge's hierarchy theory by placing the organism in the pivotal position of a bridge linking the genealogical and the ecological hierarchy (Eldredge 2008), can be mechanistically strengthened by encompassing the constitutive cohesive mechanisms that rebuild its "glue" in front of the evolutionary change in their parts—thus allowing organisms to successfully "walk the tightrope between stability and change" (Schwenk et al. 2009). For this endeavor, the epistemological premises of the reproducer account of the evolution of individuality appear promising. Such a framework might renew an organism-centered view of evolution, the task that has haunted the intellectual history of evolutionary biology ever since Fisher (Fontana and Buss 1996, 50). However, this possible "broader, organismal understanding of evolution" (Wagner 2014, 325ff.) would come at the cost of enlarging the range of what counts as a system of heredity and of the informational genealogical hierarchy in general.

5. Scaling down the Ecological Hierarchy

"Evolution is the control of development by ecology" (Van Valen 1973). An updated definition should be probably formulated in a slightly different way, such as that evolution is "ecologically driven along a developmentally favored trajectory," as shown by Kavanagh et al. (2007) at the case of the macroevolution of dentition in murine rodents.

Formally, it is quite possible to scale down the ecological hierarchy to infraorganismal ecosystems, in which "traits at various levels interact with the environment at various levels" (Korn 2002). However, scaling down the ecological hierarchy at infraorganismal levels raises several issues, such as what can be considered internal and what external in each respective microenvironment, and particularly what difference this distinction implies in causal terms.

"Normal" development requires a precise range of environmental conditions, and alteration in gene-expression patterns can depend on "regulators" that reside clearly outside the embryo, from temperature to population density to predators, as emphasized in the research field labeled eco-evo-devo (Gilbert 2000). Even the epigenetic character initiation theory holds the role of the environment as chronologically prior. Environmental change features as the initiating condition of the origination of innovations, even if it only has the role of triggering the highly structured response of the developmental system (the realizing conditions; Müller

2010). Therefore, the acknowledgment that the environment is the major source and inducer of genotypic and phenotypic variation at multiple levels of biological organization must be accompanied by the parallel recognition that it is development that acts as the major regulator, masking, releasing, and even creating new combinations of variation (Abouheif et al. 2014).

Thus the priority of environmental action does not dissolve the arguments of the developmentalist challenge. Regardless of whether the environmental input is clearly located outside or inside the respective microenvironment in the developing embryo, the way variation in environmental values is translated into variation on the level of the phenotype is nonrandom and even, according to the theory of phenotypic facilitated variation, biased to be viable (Kirschner and Gerhart 2005).

"Organisms react with one small set of phenotypic states to changes in the environment. The result of the environmental change is often integrated and detailed, rather than chaotic or random" (Wagner 2014, 160). Even the generation of mutational variation "is not unstructured with respect to survival and is neither memoryless, nor independent of the environment" (Caporale and Doyle 2013; cf. Koonin 2011, 288–89). The "hindsight" of mutations might be a product of the feedback of natural selection on the mechanisms that generate variation, reducing the uncertainty of the resulting variational response. Such feedback would allow incorporating a sort of "ghost" of environments past. In the gap between genotype and phenotype, genetic circuits, optimized by natural selection, appear to provide a reliable internal model of the environment (defined as "the time-dependent profiles of the input signals in the natural habitat of the organism"; Alon 2006, 203, 207). The modeling and anticipatory capacity of organisms would be increased by the addition (or intercalation) of regulatory layers. A recent reinstantiation of the basic tenet of hierarchy theory—the constitutive time lag between rates of processes at different biological levels—has related the filtering action of higher levels on lower-level processes to such an anticipatory capacity. As slower, coarse-grained variables at each new level extract regularities from lower level dynamics, encoding statistical information about its behavior, the addition of new level would also boost the anticipatory capacity of the whole system (Flack 2014).

Of course, no hindsight is enough to cope with massive environmental impact, and even if biological systems appear to have evolved a kind of robustness (lately described as "highly optimized tolerance") far superior than nonliving ones, they share the fragility that represents the other side of the coin of robustness—the "robust yet fragile" character that coincides with the limits of their anticipatory capacity in front of rare, and not necessarily more intense, perturbations (Carlson and Doyle 1999).

Notes

1. This view represents a partial revision of Davidson's original strict commitment to a "hardwired view" of development. See Arnone and Davidson (1997).

2. Abouheif (1997, 406), who considers four levels: genes, gene expression patterns, embryonic origins, and morphology.

References

Abouheif, Ehab. 1997. "Developmental Genetics and Homology: A Hierarchical Approach." *Trends in Ecology & Evolution* 12(10): 405–8.

Abouheif, Ehab, Marie-Julie Favé, Ana Sofia Ibarrarán-Viniegra, Maryna P. Lesoway, Ab Matteen Rafiqi, and Rajendhran Rajakumar. 2014. "Eco-Evo-Devo: The Time Has Come." In *Ecological Genomics: Ecology and the Evolution of Genes and Genomes*, edited by Christian R. Landry and Nadia Aubin-Horth, 107–25. Dordrecht: Springer.

Alderson, David L., and John C. Doyle. 2010. "Contrasting Views of Complexity and Their Implications for Network-Centric Infrastructures." IEEE *Transactions on Systems, Man, and Cybernetics* 40(4): 839–52.

Allen, Timothy F. H., and Thomas B. Starr. 1982. *Hierarchy: Perspectives for Ecological Complexity*. Chicago: Chicago University Press.

Alon, Uri. 2006. "An Introduction to Systems Biology: Design Principles of Biological Circuits." London: Chapman & Hall.

———. 2007. "Simplicity in Biology." *Nature* 446: 497.

Amundson, Ron. 2005. "The Changing Role of the Embryo in Evolutionary Thought: Roots of Evo-Devo." Cambridge: Cambridge University Press.

Arnone, Maria I., and Eric H. Davidson. 1997. "The Hardwiring of Development: Organization and Function of Genomic Regulatory Sequences." *Development* 124: 1851–64.

Auletta, Gennaro, George F. R. Ellis, and Luc Jaeger. 2008. "Top-Down Causation by Information Control: From a Philosophical Problem to a Scientific Research Programme." *Journal of the Royal Society Interface* 5: 1159–72.

Balavoine, Guillaume. 2015. "Evolutionary Developmental Biology and Its Contribution to a New Synthetic Theory." In *Handbook of Evolutionary Thinking in the Sciences*, edited by Thomas Heams, Philippe Huneman, Guillaume Lecointre, and Marc Silberstein, 443–70. Dordrecht: Springer.

Bechtel, William. 2007. "Biological Mechanisms: Organized to Maintain Autonomy." In *Systems Biology: Philosophical Foundations*, edited by Fred C.

Boogerd, Frank J. Bruggeman, Jan-Hendrik Hofmeyr, and Hans V. Westerhoff, 181–213. Amsterdam: Elsevier.

Bertolaso, Marta. 2015. "A System Approach to Cancer: From Things to Relations." In *Philosophy of Systems Biology: 5 Questions*, edited by S. Green, 13–24. Copenhagen: Automatic Press.

Bertolaso, Marta, and Silvia Caianiello. 2016. "Robustness as Organized Heterogeneity." *Rivista di filosofia Neoscolastica* (in print).

Bolker, Jessica A., and Rudolf A. Raff. 1996. "Developmental Genetics and Traditional Homology." *BioEssays* 18: 489–94.

Callebaut, Werner, Gerd B. Müller, and Stuart A. Newman. 2008. "The Organismic Systems Approach: Evo-Devo and the Streamlining the Naturalistic Agenda." In *Integrating Evolution and Development: From Theory to Practice*, edited by Roger Sansom and Robert N. Brandon, 25–92. Cambridge, MA: MIT Press.

Caporale, Lynn H., and John Doyle. 2013. "In Darwinian Evolution, Feedback from Natural Selection Leads to Biased Mutations." *Annals of the New York Academy of Sciences* 1305: 18–28.

Carlson, Jean M., and John Doyle. 1999. "Highly Optimized Tolerance: A Mechanism for Power Laws in Designed Systems." *Physical Review E* 60: 1412–27.

Carroll, Sean B. 2008. "Evo-Devo and an Expanding Evolutionary Synthesis: A Genetic Theory of Morphological Evolution." *Cell* 134: 25–34.

Carroll, Sean B., Jennifer K. Grenier, and Scott B. Weatherbee. 2001. *From DNA to Diversity: Molecular Genetics and the Evolution of Animal Design*. London: Blackwell.

Cordeschi, Roberto. 2011. "Artificial Intelligence and Evolutionary Theory: Herbert Simon's Unifying Framework." In *Logic and Knowledge*, edited by Carlo Cellucci, Emily Grosholz, and Emiliano Ippoliti, 197–215. Newcastle, UK: Cambridge Scholars Publishing.

Davidson, Eric H. 2006. *The Regulatory Genome: Gene Regulatory Networks in Development and Evolution*. San Diego: Academic Press.

———. 2011. "Evolutionary Bioscience as Regulatory Systems Biology." *Developmental Biology* 35(1): 35–40.

Davidson, Eric H., and Douglas H. Erwin. 2006. "Gene Regulatory Networks and the Evolution of Animal Body Plans." *Science* 311(5762): 796–800.

Deutsch, Jean S. 2010. "Homeosis and Beyond: What Is the Function of the Hox Genes?" In *Hox Genes: Studies from the 20th to the 21st Century*, edited by Jean S. Deutsch, 155–65. New York: Springer Science+Business Media.

Doolittle, W. Ford. 1988. "Hierarchical Approaches to Genome Evolution." *Canadian Journal of Philosophy* 14 (supplementary): 101–33.

Douboule, Denis. 2010. "The Evo-Devo Comet." *EMBO Reports* 11(7): 489.

Eldredge, Niles. 1985. *Unfinished Synthesis: Biological Hierarchies and Modern Evolutionary Thought*. New York: Oxford University Press.

———. 2008. "Hierarchies and the Sloshing Bucket: Toward the Unification of Evolutionary Biology." *Evolution: Education and Outreach* 1: 10–15.

Eldredge, Niles, and Stanley N. Salthe. 1984. "Hierarchy and Evolution." In *Oxford Surveys in Evolutionary Biology*, edited by Richard Dawkins and Mark Ridley, 1: 182–206. Oxford: Oxford University Press.

Eldredge, Niles, and Elisabeth S. Vrba. 1984. "Individuals, Hierarchies and Processes: Towards a More Complete Evolutionary Theory." *Paleobiology* 10(2): 146–71.

Erwin, Douglas H. 2011. "Evolutionary Uniformitarianism." *Developmental Biology* 357(1): 27–34.

Erwin, Douglas H., and Eric H. Davidson. 2009. "The Evolution of Hierarchical Gene Regulatory Networks." *Nature Reviews Genetics* 10: 141–48.

Flack, Jessica C. 2014. "Life's Information Hierarchy." *SFI Bulletin* 28: 13–24.

Fontana, Walter, and Leo W. Buss. 1996. "The Barrier of Objects: From Dynamical Systems to Bounded Organizations." In *Boundaries and Barriers*, edited by John Casti and Anders Karlqvist, 56–116. Reading, MA: Addison-Wesley.

Ghiselin, Michael T. 1997. *Metaphysics and the Origin of Species*. New York: State University Press.

Gilbert, Scott F. 2004. "Ecological Developmental Biology: Developmental Biology Meets the Real World." *Russian Journal of Developmental Biology* 35(6): 425–38.

Goodwin, Brian C. 2006. "Developmental Emergence, Genes and Responsible Science." In *Genes in Development: Re-Reading the Molecular Paradigm*, edited by Eva Neumann-Held and Christoph Rehmann-Sutter, 337–48. Durham: Duke University Press.

Gould, Stephen J. 2002. *The Structure of Evolutionary Theory*. Cambridge: Belknap Press of Harvard University Press.

Gould, Stephen J., and Niles Eldredge. 1977. "Punctuated Equilibria: The Tempo and Mode of Evolution Reconsidered." *Paleobiology* 3: 115–51.

Gregory, T. Ryan. 2004. "Macroevolution, Hierarchy Theory, and the C-Value Enigma." *Paleobiology* 30(2): 179–202.

Grene, Marjorie. 1987. "Hierarchies in Biology." *American Scientist* (5): 504–10.

Griesemer, James. 2000. "The Units of Evolutionary Transition." *Selection* 1: 67–80.

———. 2002. "Limits of Reproduction: A Reductionistic Research Strategy in Evolutionary Biology." In *Promises and Limits of Reductionism in the Biomedical Sciences*, edited by Marc H. V. Van Regenmortel and David L. Hull, 211–31. Chichester, UK: John Wiley & Sons.

Harrison, Peter W., Alison E. Wright, and Judith E. Mank. 2012. "The Evolution of Gene Expression and the Transcriptome-Phenotype Relationship." *Seminars in Cell & Developmental Biology* 23: 222–29.

Hoffmeyer, Jesper. 2006. "Code-Duality and the Epistemic Cut." *Annals of the New York Academy of Sciences* 901: 175–86.

Jablonka, Eva. 2002. "Information: Its interpretation, Its Inheritance, and Its Sharing." *Philosophy of Science* 69(4): 578–605.

Jablonka, Eva, and Eörs Szathmáry. 1995. "The Evolution of Information Storage and Heredity." *TREE* 10(5): 206–11.

Kavanagh, Kathryn D., Alistair R. Evans, and Jukka Jernvall. 2007. "Predicting Evolutionary Patterns of Mammalian Teeth from Development." *Nature* 449: 427–33.

Kirschner, Marc, and John Gerhart. 2005. *The Plausibility of Life*. New Haven: Yale University Press.

————. 2010. "Facilitated Variation." In *Evolution—The Extended Synthesis*, edited by Massimo Pigliucci and Gerd B. Müller, 253–80. Cambridge, MA: MIT Press.

Koonin, Eugene V. 2012. *The Logic of Chance*. Upper Saddle River, NJ: FT Press.

Korn, Robert W. 2002. "Biological Hierarchies: Their Birth, Death and Evolution by Natural Selection." *Biology and Philosophy* 17(2): 199–221.

Laubichler, Manfred D., and Günter P. Wagner. 2001. "How Molecular Is Molecular Developmental Biology? A Reply to Alex Rosenberg's Reductionism Redux: Computing the Embryo." *Biology and Philosophy* 16: 53–68.

Lewontin, Richard. 1978. "Adaptation." *Scientific American* 239(3): 213–30.

Lieberman, Bruce S., Warren D. Allmon, and Niles Eldredge. 1993. "Levels of Selection and Macroevolutionary Patterns in the Turritellid Gastropods." *Paleobiology* 19: 205–15.

Lieberman, Bruce S., and Elisabeth S. Vrba. 1995. "Hierarchy Theory, Selection, and Sorting." *BioScience* 45: 394–99.

Love, Alan C. 2006. "Evolutionary Morphology and Evo-Devo: Hierarchy and Novelty." *Theory in Biosciences* 124: 317–33.

————. 2015. "Conceptual Change and Evolutionary Developmental Biology." In *Conceptual Change in Biology: Scientific and Philosophical Perspectives on Evolution and Development*, edited by Alan C. Love, 1–54. Dordrecht: Springer Science+Business Media.

Lynch, Vincent G., and Günter P. Wagner. 2008. "Resurrecting the Role of Transcription Factor Change in Developmental Evolution." *BioOne* 62(9): 2131–54.

Minelli, Alessandro. 1998. "Molecules, Developmental Modules, and Phenotypes: A Combinatorial Approach to Homology." *Molecular Phylogenetics and Evolution* 9(3): 340–47.

Minelli, Alessandro, and Giuseppe Fusco. 2013. "Homology." In *The Philosophy of Biology: A Companion for Educators*, edited by Kostas Kampourakis, 289–322. Dordrecht: Springer Science+Business Media.

Minelli, Akessandro, and Beatrice Peruffo. 1991. "Developmental Pathways, Homology and Homonomy in Metameric Animals." *Journal of Evolutionary Biology* 4: 429–45.

Mitchell, Sandra D. 2003. *Biological Complexity and Integrative Pluralism*. Cambridge: Cambridge University Press.

Moczek, Armin P. 2012. "The Nature of Nurture and the Future of Evodevo: Toward a Theory of Developmental Evolution." *Integrative & Comparative Biology* 52: 108–19.

Morange, Michel. 2014. "From Genes to Gene Regulatory Networks: The Progressive Historical Construction of a Genetic Theory of Development and Evolution." In *Towards a Theory of Development*, edited by Alessandro Minelli and Thomas Pradeu, 174–82. Oxford: Oxford University Press.

Müller, Gerd B. 2003. "Homology: The Evolution of Morphological Organization." In *Origination of Organismal Form: Beyond the Gene in Developmental and Evolutionary Biology*, edited by Gerd B. Müller and Stuart A. Newman, 52–69. Cambridge, MA: MIT Press.

————. 2010. "Epigenetic Innovation." In *Evolution—The Extended Synthesis*, edited by Massimo Pigliucci and Gerd B. Müller, 307–32. Cambridge, MA: MIT Press.

Müller, Gerd B., and Stuart A. Newman. 1999. "Generation, Integration, Autonomy: Three Steps in the Evolution of Homology." In *Homology*, edited by Gregory R. Bock and Gail Cardew, 65–79. New York: John Wiley & Sons.

Müller, Gerd B., and Günter P. Wagner. 1991. "Novelty in Evolution: Restructuring the Concept." *Annual Review of Ecology and Systematics* 22: 229–56.

Mykles, Donald L., Cameron K. Ghalambor, Jonathon H. Stillman, and Lars Tomanekx. 2010. "Grand Challenges in Comparative Physiology: Integration across Disciplines and across Levels of Biological Organization." *Integrative & Comparative Biology* 50(1): 6–16.

Neumann-Held, Eva M., and Christoph Rehman-Sutter. 2006. Introduction to *Genes in Development: Re-Reading the Molecular Paradigm*, edited by Eva Neumann-Held and Christoph Rehmann-Sutter, 1–11. Durham: Duke University Press.

Newman, Stuart A. 2003. "Hierarchy." In *Keywords and Concepts in Evolutionary Developmental Biology*, edited by Brian K. Hall and Wendy M. Olson, 169–74. Cambridge, MA: Harvard University Press.

Niklas, Karl J. 2015. "Adaptive Aspects of Development: A 30-year Perspective on the Relevance of Biomechanical and Allometric Analyses." In *Conceptual Change in Biology: Scientific and Philosophical Perspectives on Evolution and Development*, edited by Alan C. Love, 57–76. Dordrecht: Springer Science+ Business Media.

Noble, Denis, Eva Jablonka, Michael J. Joyner, Gerd B. Müller, and Stig W. Omholt. 2014. "Evolution Evolves: Physiology Returns to Centre Stage." *Journal of Physiology* 592(11): 2237–44.

Okasha, Samir. 2006. *Evolution and the Levels of Selection*. Oxford: Oxford University Press.

———. 2012. "Emergence, Hierarchy and Top-Down Causation in Evolutionary Biology." *Interface Focus* 2: 49–54.

O'Malley, Maureen A., Orkun S. Soyer, and Mark L. Siegal. 2015. "A Philosophical Perspective on Evolutionary Systems Biology." *Biological Theory* 10: 6–17.

Oyama, Susan. 1985. *The Ontogeny of Information*. Durham: Duke University Press.

Pattee, Howard H. 1970. "The Problem of Biological Hierarchy." In *Towards a Theoretical Biology*, vol. 3, edited by Conrad H. Waddington, 117–36. Edinburgh: Edinburgh University Press.

Pigliucci, Massimo, and Gerd B. Müller. 2010. "Elements of an Extended Synthesis." In *Evolution—The Extended Synthesis*, edited by Massimo Pigliucci and Gerd B. Müller, 3–17. Cambridge, MA: MIT Press.

Rollo, C. David. 1994. *Phenotypes: Their Epigenetics, Ecology and Evolution*. London: Chapman Hall.

Roth, V. Louise. 1984. "On Homology." *Biological Journal of the Linnean Society* 22: 13–29.

———. 1994. "Within and between Organisms: Replicators, Lineages, and Homologues." In *Homology: The Hierarchial Basis of Comparative Biology*, edited by Brian K. Hall, 301–37. San Diego: Academic Press.

Salthe, Stanley N. 1985. *Evolving Hierarchical Systems*. New York: Columbia University Press.

———. 1988. "Notes toward a Formal History of the Levels Concept." In *Evolution of Social Behavior and Integrative Levels*, edited by Gary Greenberg and Ethel Tobach, 53–64. Hillsdale, NJ: Lawrence Erlbaum.

———. 2012. "Hierarchical Structures." *Axiomathes* 22: 355–83.

Sander, Klaus, and Urs Schmidt-Ott. 2004. "Evo-Devo Aspects of Classical and Molecular Data in a Historical Perspective." *Journal of Experimental Zoology: Part B* 302B: 69–91.

Savageau, Michael A. 2001. "Design Principles for Elementary Gene Circuits: Elements, Methods, and Examples." *Chaos* 11(1): 142–59.

Schaffner, Kenneth F. (under review). *Behaving: What's Genetic and What's Not*. Oxford: Oxford University Press.

Schank, Jeffrey C., and William C. Wimsatt. 1986. "Generative Entrenchment and Evolution." *Proceedings of the Biennial Meeting of the Philosophy of Science Association* 2: 33–60.

Schwenk, Kurt, Dianna K. Padilla, George S. Bakken, and Robert J. Full. 2009. "Grand Challenges in Organismal Biology." *Integrative and Comparative Biology* 49: 7–14.

Shapiro, James A. 2013. "How Life Changes Itself: The Read-Write (RW) Genome." *Physics of Life Reviews* 10(3): 287–323.

Soyer, Orkun S., and Maureen A. O'Malley. 2013. "Evolutionary Systems Biology: What It Is and Why It Matters." *BioEssays* 35(8): 696–705.

Stern, David L. 2011. *Evolution, Development, and the Predictable Genome*. Greenwood Village, CO: Roberts.

Stern, David L., and Virginie Orgogozo. 2009. "Is Genetic Evolution Predictable?" *Science* 323: 746–51.

Szathmáry, Eörs, and John Maynard Smith. 1995. "The Major Evolutionary Transitions." *Nature* 374: 227–32.

———. 1997. "From Replicators to Reproducers: The First Major Transitions Leading to Life." *Journal of Theoretical Biology* 187(4): 444–60.

Tëmkin, Ilya, and Niles Eldredge. 2014. "Networks and Hierarchies: Approaching Complexity in Evolutionary Theory." In *Macroevolution: Explanation, Interpretation, Evidence*, edited by Emanuele Serrelli and Nathalie Gontier, 183–226. Berlin: Springer.

Toussaint, Marc, and Werner von Seelen. 2007. "Complex Adaptation and System Structure." *BioSystems* 90: 769–82.

True, John R., and Eric S. Haag. 2001. "Developmental System Drift and Flexibility in Evolutionary Trajectories." *Evolution and Development* 3: 109–19.

Van Valen, Leigh. 1973. "A New Evolutionary Law." *Evolutionary Theory* 1: 1–30.

———. 1982. "Homology and Causes." *Journal of Morphology* 173: 305–12.

Vinicius, Lucio. 2010. *Modular Evolution: How Natural Selection Produces Biological Complexity*. Cambridge: Cambridge University Press.

Vrba, Elisabeth S., and Niles Eldredge. 1984. "Individuals, Hierarchies and Processes: Towards a More Complete Evolutionary Theory." *Paleobiology* 10(2): 146–71.

Vrba, Elisabeth S., and Stephen J. Gould. 1986. "The Hierarchical Expansion of Sorting and Selection: Sorting and Selection Cannot Be Equated." *Paleobiology* 12: 217–28.

Wagner, Günter P. 1989. "The Origin of Morphological Characters and the Biological Basis of Homology." *Evolution* 43(6): 1157–71.

———. 2014. *Homology, Genes, and Evolutionary Innovation*. Princeton, NJ: Princeton University Press.

———. 2015. "Reinventing the Organism: Evolvability and Homology in Post-Dahlem Evolutionary Biology." In *Conceptual Change in Biology: Scientific and Philosophical Perspectives on Evolution and Development*, edited by Alan Love, 327–42. Dordrecht: Springer Science+Business Media.

Wagner, Günter P., and Lee Altenberg. 1996. "Complex Adaptations and the Evolution of Evolvability." *Evolution* 50: 967–76.

Wake, David B., and Allan Larson. 1987. "Multidimensional Analysis of an Evolving Lineage." *Science* NS 238(4823): 42–48.

Wake, Marvalee H. 2015. "Hierarchies and Integration in Evolution and Development." In *Conceptual Change in Biology: Scientific and Philosophical Perspectives on Evolution and Development*, edited Alan Love, 405–20. Dordrecht: Springer Science+Business Media.

Walker, Sara I., Benjamin J. Callahan, Gaurav Arya, J. David Barry, Tanmoy Bhattacharya, Sergei Grigoryev, Matteo Pellegrini, Karsten Rippe, and Susan M. Rosenberg. 2013. "Evolutionary Dynamics and Information Hierarchies in Biological Systems." *Annals of the New York Academy of Sciences* 1305: 1–17.

Weber, Marcel. 2005. *Philosophy of Experimental Biology*. Cambridge: Cambridge University Press.

Weinberg, Robert A. 2014. "Coming Full Circle: From Endless Complexity to Simplicity and Back Again." *Cell* 157(1): 267–71.

Wiener, Norbert. (1950) 1989. *The Human Use of Human Beings: Cybernetics and Society*. New York: Da Capo Press.

Wilkins, Adam S. 2007. "Between 'Design' and 'Bricolage': Genetic Networks, Levels of Selection, and Adaptive Evolution." *Proceedings of the National Academy of Sciences of the United States of America* 104 (supplementary 1): 8590–96.

Wray, Gregory A. 2013. "Genomics and the Evolution of Phenotypic Traits." *Annual Review of Ecology, Evolution, and Systematics* 44: 51–72.

CHAPTER 8

Multilevel Selection in a Broader Hierarchical Perspective

Telmo Pievani and Andrea Parravicini

1. Group Selection: A Historically Contentious Concept

Evolutionary theory has been and remains centered on the causal pattern of natural selection since its early formulation by Charles Darwin (Darwin 1859). According to this perspective, selective mechanisms statistically favor traits that lead an individual to survive and reproduce better than its conspecifics in a common environment. The whole process occurs at the level of individual organisms and is not prescient: the advantageous variation must eventually reward the individual during its own life and be transmitted to the next generation (Pievani 2015a).

Yet there is a great number of evolutionary phenomena that cannot be readily explained in terms of organismal selection. As Darwin himself wrote, living beings, including humans, very often show behaviors that increase the chances of survival and reproduction of other individuals at their own expense. Some living beings sacrifice their lives and compromise their fitness when danger threatens their kin or the whole group (like the bee who commits suicide by stinging, a case-study discussed by Darwin), and millions of individuals (such as sterile workers among eusocial insects) are forced to forgo reproduction for serving the community. If natural selection favors only traits that cause individuals to survive and reproduce better than other individuals, should not such "altruistic" behaviors have gone extinct?

An attempt to explain altruistic behaviors as a product of natural selection occupied center stage in Darwinian thought. The problem of the honeybee's barbed stinger, the presence of sterile castes and slave-workers in some species of insects (Darwin 1859; Stauffer 1975), the spreading of altruism and cooperation in human societies (Darwin 1871), were all challenging issues for Darwin. He acknowledged that explaining such cooperative behaviors solely in terms of benefits to the individual was very difficult and extended the action of natural selection to include the higher level of the groups. In other words, Darwin proposed the idea that in those controversial and special cases, natural selection does not necessarily apply only to individual advantage but also applies to the community survival. If a colony of social insects produces sterile workers that devote their whole lives to assist the reproductive efforts of the queen and to benefit the group, that colony would have a competitive advantage over other colonies (Darwin 1872, 227–33; Stauffer 1975, 370–71, 510–13; Sober 2011). In such cases, natural selection would act at a higher evolutionary level of the colonies.

In *The Descent of Man*, Darwin (1871, v.1, 166) applied the same logic to the controversial case of the emergence of altruistic behaviors in human societies. In a famous passage he wrote, "A tribe including many members who . . . were always ready to give aid to each other and sacrifice themselves for the common good, would be victorious over most other tribes; and this would be natural selection."[1]

Despite some misconceptions in current literature,[2] we may confidently state that although Darwin mainly considered the action of selection at the level of individual organisms, he also recognized and included the theoretical possibility of selection acting on multiple levels (Borrello 2010; Sober 2011; Pievani 2011, 2014). Showing his typical theoretical flexibility, Darwin adopted a position that we may call "imperfect selfishness" (Pievani 2014), namely because even if he held an individualist view in his background perspective, he never adopted an explanatory "monism." On the contrary, he endorsed an explanatory pluralism about the levels of selection.

The emergence of altruism and the plurality of evolutionary levels were debated issues even after Darwin. Yet a clear formulation of these topics was not developed until the birth of the modern synthesis. Sewall Wright's shifting balance theory (Wright 1931) assigned a key role to differential migrations between demes as a mechanism to extend favorable gene combinations to the whole species. This kind of "group selection," however, was not specifically related to the problem of altruism. The theoretical models proposed by Ronald A. Fisher (1930) and John B. S. Haldane (1932) reoriented the evolutionary processes at the level of individual organisms; subsequent postsynthesis establishment of the neo-Darwinian research program defined the evolutionary process according to the population genetic model, defining it as the outcome of the accumulation of small genetic changes driven by natural selection (Borrello 2005, 2010).

Despite this general trend, a small group of biologists continued to support the idea that natural selection could act at levels above the individual organism. Vero Copner Wynne-Edwards, in his *Animal Dispersion* (1962), argued that a great number of social and altruistic behaviors, which do not increase individual fitness, could not be explained by focusing only on individual level selection. His commitment to group selection was theoretically radical. He depicted group selection as an effective mechanism of evolutionary restraint on population growth, where group-selected behaviors act as internal factors, maintaining populations below a threshold of overexploitation of food resources. He identified a wide variety of group-level adaptations, such as reproductive rate, foraging strategies, and strict population localization group, which he connected to self-regulation of population size (Wynne-Edwards 1962; see Borrello 2005, 2010). In Wynne-Edwards's "strong" version, group selection was considered as a sort of homeostatic mechanism of population self-control and internal regulation of activities at the group level.

Wynne-Edwards's theory, weakly supported by ethological and ecological evidence, received a host of severe and serious criticisms by major supporters of the modern synthesis. As Mark Borrello (2010) argued, the influence of the strong responses to Wynne-Edwards's book from some of the most prominent scholars of that time—such as David Lack, George C. Williams, William Hamilton, and John Maynard Smith—contributed not only to the rejection of Wynne-Edwards's group selection theory but also to a general denial of any idea of higher-level selection for the following twenty years. Such a denial is still widespread among evolutionary biologists. The new gene-centered standard for evolutionary studies, postulating that natural selection is always, or almost always, acting on single genes, was proposed by George Williams (1966)[3] and reached its most radical expression in Richard Dawkins's *The Selfish Gene* (1976) ten years later. In this gene-centered perspective, among-group selection was held as invariably too weak compared to within-group selection; behaviors that appear altruistic were considered but a subtle form of selfishness. The inclusive fitness (or kin selection) theory (Hamilton 1964; Maynard Smith 1964); the evolutionary game theory, including the concept of reciprocal altruism (Trivers 1971; Maynard Smith and Price 1973); and the selfish gene theory (Dawkins 1976) were all developed as alternative models to group selection.

Yet Darwin's suggestion about the possibility of a multilevel extension of evolutionary theory had not been completely forgotten. Despite the general trend, a small group of biologists carried on the research. George Price (1972; Harman 2010), David Sloan Wilson (1977, 1980), Michael Gilpin (1975), and Michael Wade (1984, 1985) provided mathematical models and early observational and experimental evidence of group selection, which began to redeem this conception among biologists. At the same time, a growing number of philosophers of biology—Elliott Sober (1981, 1984; Sober and Lewontin 1982; Wilson and Sober 1989, 1994), Elisabeth Lloyd (1988; Lloyd and Gould 1993), and David Hull (1976, 1980) among the others—reflected on the issue of units of selection and contributed robust arguments in favor of group selection as well.

A revision of the models and data that led to the earlier rejection of group selection has reversed the trend over the past twenty years. Today, there is a robust theory of multilevel selection, increasingly supported by empirical evidence, which has little to do with the earlier version of "group selection." This modern version does not involve any strong or "ontological" notion of the group; it is firmly grounded in the neo-Darwinian population genetics and behavioral ecology and has been provided with a robust mathematical framework (Nowak 2006; Tarnita et al. 2009; Nowak et al. 2010). Furthermore, the basic logic of this up-to-date version of multilevel selection theory applies to a wide range of social traits and behaviors, from the evolution of sexual reproduction and sex ratio to the evolution of articulated language (Deacon 2003) and religious beliefs (see Norenzayan 2013 for the hypothesis of a "cultural group selection"). Despite all this, the controversy continues to this day, both from an empirical and conceptual points of view, as evidenced from the different reactions to the "milestone" of the modern version of group selection—that is, Sober and Wilson's *Unto Others* (1998; see Maynard Smith 1998; Nunney 1998; Lewontin 1998; Dennett 2002).

2. Group Selection: Up-to-Date Version

David S. Wilson has long argued that his trait-group model (1980, 1983) provides an up-to-date version of the group selection theory. In his latest book, Wilson (2015) investigated the working of group-level functional organizations and their evolutionary emergence in natural world. According to Wilson, a group of organisms is functionally organized when its "parts work together in a coordinated fashion to achieve a given end" (2015, 9). Eusocial insects and human societies are typical examples of such kind of groups, and their members show a high degree of cooperative behaviors. The problem is still the Darwinian one: behaving for the good of the group does not maximize *relative fitness* within the group, and natural selection is basically based on this kind of fitness, according to which it is not important how well an organism survives and reproduces, only that it does so better than other organisms within the evolving population in the same environment. In other words, *natural selection operating within groups* tends to undermine group-level functional organizations and it consequently cannot account for the emergence of such groups. These kinds of organization could otherwise be well accounted by *natural selection between groups.*

We may take the case of a group consisting of two types of individuals in a certain proportion, the altruistic A-type (20 percent)—which provides a benefit (say 1.0) to everyone in its group, including itself, at a slight cost to itself (say 0.01)—and the selfish S-type (80 percent), which benefits from A-types but does not contribute to the public good. Although both types have increased their *absolute fitness*, the A-types present a *relative fitness* disadvantage (net benefit of 0.99) relative to the S-types (net benefit of 1.0). This means that in the next generation, the proportion of the A-types will be reduced to less than 20 percent and will progressively decrease generation after generation until they become extinct. If we include more strongly altruistic cases, the cost of providing the public good would increase, and the A-types will disappear even faster.

We may imagine now the same proportion of A- (20 percent) and S-types (80 percent) assembled in a series of groups, each group presenting different proportions of the two types. In this case, while the fitness differences among individuals *within groups* foster S-types, the fitness differences *between groups* support A-types. In other words, natural selection *within groups* and natural selection *between groups* appear to be two different processes that pull in opposite directions (so we have a plurality of antagonistic selective pressures). The outcome of the process—that is, what evolves in the total population in terms of genetic frequencies—depends on the relative strength of such opposing *levels of selection*, and it could result in the extinction of one type or in a mixture of both types in the population. Summing up, altruistic behaviors evolve whenever between-group selection prevails over within-group selection (Wilson 2015, 19–30).

In order for between-group selection to occur, it is essential that the population is subdivided into multiple groups, with variation among groups, so as to ensure the fitness differences that allow natural selection to work at the group level. Population structure is the crucial parameter, an ecological parameter. If population consists of just one group, no group selection can occur because there is no variation among groups. Otherwise, if each group is internally homogenous, there will be no selection within groups, while selection between groups will be still possible.

A host of empirical evidence has confirmed this mathematical model. A satisfying example is provided by a series of experiments carried out by David S. Wilson's research group on the variation in males' aggressiveness toward females in water striders (*Aquarius remigis*; Eldakar et al. 2009a,b; 2010a,b).

The results confirm that within-group selection weighs in favor of selfish behaviors, while between-group selection fosters cooperative behaviors. The final outcome depends on the balance between the two opposing levels of selective pressures.

The water-striders' example is just one of many cases that provide empirical support to the multilevel selection theory (individuals and groups), from the reduction of virulence in animal diseases (Bull 1994; Frank 1996) to the spread of more prudent strains of T4 coliphage virus in their bacterial host, *Escherichia coli* (Kerr et al. 2006); from the evolution of the "wrinkly spreader" strain of *Pseudomonas fluorescens* (Rainey and Rainey 2003) to the field studies on social vertebrates, like the costly cooperation of lionesses for territorial defense (Heinsohn and Packer 1995; Packer and Heinsohn 1996). In summary, evidence from studies on organisms as diverse as microbes, plants, insects, and vertebrates (Goodnight and Stevens 1997; Goodnight 2005), including a recent controversial case of the social spider *Anelosimus studiosus* (Pruitt and Goodnight 2014; followed by Grinsted et al. 2015; Gardner 2015; for a reply to objections, see Pruitt and Goodnight 2015), provide a growing empirical basis for building up a sound multilevel theory of evolution.

At the same time, a good number of critics of group selection, such as William Hamilton (1996) and George Williams (1992, 1996), have changed their minds and come to accept the role of group selection as an explanatory mechanism for some evolutionary processes, depending on the population structure (Borrello 2010; Pievani 2014). In recent contributions, even the father of sociobiology, Edward O. Wilson, has reversed (twice as of now!) his ideas exposed in his original book on "the new synthesis" (1975), where he rejected group selection in favor of a strong commitment to explaining social behaviors exclusively in terms of kin selection theory.

In Wilson and Wilson (2007), he expressed the need to refound the basic approach to sociobiology on the pluralistic principles of multilevel selection theory. More recently, he changed his mind again—and given these shifts of opinion, we suggest scholars to specify which of the three "Wilsons" they refer to when discussing his ideas on multilevel selection. The "third" Wilson (2012) argued that studies on social behavior need to reject explanations based on kin selection in favor of group selection mechanisms (raising the rugged reaction of Richard Dawkins in a *Prospect* article on May 24, 2012 [see Pievani 2014]). According to Wilson (2012), colonies of insects have evolved by standard natural selection (queen versus queen), or between-colony selection, while human-specific eusociality has emerged thanks to a strong interplay between individual and group selection. This conflict among levels of selection, Wilson argued, is likely at the origin of the ambiguity of human nature, which expresses simultaneously selfish attitudes and cooperative attitudes, aggressiveness between groups alternating with altruism within groups (for the out-group versus in-group dynamics, see Bowles 2008).

3. Toward a Pluralistic Approach

Other concepts have been proposed in order to account for the evolution of altruism and cooperative behaviors without invoking group selection, such as indirect reciprocity (Nowak and Sigmund 2005; Nowak 2006), by-product mutualism (Dugatkin 2002; Sachs et al. 2004), and costly signaling (Lachmann et al. 2001; Cronk 2005).

According to David S. Wilson and Edward O. Wilson (2007, 334–36), all these evolutionary models basically incorporate the logic of multilevel selection within their structure and, therefore, they are far from being alternative hypotheses to group selection.

First, they argued, all these models of social behavior assume the existence of multiple groups because social interactions almost invariably occur among subsets of population—for example, in kin selection theory, individuals interact with a part of the total population, with whom they share a certain degree of kinship. In this sense, kin selection is a special form of group selection.

Second, according to the concept of trait-group (D. S. Wilson 1975), "for any particular trait . . . there is an appropriate population structure that must conform to the biology of situation" (Wilson and Wilson 2007, 334), regardless the theoretical framework we are dealing with. Hence, all the abovementioned models use the same definition of groups for any particular trait. In general, the population structure and the population ecology represent firm criteria of compatibility for different patterns, such as inclusive fitness and group selection, within the framework of multilevel selection theory (Pievani 2014). Demographic and mathematical models are confirming the key role played by population structure (group size, number of groups, interactions), demography, and nonlinear interactive effects in affecting such cooperative evolutionary dynamics (Tarnita et al. 2009; Nowak 2006; Nowak and Highfield 2011).

Third, almost every cooperative and altruistic trait requires between-group selection in order to evolve, given that it is "detrimental" from the point of view of the individual selection. This was recognized by Hamilton (1975) when he reformulated his inclusive fitness theory in terms of the Price equation (Price 1972), which separates total gene frequency change into within- and between-group elements. Hamilton realized that altruistic traits evolve because of the contribution of the kin-groups with more altruists, while the same traits are disadvantageous *within* kingroups. Thus inclusive fitness theory cannot be considered as an *alternative* to group selection, as Hamilton himself came to acknowledge (Segerstrale 2013, 139–55), even though the importance of genetic relatedness remained a valid insight. The importance of kinship lies in favoring an increasing of genetic variation among groups and, therefore, in strengthening the among-group selection compared to within-group selection. In any case, genetic relatedness can be expressed by the same parameters of multilevel selection and, therefore, it could be mathematically viewed as a kind of group selection (Wilson and Wilson 2007; Sober 2011, 64).

Even in game theory, tit-for-tat and other cooperative strategies do evolve because groups of cooperators contribute more to the total gene pool than groups of noncooperators. The same multilevel dynamics can work in any N-person game theory, from the very little and ephemeral "groups"—like the pairs of socially interacting individuals in tit-for-tat strategy—to much larger and persistent groups. The underlying evolutionary dynamics in all these cases overlap with traditional group selection models: "Selfishness beats altruism within single groups. Altruistic groups beat selfish groups" (Wilson and Wilson 2007, 335).[4]

From the point of view of the selfish gene theory, group selection has to be rejected because gene is the fundamental unit of selection. As explained above, the argument is not consistent. Despite the fact that many scholars involved in the controversy (such as George Williams and Richard Dawkins) have openly acknowledged their error in addressing this point and now concede that group selection is a matter about the vehicles for genetic transmission, not dismissing the evolutionary relevance of replicators (see Sober and Wilson 1998, 92), many articles and textbooks still commonly raise this charge against multilevel selection theory. Multilevel selection models calculate the average effects of genes in population, just like any other population genetics model, and the concept of average effects in population genetics is identical to the concept of genes as "units of selection" and "replicators." Individual selection and group selection are related to the "targets"—namely, traits and behaviors—of the multilevel selection process, but the "units of selection" remain the genetic codes of the individual. Individuals and groups could be viewed as "vehicle" of transmission of functional genes even in a multilevel scenario, possibly distributed in several nested levels of organization (like in Pagel 2012).

In conclusion, modern group selection theory is consistent with many of the theories that had been treated as rival (and mutually exclusive) explanations over the past thirty years. As David S. Wilson (2015; Wilson and Wilson 2007) claimed, all these models can be considered as equivalent—that is, they all correctly predict what evolves in the total population. The models are translatable with each other, as they largely overlap, being their common denominator the population structure and the kind of traits to be considered. In the words of Elisabeth Lloyd, "game theory and kin selection models do not compete with multilevel Price-type models; rather, they are all different ways of looking at evolution in group structured populations" (Lloyd 1999, 447).

Today, there is a growing consensus in the field toward such pluralistic approach, and Martin Nowak's general theory of cooperation (2006) is a good example of this kind of integration among different selective mechanisms in a sort of "selective pluralism" within a multilevel framework (see Pievani 2014).

4. Formalizing the Multilevel Selection Theory

Samir Okasha (2006) provided a detailed analysis of the debates on multilevel selection theory (MLS) and revised the mathematical formalization of the theory on the basis of Price's equation.

According to Okasha, the levels-of-selection problem emerges directly from the core logic of Darwinism at the interaction of three distinct factors:

A. Natural selection operates on individuals that vary, reproduce differentially, and generate offspring that resemble them. In principle, natural selection can act on any entities that meet the three conditions identified by Richard Lewontin (1970; i.e., phenotypic variation, differential fitness, heritability), so natural selection is not limited to the level of individual organisms.

B. The biological world is characterized by hierarchical organization, where the lower levels are nested within the higher ones. In a nested compositional hierarchy, levels are classes and their ranks correspond to the scale of the entities that are their members. Entities at lower levels (genes to cells, cells to tissues and organs, organs to organisms) make up the higher-level entities; they are not nested within them, but the levels themselves are. Above organisms we find kin groups, colonies, demes, species, up to the higher taxa. Entities at various hierarchical levels could satisfy the condition at the point A, then, in principle, they might form populations that evolve by natural selection.

C. What is advantageous at one hierarchical level may be disadvantageous at another level, and this could generate potential conflicts among levels. Altruistic behaviors may have evolved through selection at the group level, being disadvantageous at the organism level. Carcinogenesis and many other maladaptive features of the organism as a whole may be understood by appealing to the same interlevel conflicting logic (Okasha 2006, 11).

The interaction among the three factors gives place to the long-debated question of the levels-of-selection.[5] Across the biological hierarchy, each population of entities that *simultaneously* satisfies the aforementioned three conditions—that is, character variation, associated differences in fitness and heritability—may be called, according to Okasha, a *unit of selection* that occupies a specific hierarchical *level of selection*.

In a two-levels scenario with strict nesting, for example, we shall have *particles*, as units of the lower level, which are nested in *collectives*, as units of the higher level. Collective's characters are said to be *aggregate* if they form by aggregation the characters of the particles in a collective, and then they derive directly from underlying particles' characters (as in the case of altruist/selfish groups). If all organisms in a given species are white, then white is an aggregate character of that species. Collective's characters are called *emergent* if they are not simple functions of particle-level characters, even though they usually indirectly depend on the particle-level characters. For example, sex ratio is an emergent trait of a population or species because it denotes a collective character of many organisms that cannot be characterized by having this property individually. Identical genes in every cell form a series of aggregate characters of the organism, while phenotypic characters (like coat colors or behaviors) are said to be emergent at the organismal level. Although the aggregate/emergent distinction is hard to precisely define and literature is not unanimous on this issue,[6] both kinds of characters are relevant for multilevel selection theory (Okasha 2006).

In a MLS scenario, we shall have a concept of fitness applied to entities at different levels. Particle's fitness—namely, the number of offspring that particles leave over generations—and collective's fitness, which in turn can be defined in two different ways. Collective's fitness could be the average or total fitness of the constituent particles of a collective (fitness 1) or it could be the number of new generation of collectives that a collective leaves (fitness 2). In fitness 1, the fittest collective will be the one that contributes the most offspring particles to the next generations of particles; in fitness 2, the fittest collective will be the one that provides the most offspring collectives to future generations of collectives (Okasha 2006, 53–56).

The group selection model described in section 2 deals with fitness 1 because it aims at accounting for the evolution of an individual phenotype (cooperative traits, altruism) distributed in a population differentiated into multiple groups. Conversely, for example, the controversial concept of species selection refers to fitness 2, which deals with the expected number of offspring species. This means that species selection and group selection are not analogous mechanisms, but they rely on different logical assumptions. If we are interested in the frequency of the different types of particles within a population of particles or groups of particles, then we are considering the lower level of particles as the "focal" level. If our focal units are the collectives, then we shall be interested in tracking the frequency of different particle types and collective types. Depending on whether the particles or the collectives are the "focal" units, we shall adopt two different approaches, which Damuth and Heisler (1988) have dubbed MLS1 and MLS2.

Following Okasha's analyses (2006, 62–75), which are largely based on George Price's mathematical formalisms,

- MLS1 aims at accounting for the changing frequency of an individual trait in the overall population, and it is essentially based on the group's structure. In this case, the focal units are particles and the considered fitness is fitness 1. This is the kind of approach that we have already described in relation to the evolution of altruism, where focal units were the altruist and selfish organisms, and where the final outcome depended on the balance between within-group selection, which favors selfishness, and between-group selection, which favors groups containing a high proportion of altruists. Because groups containing a high proportion of altruists contribute more individual offspring to the entire population, they are said to have a higher group fitness 1—that is, a higher average fitness of the group's members.

- MLS2 aims at addressing the changing frequency of a collective trait, with independently defines fitness at each level of selection. We may consider, for example, David Jablonski's (1987) hypothesis about the late-Cretaceous mollusk species. Among a particular mollusk clade, species with large geographic ranges became more common than those with smaller ranges. Jablonski argued that species with large ranges left more offspring species—that is, they had greater fitness 2—and geographic range was heritable. Here the focal level is the collective's level (species, in this case). Despite the fact that the species-level trait (geographic range) is likely dependent on organismic traits (like dispersal and motility), the hypothesis does not deal with the frequency of different types of organisms. Each organism's character can evolve by natural selection at the organism's level, interacting with the higher level of selection, but the fitnesses at the two levels are independently defined and selection at each level results in different kinds of evolutionary change.

The role of collectives in MLS1 lies in generating population structures that affect particles' fitness without any determined parent-offspring relations at the collective level. Differently, in MLS2, what is fundamental is that collectives "make more" collectives—namely, that they reproduce in the ordinary way standing in parent-offspring relations.

According to the supporters of multilevel selection theory, MLS1 and MLS2 are distinct processes that occur in nature. The use of the former approach rather than the latter in any particular case depends on which kind of process we are focusing on and what we wish to model.

It is clear, from this briefly sketched outline of the arguments, that group selection theory could be intended as a part of a wider and more articulated multilevel framework.

5. Multilevel Selection in a Diachronic Perspective: Major Evolutionary Transitions

The debates on a multilevel evolutionary theory are not confined within the limits of the MLS theory. Hierarchy shapes the biologic world in a series of nested levels, where the higher levels contain the lower. The MLS theory addresses a synchronic approach to the hierarchy of levels. Another kind of approach is diachronic and investigates the evolutionary origin of the biological hierarchy itself. How has this hierarchy evolved from genes to organisms, up to the species and higher taxa? How have hierarchical levels emerged? From the MLS perspective, the significant question may be how could novel entities evolve as higher units of selection? In other words, how could *groups* of individuals convert to true biological units or *individuals*?

The distinction between *individual* and *group* was strongly put into question by the cell biologist Lynn Margulis (1970), who proposed that eukaryotic cells evolved by symbiotic associations of bacterial cells (prokaryotes) rather than by an accumulation of mutational steps from bacteria. In this view, what we usually call "individual" (namely, individual organism) is actually an evolved, functionally integrated group of other individuals at a higher level, as in any holobiontic wholes (Bordenstein and Theis 2015). A series of relevant contributions (Buss 1987; Maynard Smith and Szathmáry 1995, 1999; Michod 1999, 2005; Bourke 2011; Calcott and Sterelny 2011) generalized Margulis's concept in order to account for other major transitions in evolution (MTE), ranging from groups of functionally organized molecular interactions in early living beings to the evolution of human societies, including the emergence of the first bacterial cells, multicellular organisms, and social insect colonies. As a common pattern, during each transition, a series of smaller free-living units, capable of independent replication, gave place to a larger unit, thus creating a new level of organization.

The process of emergence of a new level of biological organization may be nonselective (as in the case of Margulis's symbiogenetic theory), but the result of the process may be a new potential unit of selection at a higher level. The fundamental evolutionary problem here is how the passage from groups *of* organisms to groups *as* organisms occurs, and why lower level units sacrifice their individuality to be functional parts of a larger body. These questions immediately raise the level-of-selection issue that David S. Wilson defined as "the iron law of multilevel selection." According to this law, "adaptation at any given level of a multitier hierarchy requires a process of selection at that level and tends to be undermined by selection at lower levels" (Wilson 2015, 137). The evolution of the multicellular organisms, for example, could be seen as the outcome of a competition between selection acting at the lower level (among cells within groups) and selection at a higher level (among multicelled groups). While the former tends to disrupt the integrity of the emerging multicelled organism, the latter may favor the evolution of adaptations for suppressing internal contention. If between-group selection is able to evolve such mechanisms that suppress the potential for disruptive selection, it can become the primary evolutionary force, and groups may evolve as novel higher-level organisms by increasing their internal functional organization.

Both the MLS theory and the MTE's issue involve interactions among levels of selection. For this reason, the MLS theory is widely accepted as a theoretical framework for the study of MTE, even though the MTE's perspective considers the balance between levels of selection in a more dynamic way (i.e., balance can itself evolve) than the MLS theory (Okasha 2005, 2006; D. S. Wilson 2015).

Individuals and groups, collectives and particles, become relative terms— both historically and logically—depending on the focal level of our analysis. In the light of the MTE, multicellular organisms can be viewed as groups with a high amount of internal functional organization, despite the fact that they have been traditionally regarded as the archetypes of the notion of "individual." They actually became individuals when their functional organization was effectively able to suppress the disruptive tendencies coming from selection at the lower level (never entirely eliminated, as is evident in many cases, such as in the cellular Darwinian selfishness of cancerous growth).

At any level of the biological hierarchy, groups can evolve functional organization to the extent that selection operates at their level, and it can develop mechanisms that suppress disruptive forms of selection within groups. Higher-level organisms, such as nucleated cells, multicellular organisms, and eusocial insect colonies, prevailed everywhere over their lower-level competitors (bacterial cells, single nucleated cells, solitary insect species) in the ecological arena.

Yet such a grand view drawn by the MTE's theorists, with a progressive complexification of hierarchical units and levels in a multilevel framework, echoes older "progressionistic" evolutionary views. In this relentless increasing complexification, all these views unfailingly culminate with the evolution of humans, which, according to David S. Wilson, "are evolution's latest major transition," being the only species among primates that "crossed the threshold from groups of organisms to groups as organisms" (2015, 49). In Wilson's view, which combines MLS theory and MTE theory, human societies have developed a group-level functional organization— like that of a single body or a social insect colony—and have become able to effectively (but not totally) suppress the disruptive forms of lower-level selection due to the action of group selection. Therefore, in such view, the human species is a doubly rare case because it brings all the hallmarks of the rare event of a major evolutionary transition and, at the same time, it has evolved a kind of intelligence quite unique in the animal world (i.e., symbolic intelligence).

Despite these recurrent types of evolutionary progressionism or some revivals of radical and unwary ontological analogies between "groups" (human societies as insect colonies in David S. Wilson), MLS theory in its strictly mathematical version neither involves nor requires such kind of theoretical implications. In general, we agree with Okasha (2006; see also Borrello 2010) that connecting the issue of the levels of selection to that of major transitions may bring about a needed formal clarity, provided that at the same time we adopt a scientifically cautious and theoretically aware perspective about the slippery connections between evolution and progress.

6. Cross-Level Exaptive By-Products

As mentioned in section 4, natural selection may act on any entities that show variation, differential fitness, and heritability, and this may occur at any level of the biological hierarchy. Furthermore, processes occurring at a certain level can propagate their effects horizontally, through the same level, or vertically (upward or downward) along the hierarchy, affecting other levels and units. Therefore, a character-fitness covariance at one level may be a side effect of direct selection occurring either at the same level—such as in cases of allometry, pleiotropy, and hitchhiking, which arise as side consequences of organismic selection on other traits—or at a different level, as in the following case described by Elisabeth Vrba (1989).

In two African antelope clades, the stenotope species (i.e., characterized by ecological specialization) left more offspring species than the eurytope species (i.e., ecological generalists). Vrba argued that the differential rates of speciation between the two kinds of species were not the outcome of causal processes at the species level but side effects of selection at the organismic level, which generated more local differentiation in stenotopes than in eurytopes.

A host of other examples in literature could provide empirical evidence for the occurrence of cross-level by-products caused by natural selection acting at different levels. Natural selection acting at one level may generate effects that filter up or down the hierarchy, bringing about cross-level by-products. The possibility of this cross-level causality, centered on the agency of natural selection, is well known and widely acknowledged by the supporters of the MLS and MTE theories (see Okasha 2006, chap. 3).

Generally speaking, both theories (MLS and MTE) are almost entirely centered on the selective dynamics and the adaptationist logic typical to the population genetics models. From the point of view of the proponents of the MLS and MTE theories, phenomena related to natural selection and to adaptive dynamics are the real interesting patterns and processes that truly deserve attention in evolutionary research. As Okasha noted, "I think the concept of a cross-level by-product should be restricted to a character-fitness covariance at one level that results from *selection* at another level, rather than just some causal process or other, for the former is what matters in the context of the levels of selection" (Okasha 2006, 106, author's italic). Although this claim is acceptable in a context of multilevel *selection* process, it is nevertheless disputable that multilevel *selection* theory could cover and unify the entire range of the multiple patterns affecting the biological hierarchy. As Darwin (1859, 6) himself acknowledged, "natural selection has been the main but not exclusive means of modification."

Again on the wake of Darwinian pluralism, more than a century later, Elisabeth Vrba and Stephen Jay Gould (1986) proposed to regard natural selection as the main, but not the exclusive, causal process that falls under a more general concept of sorting. The term *sorting* describes differential birth and death among varying organisms within a population, while "natural selection" denotes *one* direct *cause* of sorting (among others). As the authors argued, "in a hierarchical world, with entities acting as evolutionary individuals (genes, organisms, and species among them) at several levels of ascending inclusion, sorting among entities at one level has a great range of potential causes. Direct selection upon entities themselves is but one possibility among many" (Vrba and Gould 1986, 217).

According to Vrba and Gould, sorting can occur both as an effect of causes at the same focal level and as an effect of causes at higher or lower levels. These sorting effects may be driven by selection at the same level or at another level above or below, but also may not. Mass extinction events caused by global ecological disruptions, for example, triggered radical changes that are random at the focal level of populations, for selective processes could not prepare for these rare, unpredictable, and highly disruptive events. Such macroevolutionary phenomena, which set the stage for new environmental conditions, produce sorting at the level of species and higher taxa, triggering, therefore, a downward cascade of propagating effects at the population level. This reshaped environmental scenario will eventually provide new conditions for future selection among individuals, and it "may generate a pervasive realignment of life's diversity" by providing "a largely fortuitous pool of exaptive potential" (Vrba and Gould 1986, 225). In such a hierarchical scenario, the exaptive mechanisms—namely, traits selected for initial usages (or without function at all) and later co-opted for their different current functions (Gould and Vrba 1982; Gould 2002; Pievani and Serrelli 2011)—become a key concept of cross-level causation and sorting because they connect one level to another across the biological hierarchy. Genetic variation (gene-level) can be considered an exaptive by-product with respect to its phenotypic effects (organism-level) because it is not directly caused by environmental pressures. On the other side of the hierarchy, the radical change among differential birth and death in populations, as in the case of a mass extinction, is an example of sorting due to external random causes acting at different levels of the biological hierarchy.

7. Multilevel Selection and the Hierarchy Theory of Evolution: An Ecological Extension

The differences between the two codiscoverers of punctuated equilibria—Stephen J. Gould and Niles Eldredge—about the hierarchical levels of evolutionary change is a good case study that highlights the gap between a standard multilevel selection theory and a hierarchical theory of evolution. As discussed above, individuals in a hierarchical framework are evolved entities themselves, which result from complex interactions among different levels. In a MLS1 scenario (Wilson's trait-groups), functionally organized groups may evolve due to selection at a higher level, even though the groups involved in this mechanism do not have such strong identity as individuals (i.e., groups are not necessarily discrete, not reproductively isolated from each other, their members are often held together mainly by pure ecological-economic relations, they can easily migrate from one group to another, groups' fitness focuses on making more particles and not more groups, and so on). Although the notion of "group" violates many of the traditional eligibility norms to be considered as an evolutionary "individual," group selection is better defined than species selection, even if "species" could show a kind of "individuality" at higher levels (see Gould 2002, 750–64; Okasha 2006, 210–12, on species selection versus avatars selection). But the species as a real and autonomous Darwinian unit is still a controversial concept, without substantial supporting evidence. "Selection" is a very demanding causal concept and it is not enough to have somehow a discrete individuality for representing a good unit of selection. Without species traits, competition, differential survival, and inheritance of some kind, no natural selection occurs.

Gould acknowledged that evolutionary hierarchy is essentially shaped by two kinds of factors: an "internal dynamic"—namely, the "genealogical context"—and the external dimension of geological history, which Gould defined as the "mover of life's patterns" (Gould to Sepkoski, August 13, 1985, in Sepkoski 2012, 382). Nevertheless, according to Gould, the hierarchical perspective is still a standard multilevel selection theory, involving "interactors with adequate modes of plurification" at different levels (Gould 2002, 642). Natural selection is multiplied at different levels. The interplay between the two different, but closely intertwined, dimensions of evolution—that is, the "internal," genealogical dimension and the "external," geological, environmental, geographical contexts—is, instead, the hallmark of Niles Eldredge's hierarchical framework.

According to Eldredge, environmental events may have different magnitude. The greater is the magnitude, the greater is the impact on ecosystems; the greater the loss of higher taxa, the more different will be the newly evolved taxa. Eldredge compared his model to "water sloshing in the bucket—the size of the sloshes depend[s] on how hard the bucket is jolted" (Eldredge 2008, 14). The sloshing bucket model predicts that the magnitude of the ecological perturbations affects the scope of any evolutionary change. *Global* mass extinctions are caused by global environmental disruptions, which affect the entire world's biotas. At this level, taxa that survived give place to new species and eventually to radiation of larger groups. *Regional* disturbances may cause extinction across different lineages due to ecosystems collapses, and they may be followed by the birth of new species and even "turnover pulse" events. At the lower ecological level, *localized* ecosystem disturbances may eliminate some groups, which are soon replaced by other conspecific populations, with little or no perceptible evolutionary change (Eldredge 1999, 2003).

According to Eldredge, the vertical dimension of evolution (i.e., changes deriving from the transmission of sorted genetic modifications) is closely interrelated with the horizontal dimension of environmental, geographical, geophysical factors, able to produce episodic ecological changes that, in turn, affect genealogical relations among organisms, populations, and species.

In Eldredge's (2008) view, this complex interplay of ecological and genealogic factors is thus better described by a double hierarchy. The *genealogical hierarchy* involves genetically based *information* systems, and it concerns reproduction or replication: the microevolutionary level of genes is nested into the higher level of organisms, which are parts of local breeding populations and species. The *ecological hierarchy* is about *matter-energy* transfer systems and refers to interactions among entities: organisms are nested into local conspecific populations, seen in this case as "economic" entities acting in their ecological niches for physical survival. They are part of local ecosystems that are nested into regional ecosystems up to encompass the whole biosphere.

Changes in ecological (interactive) dynamics affect the information stored in the genealogical hierarchy, and vice versa. Eldredge (1999, 2008) challenged the traditional claim that evolution mostly occurs because of the accumulation of microevolutionary processes within populations in a context of competition for reproductive success. Evolution takes place in wider ecological contexts, and the role of abiotic episodic events and intergenealogical patterns should not be overlooked. Lower-level phenomena penetrate and transform the higher levels, but higher-level events set the stage for the agency of lower-level processes (Lieberman et al. 2007).

Gould refused the double hierarchy proposed by Eldredge (Gould 2002, 642). Gould had steadily in mind the intellectual fight against Richard Dawkins and his gene-centered reductionism, so he thought to simply broaden the concept of replication in a hierarchical scheme of Darwinian units (genomes, organisms, groups, even species). The result is a hierarchy with standard, organism-like units (groups, species, superorganisms) intended as interactors, with the risks related to a strongly discontinuous concept of macroevolution as an independent theoretical domain. In contrast, the approach to hierarchy taken by Eldredge is a more externalist extension of the mode of speciation inherent in punctuated equilibria, with a double genealogical (time) and ecological (space) logic (Eldredge 1999). The two parallel hierarchies in Eldredge are not a mere extension of Dawkins's replicator/interactor scheme because they are two causally interdependent levels of evolutionary change. On the contrary, in Dawkins, interactors are mere vehicles for replicators, and the replicative logic is the fundamental one.

The refusal of the double hierarchy by Gould is based on two misleading arguments: useless complexity and overlapping of the elements of the two hierarchies. In Eldredge's sloshing bucket model (2008), the nested evolutionary individualities are defined as kinds of biological organization, from the points of view of genetic transmission (genealogical or evolutionary hierarchy), and exchanges of matter and energy (ecological or economical hierarchy). Thus the groups of organisms inside a species, at the same population level above organisms, could be organized in two different ways. It is not essential that replication is a necessary and sufficient criterion for individuality because the two hierarchies are not independent, but interdependent. In Eldredge's model, no faithful inheritance is required and the levels are wider units of evolutionary change (ecological and genealogical). Gould's criticism is linked to a standard way to see hierarchy in an exclusively selective way (contra Dawkins, but still within Dawkins's intellectual framework), trying to define what exactly should be an "individual." Hence, all the problems related to "species selection" (and eventually to strong versions of "group selection," as seen above) arose as well.

The novelty of the dual hierarchical approach is the *ecological extension* of the standard genealogical hierarchy. Organisms become simultaneously part of two different interacting hierarchies. As matter-energy transfer systems (interactors), organisms compete and cooperate in order to find nutrients and energy sources, being part of the ecological hierarchy. As reproducing "packages" of genetic information (replicators), they are also a part of the genealogical hierarchy. At the scale of local populations, social systems (groups) are a hybrid fusion of economic and reproductive entities, which may coincide (like in colonies) or may not. We should note that Eldredge's distinction between groups in the ecological-economic sense (avatars) and groups in the genealogical sense (demes) reflects a similar distinction proposed by David S. Wilson in his trait-group model—between "trait" groups (as the groups where individuals interact in an ecological context) and the breeding population that forms the unit within which evolutionary change occurs (e.g., D. S. Wilson 1980).

Let us see briefly some examples of the heuristic power of this dual hierarchical approach. The process of natural selection itself, within Eldredge's hierarchical framework, turns out to be a genuinely ecological process, like in Darwin's original formulation. Natural selection is not only a matter of ever-changing fitness dynamics. It is mainly because an organism or a group does well in the economic-ecological context that they could do well in their reproductive, more-making activities: "Given an economic world of finite resources, it is because organisms are both interactors and more-makers that natural selection arises as a near inevitability—a view, of course, that goes back to Darwin" (Eldredge 1986, 356–57).

At the highest levels, species and ecosystems are two very different entities and they could not be put in the same hierarchy. Species are genealogical entities, reproductive communities, and they are not part of any specifiable economic system—except in the case of a species consisting of but a single local population. Normally, species are featured in many ecosystems within their distribution area, and they do not occupy any sharply defined ecological niche (Eldredge 1986, 1999). At the lowest levels, below the organism level, the host is the environment of its endosymbionts, and the holobionts could be formalized as nested hierarchies of interacting levels.

This kind of double-hierarchy perspective fits very well with a lot of contemporary integrated field researches, in which molecular biology, paleontology, ecology, paleoclimatology, demography, population structures, and other points of views at different levels (ecological and genealogical) make evidence and patterns to converge in shaping an evolutionary scenario—for instance, in the integrated phylogenies and in paleoanthropology (Pievani 2012; Parravicini and Pievani 2015).

In conclusion, Eldredge's double hierarchy extends the multilevel selection theory in a broader ecological and evolutionary framework. Within this plurality of patterns, the MLS theory seems to cover only one, though essential, aspect of the evolutionary explanation—namely, the fundamental role played by multiple selective processes across a hierarchical evolutionary framework.

In summary, we argue that two different theories of the connections between micro- and macroevolutionary patterns are contending the field: (1) selective hierarchies, in several versions, and (2) the dual ecological-genealogical hierarchical system. The latter includes the former as a subset of processes. A dual hierarchical approach can be applied at levels above and below organism level. Darwinian natural selection is the core where the two hierarchies fuse together. The crucial role of population structure as ecological parameter in group selection versus kin selection debate (see above par. 1 and 2) confirms that multilevel selection processes depend on both ecological and genealogical interactions, like in the cases of trade-offs between antagonistic or different selective pressures. All these data suggest that the hierarchy theory of evolution could be candidate as an updated and unifying metatheory of evolutionary patterns

(Pievani 2015b). Among all contemporary sciences, the theory of evolution has one of the broadest and more diversified empirical bases, ranging from molecules to ecosystems. Such a challenging scope needs a consistent theoretical bravery.

Notes

1. See also Darwin (1871, v.1, 70–106), in particular pp. 72, 98, 103.

2. We refer in particular to Michael Ruse, who wrote that "Darwin opted firmly for hypotheses supposing selection always to work at the level of the individual rather than the group" (Ruse 1980, 615) and to Richard Dawkins, who stated in a *Prospect* article on May 24, 2012, that Darwin spoke about group selection only in "one anomalous passage."

3. See Sober (2011, chap. 2) for a detailed analysis of George Williams's arguments against the legitimacy of group selection, which he resolutely considered both an empirical mistaken and an idea at odds with the basic logic of rigorous Darwinism.

4. The phenomenon, sometimes called "equilibrium selection," could be considered as an exception to this rule. This case involves models that result in multiple local equilibria and are internally stable by definition. Here, group selection can favor the local equilibria that function best at the group level (Samuelson 1997; Gintis 2000; Wilson and Wilson 2007).

5. See Buss (1987); Sober and Wilson (1998); and Gould (2002) for a detailed account of the history of the levels-of-selection debate.

6. See, for example, Salt (1979); Vrba (1989). For a critical point of view, see Williams (1992); Damuth and Heisler (1988). Grantham (2007) provides a review of the approaches for distinguishing between emergent and aggregate traits in biological systems.

References

Bordenstein, Seth R., and Kevin R. Theis. 2015. "Host Biology in Light of the Microbiome: Ten Principles of Holobionts and Hologenomes." *PLoS Biology* 13: e1002226. doi:10.1371/journal.pbio.1002226.

Borrello, Mark E. 2005. "The Rise, Fall and Resurrection of Group Selection." *Endeavour* 29: 43–47.

———. 2010. *Evolutionary Restraints: The Contentious History of Group Selection*. Chicago: University of Chicago Press.

Bourke, Andrew F. G. 2011. *Principles of Social Evolution*. Oxford: Oxford University Press.

Bowles, Samuel. 2008. "Being Human: Conflict—Altruism's Midwife." *Nature*, 456: 326–27.

Bull, James J. 1994. "Perspective: Virulence." *Evolution* 48: 1423–37.

Buss, Leo W. 1987. *The Evolution of Individuality*. Princeton, NJ: Princeton University Press.

Calcott, Brett, and Kim Sterelny. 2011. *The Major Transitions in Evolution Revisited*. Cambridge, MA: MIT Press.

Cronk, Lee. 2005. "The Application of Animal Signaling Theory to Human Phenomena: Some Thoughts and Clarifications." *Social Science Information* 44: 603–20.

Damuth, John I., and Lorraine Heisler. 1988. "Alternative Formulations of Multi-Level Selection." *Biology and Philosophy* 3: 407–30.

Darwin, Charles. (1859) 1872. *On the Origin of Species by Means of Natural Selection, or the Preservation of Favoured Races in the Struggle for Life*, 6th ed. London: John Murray.

———. 1871. *The Descent of Man, and Selection in Relation to Sex*. 2 vols. London: John Murray.

Dawkins, Richard. 1976. *The Selfish Gene*. Oxford: Oxford University Press.

Deacon, Terrence W. 2003. "Multilevel Selection in a Complex Adaptive System: The Problem of Language Origins." In *Evolution and Learning: The Baldwin Effect Reconsidered*, edited by B. Weber and D. Depew, 81–106. Cambridge, MA: MIT Press.

Dennett, Daniel C. 2002. "Commentary on Sober and Wilson's *Unto Others: The Evolution and Psychology of Unselfish Behavior*." *Philosophy and Phenomenological Research* 65(3): 692–96.

Dugatkin, Lee A. 2002. "Cooperation in Animals: An Evolutionary Overview." *Biology and Philosophy* 17: 459–76.

Eldakar, Omar T., Michael J. Dlugos, Galen P. Holt, David Sloan Wilson, and John W. Pepper. 2010a. "Population Structure Influences Sexual Conflict in Wild Populations of Water Striders." *Behavior* 147: 1615–31.

Eldakar, Omar T., Michael J. Dlugos, John W. Pepper, and David S. Wilson. 2009a. "Population Structure Mediates Sexual Conflict in Water Striders." *Science* 326: 816.

Eldakar, Omar T., Michael J. Dlugos, R. Stimson Wilcox, and David S. Wilson. 2009b. "Aggressive Mating as a Tragedy of the Commons in a Water Strider *Aquarius remigis*." *Behavioral Ecology and Sociobiology* 64: 25–33.

Eldakar, Omar T., David S. Wilson, Michael J. Dlugos, and John W. Pepper. 2010b. "The Role of Multilevel Selection in the Evolution of Sexual Conflict in the Water Strider *Aquarius Remigis*." *Evolution* 64: 3183–89.

Eldredge, Niles. 1986. "Information, Economics, and Evolution." *Annual Review of Ecology and Systematics* 17: 351–69.

———. 1999. *The Pattern of Evolution*. New York: W. H. Freeman.

———. 2003. "The Sloshing Bucket: How the Physical Realm Controls Evolution." In *Evolutionary Dynamics: Exploring the Interplay of Selection, Accident, Neutrality, and Function*, edited by J. P. Crutchfield and P. Schuster, 3–32. Oxford: Oxford University Press.

————. 2008. "Hierarchies and the Sloshing Bucket: Toward the Unification of Evolutionary Biology." *Evolution: Education & Outreach* 1: 10–15.

Fisher, Ronald A. 1930. *The Genetical Theory of Natural Selection*. Oxford: Clarendon Press.

Frank, Steven A. 1996. "Models of Parasite Virulence." *Quarterly Review of Biology* 71: 37–78.

Gardner, Andy. 2015. "Group Selection versus Group Adaptation." *Nature* 524: E3–E4.

Gilpin, Michael E. 1975. *Group Selection in Predator-Prey Communities*. Princeton, NJ: Princeton University Press.

Gintis, Herbert. 2000. *Game Theory Evolving: A Problem-Centered Introduction to Modeling Strategic Interaction*. Princeton, NJ: Princeton University Press.

Goodnight, Charles J. 2005. "Multilevel Selection: The Evolution of Cooperation in Non-Kin Groups." *Population Ecology* 47: 3–12.

Goodnight, Charles J., and Lori Stevens. 1997. "Experimental Studies of Group Selection: What Do They Tell Us about Group Selection in Nature?" *The American Naturalist* 150 (supplementary): S59–S79.

Gould, Stephen J. 2002. *The Structure of Evolutionary Theory*. Cambridge, MA: Harvard University Press.

Gould, Stephen J., and Elisabeth S. Vrba. 1982. "Exaptation: A Missing Term in the Science of Form." *Paleobiology* 8: 4–15.

Grantham, Todd. 2007. "Is Macroevolution More than Successive Rounds of Microevolution?" *Palaeontology* 50: 75–85.

Grinsted, Lena, Trine Bilde, and James D. J. Gilbert. 2015. "Questioning Evidence of Group Selection in Spider." *Nature* 524: E1–E3.

Haldane, John B. S. 1932. *The Causes of Evolution*. London: Longmans, Green.

Hamilton, William D. 1964. "The Genetical Evolution of Social Behavior, I and II." *Journal of Theoretical Biology* 7: 1–52.

————. 1975. "Innate Social Aptitudes in Men: An Approach from Evolutionary Genetics." In *Biosocial Anthropology*, edited by Robin Fox, 133–55. London: Malaby Press. Reprinted in Hamilton 1996, 329–51.

————. 1996. *The Narrow Roads of Gene Land: The Collected Papers of W. D. Hamilton*. Evolution of Social Behavior, vol. 1. Oxford: W. H. Freeman.

Harman, Oren. 2010. *The Price of Altruism*. New York: W. W. Norton.

Heinsohn, Robert, and Craig Packer. 1995. "Complex Cooperative Strategies in Group-Territorial African Lions." *Science* 269: 1260–62.

Hull, David L. 1976. "Are Species Really Individuals?" *Systematic Zoology* 25: 174–91.

————. 1980. "Individuality and Selection." *Annual Review of Ecology and Systematics* 11: 311–32.

Jablonski, David. 1987. "Heritability at the Species Level: Analysis of Geographic Ranges of Cretaceous Mollusks." *Science* 238: 360–63.

Kerr, Benjamin, Claudia Neuhauser, Brendan J. M. Bohannan, and Antony M. Dean. 2006. "Local Migration Promotes Competitive Restraint in a Host-Pathogen 'Tragedy of the Commons.'" *Nature* 442: 75–78.

Lachmann, Michael, Szabolcs Számadó, and Carl T. Bergstrom. 2001. "Cost and

Conflict in Animal Signals and Human Language." *Proceedings of the National Academy of Sciences of the United States of America* 98: 13189–94.

Lewontin, Richard C. 1970. "The Units of Selection." *Annual Review of Ecology and Systematics* 1: 1–18.

———. 1998. "Survival of the Nicest? Review of *Unto Others* by E. Sober and D. S. Wilson." *New York Review of Books* (October 22): 59–63.

Lieberman, Bruce S., William Miller III, and Niles Eldredge. 2007. "Paleontological Patterns, Microecological Dynamics and the Evolutionary Process." *Evolutionary Biology* 34: 28–48.

Lloyd, Elisabeth A. 1988. *The Structure and Confirmation of Evolutionary Theory.* Princeton, NJ: Princeton University Press.

———. 1999. "Altruism Revisited." *Quarterly Review of Biology* 74: 447–49.

Lloyd, Elisabeth A., and Stephen J. Gould. 1993. "Species Selection on Variability." *Proceedings of the National Academy of Sciences of the United States of America* 90: 595–99.

Margulis, Lynn. 1970. *Origin of Eukaryotic Cells.* New Haven: Yale University Press.

Maynard Smith, John. 1964. "Group Selection and Kin Selection." *Nature* 201: 1145–47.

———. 1998. "The Origin of Altruism." *Nature* 393: 639–40.

Maynard Smith, John, and George R. Price. 1973. "The Logic of Animal Conflict." *Nature* 246: 15–18.

Maynard Smith, John, and Eörs Szathmáry. 1995. *The Major Transitions in Evolution.* Oxford: Oxford University Press.

———. 1999. *The Origins of Life: From the Birth of Life to the Origin of Language.* Oxford: Oxford University Press.

Michod, Richard E. 1999. *Darwinian Dynamics: Evolutionary Transitions in Fitness and Individuality.* Princeton, NJ: Princeton University Press.

———. 2005. "On the Transfer of Fitness from the Cell to the Organism." *Biology and Philosophy* 20: 967–87.

Norenzayan, Ara. 2013. *Big Gods: How Religion Transformed Cooperation and Conflict.* Princeton, NJ: Princeton University Press.

Nowak, Martin A. 2006. "Five Rules for the Evolution of Cooperation." *Science* 314: 1560–63.

Nowak, Martin A., and Roger Highfield. 2011. *SuperCooperators, Altruism, Evolution and Why We Need Each Other to Succeed.* New York: Free Press.

Nowak, Martin A., and Karl Sigmund. 2005. "Evolution of Indirect Reciprocity." *Nature* 437: 1291–98.

Nowak, Martin A., Corina. E. Tarnita, and Edward O. Wilson. 2010. "The Evolution of Eusociality." *Nature* 466: 1057–62.

Nunney, Leonard. 1998. "Are We Selfish, Are We Nice, or Are Nice Because We Are Selfish?" *Science* 281: 1619–21.

Okasha, Samir. 2005. "Multilevel Selection and the Major Transitions in Evolution." *Philosophy of Science* 72: 1013–25.

———. 2006. *Evolution and the Levels of Selection.* Oxford: Oxford University Press.

Packer, Craig, and Robert Heinsohn. 1996. "Response: Lioness Leadership." *Science* 271: 1215–16.

Pagel, Mark. 2012. *Wired for Culture: The Natural History of Human Cooperation.* New York: Allen Lane.

Parravicini, Andrea, and Telmo Pievani. 2016. "Multi-Level Human Evolution: Ecological Patterns in Hominin Phylogeny." *Journal of Anthropological Sciences* (Erice Special Issue) 94:1–16. doi:10.4436/jass.94026.

Pievani, Telmo. 2011. "Born to Cooperate? Altruism as Exaptation and the Evolution of Human Sociality." In *Origins of Altruism and Cooperation*, edited by Robert W. Sussman and C. Robert Cloninger, 41–61. New York: Springer.

———. 2012. "The Final Wave: *Homo sapiens* Biogeography and the Evolution of Language." *Rivista Italiana di Filosofia del Linguaggio*, SFL: 203–16. doi:10.4396/20120618.

———. 2014. "Individuals and Groups in Evolution: Darwinian Pluralism and the Multilevel Selection Debate." *Journal of Biosciences* 39: 319–25.

———. 2015a. "Philosophy of Selection (Natural, Sexual and Drift)." *eLS.* http://www.els.net. doi:10.1002/9780470015902.a0003461.

———. 2015b. "How to Rethink Evolutionary Theory: A Plurality of Evolutionary Patterns." *Evolutionary Biology.* doi:10.1007/s11692-015-9338-3.

Pievani, Telmo, and Emanuele Serrelli. 2011. "Exaptation in Human Evolution: How to Test Adaptive *vs* Exaptive Evolutionary Hypotheses." *Journal of Anthropological Sciences* 89: 1–15.

Price, George R. 1972. "Extension of Covariance Selection Mathematics." *Annals of Human Genetics* 35: 485–90.

Pruitt, Jonathan N., and Charles J. Goodnight. 2014. "Site-Specific Group Selection Drives Locally Adapted Group Compositions." *Nature* 514: 359–62.

———. 2015. "Pruitt and Goodnight Reply." *Nature* 524: E4–E5.

Rainey, Paul B., and Katrina Rainey. 2003. "Evolution of Cooperation and Conflict in Experimental Bacterial Populations." *Nature* 425: 72–74.

Ruse, Michael. 1980. "Charles Darwin and Group Selection." *Annals of Science* 37: 615–30.

Sachs, Joel L., Ulrich G. Mueller, Thomas P. Wilcox, and James J. Bull. 2004. "The Evolution of Cooperation." *Quarterly Review of Biology* 79: 135–60.

Salt, George W. 1979. "A Comment on the Use of the Term *Emergent Properties.*" *American Naturalist* 113: 145–48.

Samuelson, Larry. 1997. *Evolutionary Games and Equilibrium Selection.* Cambridge, MA: MIT Press.

Segerstrale, Ullica. 2013. *Nature's Oracle: A Life of W. D. Hamilton.* Oxford: Oxford University Press.

Sepkoski, David. 2012. *Rereading Fossil Record: The Growth of Paleobiology as an Evolutionary Discipline.* Chicago: University of Chicago Press.

Sober, Elliott. 1981. "Holism, Individualism and the Units of Selection." *Proceedings of the Biennial Meeting of the Philosophy of Science Association 1980* 2: 93–121.

———. 1984. *The Nature of Selection: Evolutionary Theory in Philosophical Focus.* Chicago: University of Chicago Press.

————. 2011. *Did Darwin Write the Origin Backwards? Philosophical Essays on Darwin's Theory*. Amherst, NY: Prometheus Books.

Sober, Elliott, and Richard Lewontin. 1982. "Artifact, Cause, and Genic Selection." *Philosophy of Science* 49: 157–80.

Sober, Elliott, and David S. Wilson. 1998. *Unto Others: The Evolution and Psychology of Unselfish Behavior*. Cambridge, MA: Harvard University Press.

Stauffer, Robert C., ed. 1975. *Charles Darwin's Natural Selection: Being the Second Part of His Big Species Book Written from 1856 to 1858*. Cambridge: Cambridge University Press.

Tarnita, Corina E., Tibor Antal, Hisashi Ohtsuki, and Martin A. Nowak. 2009. "Evolutionary Dynamics in Set Structured Populations." *Proceedings of the National Academy of Sciences of the United States of America* 106: 8601–4.

Trivers, Robert L. 1971. "The Evolution of Reciprocal Altruism." *Quarterly Review of Biology* 46: 35–57.

Vrba, Elisabeth S. 1989. "Levels of Selection and Sorting with Special Reference to the Species Level." *Oxford Surveys in Evolutionary Biology* 6: 111–68.

Vrba, Elisabeth S., and Stephen J. Gould. 1986. "The Hierarchical Expansion of Sorting and Selection: Sorting and Selection Cannot Be Equated." *Paleobiology* 12: 217–28.

Wade, Michael J. 1984. "Changes in Group Selected Traits When Group Selection Is Relaxed." *Evolution* 38: 1039–46.

————. 1985. "Soft Selection, Hard Selection, Kin Selection and Group Selection." *American Naturalist* 125: 61–73.

Williams, George C. 1966. *Adaptation and Natural Selection: A Critique of Some Current Evolutionary Thought*. Princeton, NJ: Princeton University Press.

————. 1992. *Natural Selection: Domains, Levels, Challenges*. New York: Oxford University Press.

————. 1996. Preface to the reprint edition of *Adaptation and Natural Selection: A Critique of Some Current Evolutionary Thought*, ix–xiv. Princeton, NJ: Princeton University Press.

Wilson, David S. 1975. "A Theory of Group Selection." *Proceedings of the National Academy of Science* 72: 143–46.

————. 1977. "Structured Demes and the Evolution of Group-Advantageous Traits." *American Naturalist* 111: 157–85.

————. 1980. *The Natural Selection of Populations and Communities*. Menlo Park, CA: Benjamin/Cummings.

————. 1983. "The Group Selection Controversy: History and Current Status." *Annual Review of Ecology and Systematics* 14: 159–87.

————. 2015. *Does Altruism Exist? Culture, Genes, and the Welfare of Others*. New Haven: Yale University Press.

Wilson, David S., and Elliott Sober. 1989. "Reviving the Superorganism." *Journal of Theoretical Biology* 136: 337–56.

————. 1994. "Reintroducing Group Selection to the Human Behavioral Sciences." *Behavioral and Brain Sciences* 17: 585–654.

Wilson, David S., and Edward O. Wilson. 2007. "Rethinking the Theoretical Foundation of Sociobiology." *Quarterly Review of Biology*, 82: 327–48.

Wilson, Edward O. 1975. *Sociobiology: The New Synthesis*. Cambridge, MA: Harvard University Press.

———. 2012. *The Social Conquest of Earth*. New York: W. W. Norton.

Wright, Sewall. 1931. "Evolution in Mendelian Populations." *Genetics* 16: 97–159.

Wynne-Edwards, Vero C. 1962. *Animal Dispersion in Relation to Social Behavior*. Edinburgh: Oliver and Boyd.

CHAPTER 9

Systems Emergence

The Origin of Individuals in Biological and Biocultural Evolution

Mihaela Pavličev, Richard O. Prum, Gary Tomlinson, and
Günter P. Wagner

Introduction

A key problem in recent thinking on the origin and evolution of life involves the concept of emergence and the challenges it might present to traditional models of causality (Mitchell 2009; Ellis et al. 2012; Love 2012; Okasha 2012). The question might be succinctly put as follows: granted that there are systems whose properties might not be *predicted* from the properties of their component parts alone, are there systems whose properties cannot be *explained* by the more or less straightforward interaction of those parts? Aggregates of water molecules, to cite the classic example of simple emergent complexity, show properties that neither oxygen nor hydrogen atoms show in isolation, but the properties admit of causal explanation according to the nature of the interaction of those atoms. Do some biological systems go further, challenging our heuristic models and eluding causal explanation of the usual sorts? To pose these questions is to raise large and difficult issues concerning the nature of generalization and the role of laws in biological thought, the possibility of modes of causation not yet encountered in today's science, and more.

Emergence and the kinds of interactions that contribute to emergent properties have been addressed in many living and nonliving systems, and various types and degrees of emergence—horizontal and vertical, strong and weak, and more—have been recognized (Noble 2012). Terrence Deacon has introduced a typology of three orders of emergence of increasing complexity (Deacon 2003). His first order involves the properties of certain aggregates—for example (to return to instance of water), liquidity. Second-order emergence, in addition, shows local, short-term contingency—for instance, the bifurcations during the crystallization of a snowflake, its final form dependent not only on the initial conditions but also on conditions reset anew with each stage of its growth. Third-order emergence furthermore involves some kind of coded information or *memory* that enables contingency to be communicated across generations—in other words, heritability. This is a property of living systems (and of some things a few of them have created). Emergence in living systems is particularly intriguing because evolutionary processes that often involve gradual changes are so important in their origin, and these processes can appear to conflict with the idea of emergent properties.

We take, for now at least, an agnostic position on debates about causality and emergent properties, and we begin instead with the apparent connection between emergent phenomena characteristic of living things and complex systems. We perceive such systems across the ontological span of life, on scales from the intracellular and molecular to the sociocultural, as well as across the whole phylogenetic history of life. And we propose that the most constructive way to tackle the problem of emergence in biology will start from the analysis of specific instances of such complex systems.

We offer analyses of two cases from opposite ends of both the ontological and the phylogenetic or historical spectra. The first, no doubt evolutionarily ancient and hugely dispersed, is the case of epigenetic *character identity networks* (ChINs), systems of molecular interaction by which phenotypic characters are maintained across species through phylogeny and proliferated. The second, recent in development and restricted to a tiny corner of the biosphere, is the case of special cultural systems or *cultural epicycles* that arose late in hominin evolution and that help to explain the peculiarities of that evolution. In each of these cases, we will be concerned to describe features that are similar despite the vast differences between the systems themselves: the stability or *robustness* of the systems, the tendency of the systems toward *self-maintenance* of this robustness, the central role of *feedback* (and feed forward) dynamics in this maintenance, and the systemic *autonomy* thus gained. These are all issues that concern the nature of these emergent systems, but they do not address the difficulty of explaining their origin. What is it that makes the explanation of emergent systems so difficult? How can this difficulty be overcome in a rational and scientific way? These issues will be discussed in the last part of this chapter.

Emergence Is the Origin of Novel Individuals

In this section, we argue that Deacon's third level of emergence—namely, the origin of contingent structures with coded information or memory—is connected to the origin of novel individuals (Deacon 2003). Deacon's notion of coding or memory implies that there is some kind of diachronic continuity—that is, some mechanism that "represents" the memory across time, this representation being the delimited temporal extension we call heritability. This mechanism tends to be local—that is, it is a process that is spatially delimited within which memory can be actualized and heritability can occur. We argue that such spatiotemporal localization is fundamental to all cases of Deacon's "third-order" emergence, and it helps to define what we mean by *individuals*.

The origin of life cannot be fully explained by the emergence of "memory molecules" (RNA and DNA) but critically depends on the origin of cells. Cells are the entities that are capable of self-replication, scaffolding DNA replication within them, and they are necessary both for physiological/metabolic integrity and for the possibility of natural selection of genomes larger than very short polynucleotides (Maynard-Smith and Szathmáry 1995). Cells house mechanisms for living third-order emergence. Hence the emergence of life was tied to the origin of a novel kind of individual, the cell. Later, the life of animals and other multicellular organisms became possible through the integration of cells into higher-level individuals (Buss 1987). In general, biological diversity is increasingly thought of as the individuation of populations from other similar populations. Species are such populations or groups of populations that are genetically isolated from other similar populations and are thus independent units of evolutionary change (Mayr 1942).

Extending this thinking, we hypothesize that any major transition in the evolution of life is accomplished through the origin of novel *kinds* of individuals. Individual in this context refers to a distinct entity, which has an origin, a continuation in time over generations or species even, and potentially an end. An important consequence of emphasizing the individual as a historical entity is that its extant features are often misleading, as will be shown in examples below. Individuals thereby are not limited to a particular organizational level; rather, they can be morphological or behavioral traits, cell types, as well as higher-level entities involving multiple organisms. Similarly, the emergence of human modernity over the last two hundred to three hundred millennia can also be recounted as a history of the origin of novel kinds of systems of human activity, which in turn gave rise to music, language, and all forms of sociality and institutions characteristic of *Homo sapiens* today (Tomlinson 2015). Here we will discuss two kinds of emergence of novel kinds of individuals. One involves suborganismic individualities, in biology variously called homologues, characters, or organs and body parts (Wagner 2014). The other involves systems of human cultural coalescence. Our purpose is to suggest structural similarities between those quite disparate kinds of individuals: one strictly biological, the other biocultural.

Suborganismal Organic Individuals

The biological literature on the emergence of individuality has focused on a small handful of emergent units that mark in their appearance the major transitions in the history of life (see, for instance, Maynard-Smith and Szathmáry 1995). These include the origin of life and cells, the origin of multicellularity, and the origin of eusocial aggregates of organisms, as in the case of social insects (Wilson 1975; Rajakumar et al. 2012) or physiologically integrated colonies of animals (with syphonophores as the classic example; Haeckel 1869; Dunn and Wagner 2006). Whereas these cases certainly represent dramatic increases in biological complexity, the vast majority of evolutionary novelties contributing to the emergent complexity of life forms are due to a less well-investigated kind of emergence, the differentiation of Darwinian individuals into smaller subsystems. By "Darwinian individuals," we mean members of the population with capacity to undergo evolutionary change (Godfrey-Smith 2011). These subsystems nevertheless acquire a degree of independence that betrays their status as emergent individuals. In the following paragraphs, we will briefly discuss a few examples and then reflect on the biological mechanisms and evolutionary causes for their individuation.

 If one is asked to rank intuitively a set of animal species according to their morphological complexity, the resulting classification will reflect the degree of differentiation of the body into quasi-independent body parts or organs. Animals with few tissue types and no organs will occupy the lower end of the spectrum. Examples are ctenophores and cnidarians, with the (apparently simple) hydra as exemplar. On the other end of the spectrum will be vertebrates with hundreds of anatomically defined subparts, from eyes, muscles, bones, and the many parts of the brain to the many different skin differentiations that make up the surface of the body. Organismal complexity, at least in this sense of the word, is proportional to the number of differentiated parts an organism is composed of (McShea 1996). This increase in complexity occurs without the creation of new aggregates, which is instead the case in the origin of multicellular organisms and the origin of social insect colonies. They have received relatively less attention, since they seem trivial from the functionalist point of view: division of labor and functional specialization seem like "good ideas" to be expected as outcomes of natural selection. Before we discuss the likely mechanisms for the origin of individualized body parts, we want to mention two other examples of "suborganismal" individualities: cell types and worker casts in social insects.

Animals of different degrees of complexity differ not only in the number of differentiated aggregates of cells—that is, organs or body parts (brains, livers, muscles, etc.)—but also in the number of specialized and individuated cellular elements, so-called *cell types*. Cell types are usually recognized through the specializations they acquired to fulfill different functions as muscle cells, nerve cells, secretory cells, and so on. This functional perspective is useful as long as we focus on one species (e.g., *Homo sapiens*) or on a group of sufficiently related species (e.g., mammals). When comparing cells from more distantly related animals, however, the usefulness of the function-based classification diminishes. This is principally due to the fact that during evolution cell types change function in a variety of ways: one frequent pathway for this change is the division of function, in which two or more functions are performed by one ancestral cell type, but in derived lineages, one finds two or more specialized cell types each performing a subset of the functions of the ancestral cell type (Arendt 2008; figure 9.1). Another pathway is the acquisition of novel functions, as in the case of the evolution of endometrial stromal cells to accommodate an invasive placenta in the uterus (Wagner et al. 2014). In general, in a broad evolutionary context, cell types are not well defined by a specific function; they are better defined as historical individuals that can and do perform different sets of functions in different lineages. Nevertheless, cell types retain their identity in the historical process of transformation and adaptation to different

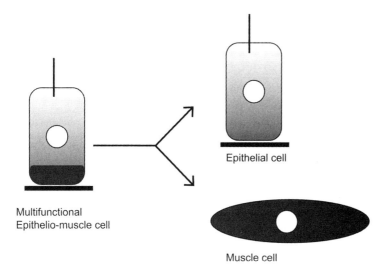

FIGURE 9.1 Example of the origin of a novel cell type through the specialization of an ancestral multifunctional cell: in many lower animals epithelial cells also function as contractile cells (muscle cells) where they have a contractile apparatus at the base of the cell (black area). In more derived animals, these two functions are usually performed by specialized cell types: epithelial cells and muscle cells. This is an example of the origin of a new cell type without a novel function, because the two functions are already both performed by the ancestral multifunctional epithelia-muscle cell.

needs. How this is possible will be discussed below. But first let us consider another example of individuation through differentiation: the evolution of worker casts.

Eusociality evolved in many lineages of animals, from ants, bees, and termites to naked mole rats and various forms of colonial marine invertebrates (Crespi and Yanega 1995; Wilson and Hölldobler 2005). In each case an aggregate of individuals performs the functions of an organism by dividing somatic functions (growth, nutrition, defense, etc.) and reproductive functions among the individuals (bee queens, reproductive polyps, etc.). The individuals specialized for somatic functions "give up" their long-term reproductive chances but support the reproduction of a relative. This situation is structurally identical to that of a multicellular animal, where somatic cells forgo their chance of leaving offspring in favor of a genetically related subpopulation of cells that do almost nothing but reproduce (germ line cells giving rise to gametes). As mentioned above, after the establishment of the soma-germline differentiation, further complexity arises through the differentiation of somatic cells. The same happens with respect to sterile worker classes in ants (Hölldobler and Wilson 1990). Examples are the differentiation of workers into "minor workers" and "soldiers." Workers do what workers do—namely, gather food, care for the larvae, and build the nest. Soldiers defend the nest and are sometimes recruited for specialized tasks, like cracking seeds. The existence of different worker casts is structurally identical to the case of cell type differentiation in multicellular organisms, where also nonreproductive members of the colony differentiate in order to perform specialized functions.

Given that suborganismal differentiation is a major mode of increasing complexity in organic evolution, two questions are raised: (1) Why would specialized parts of nonreproductive entities arise at all—that is, why would natural selection favor differentiation in some lineages but not in others? (2) How, mechanistically, is the individuality of these parts achieved? These two questions are two sides of the same coin, since the answer to the first also tells us something about the nature of the second.

New functions do not necessarily lead to individualized structures, specialized to these distinct functions. That has been demonstrated abundantly by centuries of comparative biology, from cells that perform a variety of functions to ant colonies where only one worker class exists that performs all the functions from food gathering to defense, without any differentiation of the workers. Mathematical analysis of soma-germ line differentiation (Michod 1999) as well as more general models (Rueffler et al. 2012) suggest that a key requirement for the evolution of specialized structures is that performance of multiple functions is consistently worse than specializing on only one function; that is, there is a tradeoff that is strong enough to favor differentiation of general-purpose elements into specialized structures. A corollary of this insight is that sharp distinction between different specialized structures is favored if intermediate phenotypes are worse in performing either function than the specialized cases (Rueffler et al. 2012). That this is not always the case is likely, given the many cases of less differentiated forms of life. Nevertheless, the pressure to avoid intermediate phenotypes gives us a hint how, mechanistically, the individuality of structures (cell types, organs, or worker casts) is realized.

Any biological structure has some degree of plasticity, meaning that it can assume different phenotypes in spite of executing the same developmental program (West-Eberhard 2003). The final phenotype is an interaction between the developmental program active in a body part and the environment. Plastic responses to intra- or extraorganismal environments are usually gradual and thus generically produce a graded distribution of

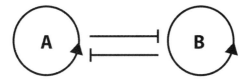

FIGURE 9.2 The logic of core networks. These are gene regulatory networks of regulatory genes (transcription factor genes) that feed back positively on themselves, stabilizing two alternative states (A or B) and at the same time inhibit the expression of the alternative gene regulatory network state.

phenotypes in response to a graded distribution of environmental signals. If, however, intermediate phenotypes along this smooth distribution are disadvantaged, a switch mechanism is favored—that is, a mechanism that, at a certain threshold, leads to the development of alternative and robust phenotypes.

There are two well-understood molecular mechanisms that can produce switch-like responses to a graded signal: (1) positive feedback loops among genes with mutual inhibitory interactions among alternative feedback loops (2) and cooperativity among gene products. Many examples have been described where the differentiation of cell types or regions in the embryo are caused by small gene regulatory networks that have a positive feedback loop configuration, variously called core networks, kernels, or character identity networks (Davidson and Erwin 2006; Wagner 2007; Graf and Enver 2009; figure 9.2). Genes of a core network activate each other's expression and are engaged in a "co-regulatory embrace," in Eric Davidson's words (Davidson and Erwin 2006). In enhancing each other's expression, the members of a core network also inhibit the expression of alternative core network genes (figure 9.2).

The existence of positive feedback loops as core networks has a number of important consequences. Positive feedback loops that cause alternative stable states of gene expression enable the expression of different developmental programs. This is due to the fact that these alternative core networks consist of regulatory genes, mostly transcription factor genes, and they can regulate not only each other but also any set of target genes. Which set of target genes they regulate is not determined by the nature of these transcription factor genes but by the cis-regulatory elements of the target genes themselves.

The ability to control different developmental or physiological pathways is inherent in the existence of core regulatory networks. As a further consequence, the ability to express different sets of genes leads to evolutionary individuation, since mutations can affect cells or characters differentially if they have different core networks activated. The ability to express different genes entails partial variational individuality and evolutionary individuality (Wagner 2015). Finally, at the population genetic level, positive core regulatory feedback loops lead to variational independence of gene expression (Pavličev and Widder 2015).

A final note, one should remember that the logic of core networks implies that they are molecular devices that enable phenotype differentiation but do not determine the nature of the difference. This is the case because the set of transcription factor genes composing a core network can regulate any set of target genes, as long as the target gene has an enhancer that binds the transcription factors of the core network. Hence the same core network, in different species, can control the development of different phenotypes and functional specializations, and the set of target genes affected by a core network can change without changes to the core network. This fact gives mechanistic meaning to the notion of "the same organ in different organisms regardless of form and function" (Owen's definition of homology, 1848; Wagner 2014). Sameness here does not mean sameness or similarity of phenotype or function but *historical continuity of the existence of a localized individual* with possibly changing functional commitments. Hence sameness (homology) of body parts gestures toward the existence of a historical individual (Ghiselin 2005; Wagner 2007, 2014).

The other molecular mechanism that can contribute to sharply delineated alternative phenotypes is cooperativity of transcription factor proteins. This mechanism is well documented in many systems but rarely discussed in its evolutionary developmental consequences (see, for instance, Wagner and Lynch 2010). Cooperativity among transcription factor proteins means that the regulation of a target gene strongly depends on the simultaneous presence of two or more transcription factor proteins in order to induce the expression of a target gene. This mechanism ensures that certain target genes are expressed only when the right combination of transcription factor proteins is present in a cell. Cooperativity is usually due to a physical protein-protein interaction in which the interaction partners reveal each other's activation domains. Examples are the interactions that evolved among the MADS box genes underlying the flower organ identity of eu-angiosperm flowers (Wang et al. 2010) and the interactions between the general transcription factor FOXO1 with either CEBPB (Lynch, May, and Wagner 2011) and/or HoxA11 (Nnamani et al. 2015). In the case of flower development, transcription factor cooperativity ensures that greatly different flower organs (petals, stamens, etc.) can develop from the relatively crowded space of a flower meristem.

Cell and organ identity is subscribed by cooperative molecular mechanisms (feedback loops, transcription factor cooperativity) that enable sharply different developmental and physiological readouts in response to otherwise graded signals from the intraembryonic or external environment.

Epicyclic Individuals in Late Hominin Evolution

If several new kinds of individuals mark the major biological transitions in the history of life, at least two more mark, among a small group of mammalian and avian species, *biocultural* transitions. Here we define *culture* in a broad way, as the passing of information learned in a lifetime to future generations (Richerson and Boyd 2005). Cultural animals are, in this view, those that can learn from past experience and transmit this learning through active or passive mimetic or other pedagogies to progeny or other conspecifics. By this definition there are many cultural animals in the world today in addition to humans, including at least some primates and birds. Historically, hominins have undoubtedly been cultural animals throughout the career of the genus, and it is to the late development of this genus—*Homo heidelbergensis*, *Homo neanderthalensis*, and *Homo sapiens*, at least, and perhaps lingering erectines (the shadowy *Homo floresiensis*) and other groups such as the Denisovans—that we will devote our attention. Our objective, again, will be to point out some structural affinities between the intracellular mechanisms discussed above and mechanisms operating at the other end of the ontological and phylogenetic spectra of earthly life.

Here Deacon's third-order emergence, with which we began, operates on a scale vastly different from the intracellular, molecular processes of core networks and character identity networks. The systems of encoded information or memory mechanisms at this end of the scale are "packages" or *archives* of cultural materials transmitted across generations, with more or less fidelity, through cooperative activity in social groups (Sterelny 2012). Such cultural transmission represents the first of the two major biocultural transitions mentioned above. We may think of these archives as akin to a hypothetical incipient core network within the cell. Like loosely organized arrays of molecules and genes not yet interacting in robust systems, cultural archives in hominin history gathered together arrays of gestures involving information and interaction with material affordances of the environment. An example of such an array is the loose collection of operations—the *chaîne opératoire* or operational sequence, as archaeologists call it (Leroi-Gourhan 1993)—by which several or many species of hominins produced symmetrical hand axes of similar design for well over a million years. Such cultural archives had considerable power to alter the interactions of hominin groups with their environments and ultimately, through the feedback operations of niche construction, to shift the selective pressures at work on the species involved (Odling-Smee et al. 2003; Godfrey-Smith 2014).

Across late hominin evolution, it is clear that cultural archives accumulated information at ever-deeper levels and transmitted it with increasing fidelity (Gamble 1999; Sterelny 2012). As the archives grew more complex, they began to take on a more systematic form; this systematization represents the second major biocultural transition. If shallow archives like the hand-axe archive resemble loosely organized molecular and genetic arrays, these more systematic cultural arrays resemble fully formed core or character identity networks (we elaborate on this resemblance below). Their systematic organization, variously involving material, behavioral, and informational ingredients, gave them a self-conserving structural integrity, and this robustness redoubled the force of culture in reshaping the niches of later hominins. Moreover, their conserved structures enabled them to step outside the feedback cycles of normal niche construction, to one degree or another; they acted then not exactly as elements in those cycles but as somewhat autonomous systems feeding *forward* into them (see figure 9.3). To characterize this difference, they deserve their own name: they are not so much cycles but *epicycles* in biocultural evolutionary dynamics (Tomlinson 2015).

Cultural epicycles help to explain the pace of cultural innovation across late hominin evolution, its sporadic burgeoning for several hundred thousand years, and its explosive acceleration, especially among sapient humans, across the last hundred millennia or so. The epicycles, like the archives, acted not merely as agents of cultural change. They are fully biocultural systems, which, as they altered hominins' capacities to reshape their environments and the selective pressures these exerted, entered into the dynamics of natural selection. Two brief examples of epicycles will suffice here, both drawn from the evolution of communication systems starting from about five hundred thousand years ago (the period in Europe of

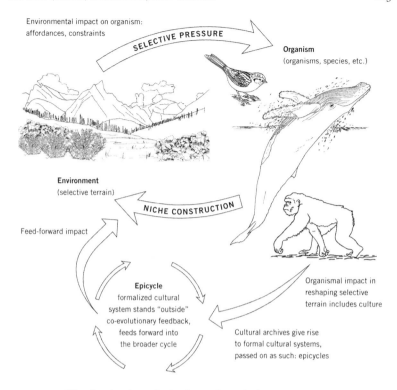

Environmental impact on organism:
affordances, constraints
SELECTIVE PRESSURE

Organism
(organisms, species, etc.)

Environment
(selective terrain)

NICHE CONSTRUCTION

Feed-forward impact

Epicycle
formalized cultural
system stands "outside"
co-evolutionary feedback,
feeds forward into
the broader cycle

Organismal impact in
reshaping selective
terrain includes culture

Cultural archives give rise
to formal cultural systems,
passed on as such: epicycles

FIGURE 9.3   Biocultural cycles and epicycles: a schematic view.

Homo heidelbergensis and in Africa of the still mysterious progenitors of *Homo sapiens*) and reaching down to modern human language and music (for these and more instances, see Tomlinson 2015).

The first example hovers close to the cultural archives and niche constructive feedback from which it springs; it is less fully autonomous, less clearly epicyclic, than the second. As hominins' cognitive capacity for shared attention grew—a growth attested in archaeologists' reconstruction of social circumstances already half a million years ago (Gamble 1999)—the coordinated turn-taking of modern human discourse could begin to form (Tomasello 2008). The kinds of gestures and calls that must have long characterized *protodiscourse*, partly innate and involving pointing signification, came to be more firmly tied to this coordination (Burling 1993; Bowie 2008), and this solidified linkage altered vocalized interactions on the landscape in new, advantageous ways. The value of intensified vocal interaction and collaboration drove selection for greater control and variety in vocalization. Such selection, however, in turn demanded not merely new capacities for production of the new repertories of calls but new capacities for their perception and interpretation as well. The selection, in other words, was not only for communicative gesture but also for the capacity to attend to it—the very shared attention that set off the communicative changes in the first place. A feedback cycle was closed, one manifesting a noteworthy degree of autonomous structuring and interaction among its elements, and communicative complexity grew.

The second example involves the separate pathways from protodiscourse toward modern human language and music and in particular the different but parallel shifts in each from graded, analog communicative scales to discrete, digital ones. (These might rightly be thought of as two distinct but interacting epicycles.) With increasing social and communicative complexity came the proliferation of diverse vocalizations and along with it sharpening distinctions of the intonations and timbres the hominin vocal tract could produce (Morley 2013). These had occupied positions along graded spectra in the earlier era of protodiscourse—think of the gradation from a giggle to a laugh to a guffaw. Now, however, the proliferation of vocalized signals put pressure on these analog spectra as distinct signals crowded together on them. In order to maintain the distinctness of the signals, a move to discreteness or digitalization of signaling arose, perhaps as a threshold switch depending on new capacities for hierarchized cognition (also attested in the archaeological record; Tomlinson 2015). This shift to discreteness must have shifted the selective environment also as the social and survival value of precise communication manifested itself more and more in its own, niche-constructive loop. Under the influence of different social uses to which vocalization could be put, meanwhile, the modes of digitalization bifurcated. On the one hand was the discretization of timbre or tone color, retained today in the small arrays of discrete vowel sonorities characteristic of the phonemes of any language (Tomlinson 2015); on the other was the weaning of the perception of discrete pitch levels from the smooth intonational contours of the earlier calls. These two modes of discreteness did not simply supersede earlier analog spectra of vocalization; graded intonations and timbres remain basic to both language and music today. Instead they overlaid on top of these graded spectra discrete arrays of sounds, thus redoubling communicative complexities and adding new kinds of elements (and percepts) to them. These new elements could lead to the distinct combinatorial systems of modern language (discretized phonemes built into morphemes and words) and music (discretized pitches as the gravitational centers of melodies).

How far can we push the analogies of structure between our intracellular molecular networks and biocultural epicycles? We can answer the question under several categories we have discerned in the former, isolating structural similarities to these in the latter. First, in each case, we have identified *historical lineages of structures* that define new kinds of individuals through the dynamics of natural selection, whether biological or biocultural. Second, these dynamics involve as their fundamental feature relations of both *cooperativity* and *feedback*. Indeed, in cultural archives, already we see the cooperativity of basic if varied elements. In the hand-axe archive, for example, the loose array of gestures involves, minimally, material affordances of flint and other minerals, ergonomic possibilities of hand and arm, and social information transmitted through mimesis or active pedagogy. These loose modes of cooperation tighten in the epicycles, linked in highly ordered feedback (and feed forward) relations, much as they are linked in core networks; these systems of feedback relations then feed forward into the broader cycles of biocultural coevolution. Third, the molecular networks and cultural epicycles both take on a *self-maintaining robustness* through the reciprocity of their feedback interactions. In the first epicycle described above, for example, shared attention (on the one hand) and increasing versatility and proliferation of vocalization (on the other) are not only enhanced but stabilized by their mutual relation, so that communicative efficacy in general comes to depend on both. The value of communicative sociality all told, then, becomes a force militating for the maintenance of the epicycle. This points to a fourth congruency between our molecular and cultural levels: in each case the feedback dynamics that enhance the stability of the systems involve not only the internal workings of the system itself but its relation to its broader environment (environments extending, in our examples, from the intracellular to the ecosystemic and sociocultural). Robustness is a product of the historical structuring of *relations between system and environment* or context.

There is, finally, a fifth, more speculative structural congruency between these two kinds of systems. We have suggested that alternative molecular networks or developmental programs might have come to be stabilized by virtue of the disadvantage (or weaker advantage) offered by programs intermediate between them. Across evolutionary time scales, a graded distribution of phenotypes could thus result in discrete (rather than graded) systems, and this might occur through the activation of switch-like bifurcations at certain threshold points. A mechanism of this sort can be seen, at the biocultural level, in the second epicycle described above.

Protodiscourse and early hominin sociality in general provided an environment of graded signals—analog signals, as we have characterized them.
From this environment arose—at a point rather late in hominin evolution,
perhaps involving only *Homo sapiens*, perhaps *Homo sapiens* and Nean-
derthals—a new kind of discreteness in the production and perception of
social signals, with huge consequences for the various modes of hierarchized
combinatoriality in human communication today. The threshold at which
this analog/digital shift came about remains mysterious, but it probably required a switch-like change, as intermediate positions along crowded analog spectra became indistinct in their signaled meanings and therefore untenable. And it probably arose from complex mutual enhancements of
shared attention, capacities for hierarchic cognition, proliferation of vocal
and gestural signals on the social landscape, and more. (In this arena the
distinct, emergent social uses of different kinds of signaling—what would
later coalesce as the different realms of language and music—no doubt
played an important role, as intermediaries between the two became less
efficacious in comparison to the extremes; see Tomlinson 2015.) It is important to notice here that the switch-like change can be explained without recourse to magic "language genes" and the like as the explanatory
fodder. The *move to switch-like thresholds*, in other words, may under certain environmental circumstances be a feature of systemic complexity in
living systems from molecular to biocultural, epicyclic levels.

What Do Robustness, Context Sensitivity, and Plasticity Tell Us about the Process of Individuation?

As illustrated by the two examples, the individuation of systems involves
two characteristics: (1) the acquiring of an ability to display distinct ranges
of possible functions or phenotypes that is new relative to other individuals and (2) a certain stabilization of this "phenotypic" range of possibilities and its persistence in the face of variability of input. How do these
properties arise? How are they related? Does the individuation in terms of
outputs (function) depend on individuation of the influences of the context
or environment?

    To explore these questions, we will briefly consider robustness of ex-
isting traits, that is, individuals at the suborganismal level. As indicated in
the section on suborganismal individuals, the autonomy of a morphological
unit is rooted in the autonomy of its development. The chicken leg offers
a good example. It develops in fairly constant intraorganismal and extra-

organismal environments, and most of the time it develops into a normal leg, distinct from ours and distinct from the forelimbs (wings). For instance, the two bones of the lower leg, the tibia and fibula, are connected by tibio-fibular syndesmosis, and only the tibia is involved in the joint to the tarsus, whereas the fibula is shorter (Müller and Streicher 1989; Müller 2003). This is a feature that appears characteristic for the whole bird clade and for theropod dinosaurs as well. Upon a simple perturbation during development (such as paralysis preventing muscular pull by the ilio-fibularis muscle), however, the syndesmosis tibio-fibularis does not develop; instead the two bones stay separate. The proper development, resulting in the wild type trait, thus requires an input from the developmental environment and is not robust to its removal.

What is surprising about this example is that even such a widely conserved part can still be dependent on an environmental factor. One possible evolutionary explanation for this lack of trait robustness involves the likelihood that this perturbation occurs. It seems likely that extreme perturbations will exceed the ability of any trait of an individual to compensate (in the extreme, take the activity of the ilio-fibularis muscle away and the syndesmosis will not develop). However, given a reliable enough environment (a normal chick always pulls on the fibula with that muscle), there is no *need* to have a trait that is robust to these perturbations. Thus the development of the leg is genetically determined only to the extent that it makes a similar leg under the range of environments it encounters. It is important to note that this does not mean that the particular gene variants making the leg in a particular environment have no role. It means that they "negotiate" a phenotype in interaction with the environment.

In the chicken limb example above, as long as the developmental context for the connection between tibia and fibula remains constant, the other aspects of the leg can evolve gradually, even though the syndesmosis fully depends on a particular environmental input. Thus eventually all the particular novel functions that this trait can perform depend on the particular stable environmental input. It is likely that evolution of most *individuals* (as defined above) involves some stable environmental aspects, and hence it seems likely that robustness to environmental changes arises secondarily rather than being intrinsic to the trait as it first arises. Therefore as the individual is not entirely context insensitive, it is likely that a change in either external or internal environment will eventually cause a substantial change in the individual's phenotype incompatible with fulfilling the original functions. In the above example, such substantial change may occur in an internal environment, consist of the genes shared with wings that come under a selective pressure due to a novel habitat, and lead to pleiotropic changes in the leg or an external environment that might no longer contain a source of a particular compound necessary for the wild type phenotype. There appear to be two options, to either preserve the original function or adopt a new function. These correspond to overcoming the contingency or building it into the new individual.

In order to preserve the ancestral function in the face of changing contexts, the individual may evolve to *compensate* for the external change. The individual may evolve its internal structure in response to the changing context while the function of the part is maintained. This evolutionary pathway leads to stability and perhaps leads to emergence—that is, robust individuality. The architecture of the unit evolves in order to strengthen its independence from those environmental factors that affect its function, leading to distinctiveness based on evolving character identity network in suborganismal sense or, at the biocultural level, to an epicycle. In the second option a new function appears in response to a changing environment. The changed environment reveals a novel behavior that may in fact be favored, and a novel kind of individual originates. This option is not entirely independent of the first because, in order to maintain some stability of the new behavior, the new individual must eventually become to some extent insensitive to the changes of the newly arising contexts by stabilizing interactions specific to those contexts.

In this view, only a hypothetical context-insensitive structure would not be able to manifest emergent properties. Such a structure might have a very reliable function, but one that is completely inert and incapable of novelty. In contrast, any entity that is context sensitive will produce emergent individuals either to maintain the status quo or to stabilize a new one.

Is the Topology of Interactions Sufficient?

Finally, it is fair to ask whether topology of interactions among parts, such as feedback, is sufficient to explain emergent systems. We can think of these topologies at every level we have described: in the interactions of proteins and genes, of developing bones and connective tissues, and of material affordances, social behaviors, and cultural information. We take the fact that emergent properties pertain to whole systems to imply that these properties derive from aspects of structure or relations (interactions, topology) between parts. If relational structure is fundamental, then we might expect that we can capture the crucial properties of emergent systems by scrutinizing their networks of interactions, regardless of the kinds of parts making up the systems. It is to broach this possibility that we have compared the network structures of systems as distant as the developmental genetic, ecological, and social systems (Newman et al. 2006). But are the properties of the components of these systems irrelevant and independent of the interactions they are engaged in? Or do these common structural motifs also involve parts with common features?

Ecologist Oswald Schmitz addressed this question explicitly by asking how much detail in an ecological system we need to know in order to predict its behavior (Schmitz 2010). Schmitz's extensive empirical work on a reasonably tractable system consisting of spiders, grasshoppers, and plants in an old field led him to demonstrate that the degree of allowable abstraction depends on how closely the levels of organization are coupled. Schmitz focused both on the topology of interactions between species and on identification of relevant features of interacting species, in particular their reaction norms. Through detailed analyses of species' roles, he was able to recognize an important aspect of ecological complexity: not only do species' effects in the ecosystem combine in nonlinear ways (the topology of their interactions is important for this), but the type of emergent behavior of the ecosystem depends on the properties of interacting species. Specifically, the consequences of interaction depend on the functional repertoires of the species involved. For example, the presence of multiple predators can lower the pressure on their prey, heighten it, or keep it the same. Rather than on the number of competitors, the actual interaction depends crucially on the ability of predatory and prey species to switch to alternative prey, alternative habitats, and so forth. Depending on their flexibility, species' effects on other species may be synergistic or antagonistic, enhancing or diminishing each other's contribution in the system. The topology of interactions is certainly crucial, but so are the reaction norms of the species in defining the network of interactions.

Conclusions

The evolutionary processes described here lead to individuation of distinct phenotypic systems that on the one hand are robust to changes in their context and on the other influence the future evolutionary potential of the organisms that carry them. These systems show what we call emergent properties in the sense of their nondecomposability, but we have sought to understand them not in their present structures but rather in the historical contingencies that created them. We have thus distinguished between emergence as a structural property of nondecomposability and emergence as a process, and we suggest that emergent properties in living systems can be best illuminated in this historical light.

The second aspect we have emphasized is the role of the interacting units and their properties. Focusing on patterns of network topology in vastly different systems points to the broad similarities of design of the systems, and we have emphasized these similarities in order to exemplify their existence across vastly divergent phenomena of biological systems. But this emphasis can create the appearance that the interacting parts of the systems are not important. Instead, we argue that the way in which the component parts of systems react to changing interaction (among themselves and with their environment) is crucial and depends on the features of the parts themselves. The phenotypic plasticity of these parts—whether proteins, connective tissue, or signifying vocalizations—seems to play an important role in determining the type of emergence generated.

Finally, we have argued that emergence, when it is most striking—Deacon's third kind of emergence, in life and its products such as culture—is always associated with the origin of novel kinds of individuals. Understanding the origin, stability, robustness, and nature of individuals is a crucial perspective in integrating the various aspects of emergence into a coherent understanding of biological and biocultural histories. In emphasizing the structurally similar properties of differing emergent individuals—positive feedback loops, emergent stability, switches, and so forth—we wish to encourage attempts to forge an abstract theory of the behavior of emergent systems and to build models of their origins.

References

Arendt, Detlev. 2008. "The Evolution of Cell Types in Animals: Emerging Principles from Molecular Studies." *Nature Reviews Genetics* 9: 868–82.

Bowie, Jill. 2008. "Proto-Discourse and the Emergence of Compositionality." *Interaction Studies* 9: 18–33.

Burling, Robbins. 1993. "Primate Calls, Human Language, and Nonverbal Communication." *Current Anthropology* 34: 25–53.

Buss, Leo W. 1987. *The Evolution of Individuality*. New York: Columbia University Press.

Crespi, Berard J., and Douglas Yanega. 1995. "The Definition of Eusociality." *Behavioral Ecology* 6: 109–15.

Davidson, Eric H., and Douglas H. Erwin. 2006. "Gene Regulatory Networks and the Evolution of Animal Body Plans." *Science* 311: 796–800.

Deacon, Terrence. 2003. "The Hierarchic Logic of Emergence: Untangling the Interdependence of Evolution and Self-Organization." In *Evolution and Learning: The Baldwin Effect Reconsidered*, edited by Bruce H. Weber and David J. Depew, 273–308. Cambridge, MA: MIT Press.

Dunn, D. Casey, and Günter P. Wagner. 2006. "The Evolution of Colony-Level Development in the Siphonophora (Cnidaria: Hydrozoa)." *Developmental Genes and Evolution*. doi:10.1007/s00427-006-0101-8.

Ellis, George F. R., Denis Noble, and Timothy O'Connor. 2012. "Top-Down Causation: An Integrating Theme within and across the Sciences? Introduction." *Interface Focus* 2 (1): 1–3.

Gamble, Clive. 1999. *The Palaeolithic Societies of Europe*. Cambridge: Cambridge University Press.

Ghiselin, Michael T. 2005. "Homology as a Relation of Correspondence between Parts of Individuals." *Theory in Bioscience* 124: 91–103.

Godfrey-Smith, Peter. 2011. *Darwinian Populations and Natural Selection*. Oxford: Oxford University Press.

————. 2014. *Philosophy of Biology*. Princeton, NJ: Princeton University Press.

Graf, Thomas, and Tariq Enver. 2009. "Forcing Cells to Change Lineages." *Nature* 462: 587–94.

Haeckel, Ernst. 1869. "Zur Entwicklungsgeschichte der Siphonophoren." *Natuurk. Verh.* 6: 1–120.

Hölldobler, Bert, and Edward O. Wilson. 1990. *The Ants*. Cambridge, MA: Harvard University Press.

Leroi-Gourhan, André. 1993. *Gesture and Speech*. Cambridge, MA: MIT Press.

Love, Alan C. 2012. "Hierarchy, Causation and Explanation: Ubiquity, Locality and Pluralism." *Interface Focus* 2 (1): 115–25.

Lynch, Vincent J., Gemma May, and Günter Wagner. 2011. "Regulatory Evolution through Divergence of a Phosphoswitch in the Transcription Factor CEBPB." *Nature* 480 (7377): 383–86.

Maynard-Smith, John, and Eörs Szathmary. 1995. *The Major Transitions in Evolution*. Oxford: Oxford University Press.

Mayr, Ernst. 1942. *Systematics and the Origin of Species*. New York: Columbia University Press.

McShea, Daniel W. 1996. "Metazoan Complexity and Evolution: Is There a Trend?" *Evolution* 50: 477–92.

Michod, Richard E. 1999. *Darwinian Dynamics: Evolutionary Transitions in Fitness and Individuality*. Princeton, NJ: Princeton University Press.

Mitchell, Sandra D. 2009. *Unsimple Truths: Science, Complexity, and Policy*. Chicago: University of Chicago Press.

Morley, Ian. 2013. *The Prehistory of Music: Human Evolution, Archaeology, and the Origins of Musicality*. Oxford: Oxford University Press.

Müller, Gerd B. 2003. "Embryonic Motility: Environmental Influences and Evolutionary Innovation." *Evolution and Development* 5 (1): 56–60.

Müller, Gerd B., and Johannes Streicher. 1989. "Ontogeny of the Syndesmosis Tibiofibularis and the Evolution of the Bird Hindlimb: A Caenogenetic Feature Triggers Phenotypic Novelty." *Anatomy and Embryology* 179: 327–39.

Newman, Mark, Albert-L. Barabási, and Duncan J. Watts. 2006. *The Structure and Dynamics of Networks*. Princeton, NJ: Princeton University Press.

Nnamani, Mauris C., Soumya Gangyly, Vincent J. Lynch, Laura S. Mizoue, Yingchun Tong, Heather Darling, Monika Fuxreiter, Jens Meiler, and Günter P. Wagner. 2015. "Evolution of Conditional Cooperativity between HOXA11 and FOXO1 through Allosteric Regulation." *bioRχiv*. http://dx.doi.org/10.1101/014381.

Noble, Denis. 2012. "A Theory of Biological Relativity: No Privileged Level of Causation." *Interface Focus* 2 (1): 55–64.

Odling-Smee, F. John, Kevin N. Laland, and Markus W. Feldman. 2003. *Niche Construction: The Neglected Process in Evolution*. Princeton, NJ: Princeton University Press.

Okasha, Samir. 2012. "Emergence, Hierarchy and Top-Down Causation in Evolutionary Biology." *Interface Focus* 2 (1): 49–54.

Pavličev, Mihaela, and Stefanie Widder. 2015. "Wiring for Independence: Positive Feedback Motifs Facilitate Individuation of Traits in Development and Evolution." *Journal of Experimental Zoology Part B: Molecular and Developmental Evolution* 324 (2): 104–13.

Rajakumar, Rajendhran, Diego San Mauro, Michiel B. Dijkstra, Ming H. Huang, Diana E. Wheeler, Francois Hiou-Tim, Abderrahman Khila, Michael Cournoyea, and Ehab Abouheif. 2012. "Ancestral Developmental Potential Facilitates Parallel Evolution in Ants." *Science* 335 (6064): 79–82.

Richerson, Peter J., and Robert Boyd. 2005. *Not by Genes Alone: How Culture Transformed Human Evolution*. Chicago: University of Chicago Press.

Rueffler, Claus, Joachim Hermisson, and Günter Wagner. 2012. "Evolution of Functional Specialization and Division of Labor." *Proceedings of the National Academy of Sciences of the United States of America* 109 (6): E326–35.

Schmitz, Oswald J. 2010. *Resolving Ecosystem Complexity*. Princeton, NJ: Princeton University Press.

Sterelny, Kim. 2012. *The Evolved Apprentice: How Evolution Made Humans Unique*. Cambridge, MA: MIT Press.

Tomasello, Michael. 2008. *Origins of Human Communication*. Cambridge, MA: MIT Press.

Tomlinson, Gary. 2015. *A Million Years of Music: The Emergence of Human Modernity*. New York: Zone Books.

Wagner, Günter P. 2007. "The Developmental Genetics of Homology." *Nature Reviews Genetics* 8: 473–79.

———. 2014. *Homology, Genes and Evolutionary Innovation*. Princeton, NJ: Princeton University Press.

———. 2015. "What Is Homology Thinking and What Is It for?" *Journal of Experimental Zoology Part B: Molecular and Developmental Evolution*, in press.

Wagner, Günter P., Koryu Kin, Louis Muglia, and Mihaela Pavličev. 2014. "Evolution of Mammalian Pregnancy and the Origin of the Decidual Stromal Cell." *International Journal of Developmental Biology* 58: 117–26.

Wagner, Günter P., and Vincent J. Lynch. 2010. "Evolutionary Novelties." *Current Biology* 20 (2): R48–52.

Wang, Yong-Qiang, Rainer Melzer, and Günter Theißen. 2010. "Molecular Interactions of Orthologues of Floral Homeotic Proteins from the Gymnosperm Gnetum gnemon Provide a Clue to the Evolutionary Origin of 'Floral Quartets.'" *Plant Journal* 64 (2): 177–90.

West-Eberhard, Marie J. 2003. *Developmental Plasticity and Evolution.* Oxford: Oxford University Press.

Wilson, Edward O. 1975. *Sociobiology: The New Synthesis.* Cambridge, MA: Harvard University Press.

Wilson, Edward O., and Bert Hölldobler. 2005. "Eusociality: Origin and Consequences." *Proceedings of the National Academy of Sciences of the United States of America* 102 (38): 13367–71.

PART 3

Biological Hierarchies and Macroevolutionary Patterns

Ecology and Evolution:
Neither Separate nor Merged

Emanuele Serrelli and Ilya Tëmkin

Darwin, Haeckel, and the Economy of Nature

Ecology was coined for the first time by Ernst Haeckel in the 1860s and 1870s and defined in the following passage: "By ecology we mean the body of knowledge concerning *the economy of nature*, *the total relations* of the animal to both its inorganic and organic environment; including its friendly and inimical relations with those animals and plants with which it comes into contact. In a word, *all the complex relationships* referred to as *the struggle for existence*" (Haeckel 1870, our emphases).

Haeckel's notion of the economy of nature derived directly from the work of Darwin, and the two scholars were exponents of a mature, unified view of the economy of nature, the struggle for existence, adaptation, and descent with modification, a view oriented to the most complete consideration of relationships in nature.

In the following years since Darwin and Haeckel, a disciplinary gap between ecology and evolution began to widen, to the point of development of two isolated scientific fields that shared little in terms of academic objectives and methodologies.

Botanists and the Disciplinary Entrenchment of Ecology at the Turn of the Century

At the turn of the twentieth century, ecology began to evolve into an autonomous discipline. Major protagonists of this development were European botanists, such as Eugenius Warming, Andreas F. W. Schimper, and Carl G. O. Drude. These pioneers studied biogeography—that is, the geographic distribution of organisms with respect to soils and climatic conditions, emphasizing the way in which organisms and abiotic environments influence each other at a large scale.

The field was developed further in the 1910s, when, for example, American biologist Frederic E. Clements devised methods to characterize assemblages of organisms for analyzing historical conditions of local environments, the work that was instrumental in defining important ecological concepts such as "community" and "ecological succession."

The plant biologists were the first to define the line of research that would inherit the name "ecology." In the same years, important chemists-physicists and mathematicians, such as Ludwig Boltzmann or Alfred J. Lotka, were discussing evolution within their tentative, general dynamic and thermodynamic view of life. For these scientists, evolution was a fundamental assumption requiring mathematical formulation. For Boltzmann, life was a thermodynamic phenomenon, where biological systems were regarded as open to energy flows. Along the same lines, Lotka argued that "the fundamental object of contention in the life-struggle, in the evolution of the organic world, is available energy. . . . The first effect of natural selection thus operating upon competing species will be to give relative preponderance (in number or mass) to those most efficient in guiding available energy. . . . Primarily the path of the energy flux through the system will be affected" (1922, 147).

The Evolutionary Synthesis and Fluxes of Traits

Whereas Boltzmann and Lotka conceptualized evolution in terms of equilibrium and change in energy flow, the proponents of the developing evolutionary synthesis (or modern synthesis) turned their attention to the particulate dynamics of alleles and traits within populations. The dominant way of studying evolution thus became the observation and modeling of the transmission and transformation of traits and alleles. After all, organismal traits could be seen as the repositories of lasting traces of energy-matter dynamics; on the other hand, the energy-matter dynamics were constrained and channeled by the inherited traits of the evolved populations of organisms. But neo-Darwinian trait-centered phenomena—such as heritability, variation, natural selection, drift, and speciation—and the resulting phylogenetic patterns were essentially *genealogical*. The powerful framework of the modern synthesis had brought into the sharpest focus vertical genealogical patterns—that is, those that can be followed and fully captured by tracing "bloodlines," related lineages, and their common ancestry.

Ecology: A Science of Ecosystems

Whereas the focus of modern synthesis on traits and genealogies exacerbated the divorce of evolution from ecology, ecology itself was further elaborated and became the "science of ecosystems." The concept of ecosystem was coined by another botanist, Arthur Tansley (1935). Tansley praised the term *biome* that Clements had coined to indicate the whole complex of organisms inhabiting a given region. However, Tansley regarded the concept of biome too restrictive, as it focused exclusively on the living entities:

> the more fundamental conception is, as it seems to me, the whole system (in the sense of physics), including not only the organism-complex, but also the whole complex of physical factors forming what we call the environment of the biome—the habitat factors in the widest sense. Though the organisms may claim our primary interest, when we are trying to think fundamentally we cannot separate them from their special environment, with which they form one physical system. It is the systems so formed which, from the point of view of the ecologist, are the basic units of nature on the face of the earth. Our natural human prejudices force us to consider the organisms (in the sense of the biologist) as the most important parts of these systems, but certainly the inorganic "factors" are also parts—there could be no systems without them, and there is constant interchange of the most various kinds within each system, not only between the organisms but between the organic and the inorganic. These ecosystems, as we may call them, are of the most various kinds and sizes. (Tansley 1935, 299)

The concept of ecosystem was an expression of a reasoning style, sometimes defined as "holistic" or "systemic," that had characterized the work of pioneer ecologists since the beginning. Ecosystems had their peculiar patterns and processes in which organisms did not possess a privileged status, least of all did their traits.

Subsequently, ecology further split into *functional* ecology, studying species' distributions and their causes, and *community* ecology, dealing with community structure, variation, and change. By the mid-1950s, many ecologists believed that they could provide a general theory of ecology without recourse to evolutionary thinking and concepts (Orians 1962).

Ecological Stage and Evolutionary Play

In the 1960s, ecologist G. E. Hutchinson (1965) promoted the idea that the outcomes of ecological interactions among species are contingent on their evolutionary histories. He did so by famously proposing the simile between ecological systems and theater, where the actors (i.e. species or individuals) have roles prescribed by their evolutionary history. The acts are played out in an unscripted fashion, contingent on the environmental setting (i.e. the local theater). The causal agents are clear: evolutionary histories and contingency. Ecological assemblies, regarded as aggregations of time-extended segments from multiple of genealogies, are characterized by secondary, derived patterns: they become the "stage" for the evolutionary play. As the twentieth century progressed, the study of genealogies steadily came to the forefront of evolutionary research, first empowered by the development of phylogenetic systematics in the 1960s and then by advances in molecular biology.

Since the 1970s, frequent attempts were made to bridge the gap between ecology and evolution. Interestingly, their principal protagonists were paleobiologists, who tried to establish a unified perspective to encompass evolution, ecology, and biogeography.

Merging Evolution with Ecology: Van Valen's "Law of Extinction"

Leigh Van Valen (1973) was a pioneer in establishing causal connections between ecological processes and macroevolutionary phenomena. By applying the survivorship curve, a standard technique derived from population ecology, to higher-level genealogical units (species, genera, and families) in the fossil record, Van Valen derived the "law of extinction," according to which the rate of extinction within any such group is constant and uncorrelated with the group's age. The best predictor of the extinction rate was, according to Van Valen, the "adaptive zone," a concept typical in ecology: extinction is stochastically constant within an "adaptive zone"— for example, in benthic versus pelagic marine organisms. According to Van Valen's famous Red Queen hypothesis, the constancy of extinction in an adaptive zone is due to coevolution, the process by which a species evolves as a result of changes in species it interacts with. For Van Valen, a tendency of all species to increase fitness eventually results in an arms race between species who occupy the same environment, which, in turn, might account for both coextinction and coevolution of species as well as more inclusive genealogical units. These ideas eventually led Van Valen to introduce his ecological species concept in 1976, according to which species are defined by ecological parameters.

For Van Valen, ecological reasons explain observations that are hardly accommodated by the modern synthesis, such as the constancy of rates of protein evolution discovered by Tomoko Ohta and Motoo Kimura: it is the constancy and diversity of environmental stresses that, over long intervals, cancel out all irregularities in molecular rates. Furthermore, Van Valen was a strong advocate of an ecological perspective of evolution that culminated in his seminal 1989 proposal of the "energetic paradigm" of evolution (potentially alternative to the "informational"), according to which "evolution is the control of development by ecology" (Van Valen 1989). In this unconventional ecological view of the nature of fitness, the evolution of biotas can be approached through changes in patterns of energy flow and their control, as trophic energy is a major controller of evolution and ecology.

Niche Construction Models

Paleobiological approaches were most fruitful in applying an ecological causal framework to genealogical patterns at the levels of species and clades but had limited success in integrating ecology with processes at population/ avatar level. The idea of "niche construction" by John Odling-Smee, elaborated by Kevin Laland, Marcus Feldman, and colleagues in the mid-1990s, was an attempt to such integration, with hope of broader application to "both population-community ecology and ecosystem-level ecology," reconciling them "under the rubric of an extended evolutionary theory" (Laland et al. 1999, 10247; cf. Odling-Smee 2010). The idea of niche construction is a revival of Darwin's theme that organisms dramatically impact their environment. But despite rolling claims, the idea of niche construction was introduced into evolutionary biology in the form of modified population genetics models. These were built to explore the feedback effects among loci, mediated by the availability of environmental resources: niche construction "analysis assumes that a population's capacity for niche construction is influenced by the frequency of alleles at a first locus, and that this niche construction is expressed as a change in the frequency of a resource in the population's environment. The amount of this resource in the environment then determines the pattern and strength of selection acting at the second locus" (Laland et al. 1996, 306).

An ecological evolutionary theory is, perhaps, more within reach today, due to a renewed attention to the organism-level processes, and to the rapid development of quantitative genetics and evolutionary developmental biology.

Making Ecology Evolutionary

Since the 1950s, many ecologists lamented the lack of evolutionary perspective in ecological research program. Gordon H. Orians described the difficulty of ecologists to take evolution and natural selection into account: "Even if one strongly believes in the action of natural selection it is exceedingly difficult, as Darwin pointed out, to keep it always firmly in mind. Neglect of natural selection in ecological thinking is, therefore, understandable though regrettable. However, its deliberate exclusion in these years following the Darwin centennial would seem to be exceedingly unwise" (Orians 1962, 262).

At the same time, Orians argued that evolution is essential for addressing any question in biology: "It is becoming increasingly apparent that a complete answer to any question should deal with physiological, adaptational and evolutionary aspects of the problem. The evolutionary process of becoming yields the most profound understanding of biological systems at all levels of organization. The non-evolutionary answer to the question of why an animal is abundant in some parts of its range and rare in others is of necessity incomplete" (261).

Orians described the state of ecology in the 1950s as divided into functional and evolutionary, with different ideas on what the main objective of ecological research program should be. Functional ecology deals with geographical distribution and abundance, and their change through time.

Such aspects are explained by the organisms' physiology and behavior, and by physical and climatic features of the environments. Functional ecology makes no appeal to evolutionary concepts: "It is the job of the ecologist to count auger-holes and so to predict the number of bees to be found and the job of the evolutionist to measure genetic change as a result of competition" (Orians 1962, 258).

"Evolutionary ecology" was not yet recognized as a distinct field of study until it was baptized by Orians in the same paper (although Darwin is often considered to be the first evolutionary ecologist before the split described above). Contrary to the approach taken by functional ecologists, evolutionary ecologists posited that current adaptations cannot be understood without reference to past evolutionary processes.

For Orians and evolutionary ecologists, evolutionary phenomenology had to be integrated into the theory of ecology rather than being haphazardly invoked as an accessory explanatory framework. Dramatically, for Orians, "evolution would seem to be the only real theory of ecology [and of other fields] today" (262). There was no chance that some functional ecologists claimed that "no general theory of ecology is possible and that each case must be considered individually" (261) while at the same time they maintained "that a general and satisfying theory of ecology can and should be constructed without recourse to evolutionary thinking and concepts" (258).

Orians characterized functional ecologists as overly focused on single organisms and consciously avoiding "distractions" from other organisms they interact with. This is particularly in contrast with community studies, based on "the hope that predictable relationships between the relative abundance and interactions of species can be discovered leading to insights into community structure" (258).

In the 1980s, ecosystem science began to be progressively concerned with processes at higher ecological levels, including the entire biosphere at the global scale, leading to (or coinciding with?) the birth of Earth system science, global ecology, and geophysiology. All these scales added further nontrivial relationships to evolution. The journal *Evolutionary Ecology* was founded in 1987. In the editorial, Michael Rosenzweig described evolutionary ecology as a hypothetical-deductive field and "the most intractable of all forms of ecology": "It is a wonderment that we do not all go insane. It is no surprise that many of us seem to, that others leave science altogether or convert to other branches. Some, however, survive the terrible realization that they are practitioners of the most weirdly complex and interactive branch of population biology" (Rosenzweig 1987).

Implementing Evolution into Ecological Models

Mathematical models have long been among fundamental approaches in the methodological arsenal of ecological research, but they have become particularly significant in recent decades due to the development of sophisticated computational methods, powerful digital technology, and vast, ever-increasing data sets. Patterns and regularities uncovered by mathematics and simulation guide data collection, field and lab work, and inferences and predictions. For this reason, it is particularly interesting to examine the growing trend among modelers to incorporate evolutionary processes into the established body of ecological models. This tendency is a realization of Orians's aspiration to achieve a greater unity within an evolutionary-ecological theory.

Current evolutionary ecological models can accommodate a wide array of evolutionary processes, such as the effect of a phenotypic evolution on an ecosystem function—for example, primary productivity. Different selection regimes correspond to peculiar phenotype distributions. Shifts in phenotype distributions affect the strength of trophic interactions, or "links" (between, for example, individual predators and prey of different sizes), and the ecological role of different species. The effect of hypothetical scenarios on the structure and function of a food web and its effect on the entire ecosystem function can be examined by differential weighting of the links between the predator population and the prey (cf. Matthews et al. 2011).

Another example of integrating evolution into ecosystem models, dealing with temporal distribution of energy and matter, includes biological invasions (Kylafis and Loreau 2008). Producers, consumers, and their resources are modeled in a condition of cyclic equilibrium. Producers and consumers can be allowed to adapt to the environment, and the possible invasion of new kinds of organisms is modeled too.

The idea that ecology and evolution are domains that exist in incommensurable time scales can be considered overcome. Instead, we find the idea that ecological-evolutionary phenomena of different magnitudes are at work. A key point to integrate ecology and evolution is that integration must happen differently at different time scales. A general theoretical framework should encompass all this complexity.

The Dual Hierarchy Approach

Following Van Valen's lead, Niles Eldredge pointed out that the modern synthesis had been "unfinished" (Eldredge 1985). He developed punctuated equilibria (Eldredge and Gould 1972) in a macroecological sense: punctuations and stasis are "coordinated" across many lineages; therefore, they have evident ecological causes and triggers.

It was Eldredge who brought forth more explicitly the fact that evolutionary theory had been concentrating on genetic processes and formation of genetic lineages. But—Eldredge pointed out—evolution does not occur in a vacuum; rather, it is what takes place inside matter-energy transfer systems that determines, in large measure, the patterns of stability and change in genetic systems that we call "evolution."

Eldredge started to propose the dual hierarchy of evolution. He observed that "organisms seem to be both energy conversion machines and reproducing 'packages' of genetic information. As such they are integrated simultaneously into two largely separate, but interacting kinds of general systems" (Eldredge 1986, 351).

Evolution happens in the economic hierarchy, a set of nested ecological systems kept together by circulation of matter and energy. The evolution of lineages is a hierarchically organized retention of genealogical information (the evolutionary hierarchy) interacting with, and largely dependent on, the dynamics of the economic hierarchy.

The dual hierarchy was also a proposal to reformulate even the most basic concepts, such as natural selection. Rather than a modification of gene frequencies, natural selection is "a very clear example of the effect of the ecological hierarchy on elements of the genealogical hierarchy" (Eldredge 1985, 183) that yields the origin, maintenance, and modification of adaptation. The genealogical hierarchy provides variation to ecological dynamics.

Eldredge never stopped reproposing and elaborating the idea as a radical restructuring of the modern synthesis. In 2003 he proposed a very general "sloshing bucket" theory of evolution according to which the genealogical and the ecological hierarchies are like the walls of a bucket filled with some water. If we carry around the bucket, we will see the water sloshing from one wall to the other. If we are very careful, this movement will be slight, but if we stumble, then the water will go high on one wall (for Eldredge, the ecological wall) and bounce back on the other wall (the genealogical one): these, for Eldredge, are the most remarkable episodes in the history of life, where ecological events trigger extinctions and evolutionary novelties.

Ecology in Hierarchy Theory

It may be said that Eldredge's way of presenting things never became mainstream. But attempts at building a unified, thermodynamic-informational evolutionary theory did not break through either (Brooks and Wiley 1988), and all attempts at "extending the modern synthesis" did not touch the heart of the problem of conceiving a unitary framework for laying down all the best theories and models from ecology and evolution. Hierarchy theory—conveniently acknowledged, generalized, expanded in time and taxonomical scope, and modified—is a standing candidate framework for the multiscale integration of ecology and evolution here evoked.

What is ecology in hierarchy theory? Ecology is certainly what it has been since the end of the nineteenth century: a science of physical and chemical flows and cycles, a science of energy and matter transfers, a science of codetermination between biotic and abiotic (weather, soil, nutrients, geology) factors. In the dual hierarchy framework, ecology is, more generally, the science of interactions. A key methodological innovation in conceiving these interactions is network theory (Proulx et al. 2005; Tëmkin and Eldredge 2015). The hierarchy perspective integrates ecology and evolution by studying and modeling how interactions determine a change in information content (the evolutionary hierarchy).

References

Brooks, Daniel, and E. O. Wiley. 1988. *Evolution as Entropy: Toward a Unified Theory of Biology*. Chicago: University of Chicago Press.
Eldredge, Niles. 1985. *Unfinished Synthesis*. New York: Columbia University Press.
———. 1986. "Information, Economics and Evolution." *Annual Review of Ecology and Systematics* 17: 351–69.
Eldredge, Niles, and Stephen J. Gould. 1972. "Punctuated Equilibria: An Alternative to Phyletic Gradualism." In *Models in Paleobiology*, edited by Thomas J. M. Schopf, 82–115. San Francisco: Freeman, Cooper.
Eldredge, Niles, and Stanley Salthe. 1984. "Hierarchy and Evolution." In *Oxford*

Surveys in Evolutionary Biology, edited by Richard Dawkins and M. Ridley, vol. 1, 182–206. Oxford: Oxford University Press.

Haeckel, Ernst. 1866. *Generelle Morphologie der Organismen. Allgemeine grundzü geder organischen formen- wissenschaft, mechanisch begrü ndet durch die von Charles Darwin reformirte descendenztheorie.* Berlin: Verlag von Georg Reimer.

———. 1870. "Über Entwicklungsgang und Aufgabe der Zoologie." *Jenaische Zeitschrift für Medizin und Naturwissenschaften* 5: 352–70. Translated in W. C. Allee, A. E. Emerson, O. Park, T. Park, and K. P. Schmidt. 1949. *Principles of Animal Ecology.* Philadelphia: Saunders.

Hutchinson, G. Evelyn. 1965. *The Ecological Theater and the Evolutionary Play.* New Haven: Yale University Press.

Hutchinson G. E., and Robert H. Macarthur. 1959. "A Theoretical Ecological Model of Size Distributions among Species of Animals." *American Naturalist* 93 (869): 117–25.

Kylafis, Grigoris, and Michel Loreau. 2008. "Ecological and Evolutionary Consequences of Niche Construction for Its Agent." *Ecology Letters* 11 (10): 1072–81.

Laland, Kevin N., F. John Odling-Smee, and M. W. Feldman. 1996. "The Evolutionary Consequences of Niche Construction: A Theoretical Investigation Using Two-Locus Theory." *Journal of Evolutionary Biology* 9 (3): 293–316.

———. 1999. "Evolutionary Consequences of Niche Construction and Their Implications for Ecology." *Proceedings of the National Academy of Sciences of the United States of America* 96 (18): 10242–47.

Levine, L., and Leigh Van Valen. 1964. "Genetic Response to the Sequence of Two Environments." *Heredity* 29: 734–36.

Lotka, Alfred J. 1922. "Contribution to the Energetics of Evolution." *Proceedings of the National Academy of Sciences of the United States of America* 8: 147–51.

Matthews, Blake, Anita Narwani, Stephen Hausch, Etsuko Nonaka, Hannes Peter, Masato Yamamichi, Karen E. Sullam, Kali C. Bird, Mridul K. Thomas, Torrance C. Hanley, and Caroline B. Turner. 2011. "Toward an Integration of Evolutionary Biology and Ecosystem Science." *Ecology Letters* 14 (7): 690–701.

Odling-Smee, F. John. 2010. "Niche Inheritance." In *Evolution: The Extended Synthesis*, edited by Massimo Pigliucci and Gerd B. Müller, 175–207. Cambridge, MA: MIT Press.

Orians, Gordon H. 1962. "Natural Selection and Ecological Theory." *American Naturalist* 96 (890): 257–63.

Pearce, Trevor. 2010. "'A Great Complication of Circumstances'—Darwin and the Economy of Nature." *Journal of the History of Biology* 43 (3): 493–528.

Proulx, Stephen R., Daniel E. L. Promislow, and Patrick C. Phillips. 2005. "Network Thinking in Ecology and Evolution." *Trends in Ecology and Evolution* 20 (6): 345–53.

Rosenzweig, Michael L. 1987. Editorial. *Evolutionary Ecology* 1: 1–3

Tansley, A. G. 1935. "The Use and Abuse of Vegetational Concepts and Terms." *Ecology* 16 (3): 284–307.

Tëmkin, Ilya, and Niles Eldredge. 2015. "Networks and Hierarchies: Approaching Complexity in Evolutionary Theory." In *Macroevolution: Explanation, Interpretation and Evidence*, edited by Emanuele Serrelli and Nathalie Gantier, 83–226. Switzerland: Springer.

Van Valen, Leigh. 1973. "A New Evolutionary Law." *Evolutionary Theory* 1: 1–30.

———. 1976. "Ecological Species, Multispecies, and Oaks." *Taxon* 25 (2/3): 233–39.

———. 1989. "Three Paradigms of Evolution." *Evolutionary Theory* 9: 1–17.

Summaries for Part 3

This section deals with the area in which hierarchy theory originated and then was most used with important results: the area commonly labeled "macroevolution." Key patterns are identified that demonstrate the co-emergence of evolution in the dynamical interaction between ecology and lineages, especially at the macroscopic scale.

William Miller III proposes a "metatheory" of evolution including not only macroevolution but also the macroecological aspect of evolution. The "foundation piece" of this metatheory is, for the author, punctuated equilibria, the theory that explains patterns of speciation and stability in the fossil record, such as species selection, coordinated stasis, and turnover pulse. The "metatheory" proposed by Miller is named "macroevolutionary consonance theory." It postulates a causal correlation between, on the one hand, the developmental history of ecosystems and, on the other hand, adaptive speciation. Environment is the driver of turnovers, and the theory allows for a classification of the possible roles of species in regional turnovers as well as for explanation of origin and persistence of rare species (that are viewed as "leftovers" of previous time intervals of regional ecosystems). Microevolutionary processes are embedded in all this theory, but they don't allow for long-range explanatory extrapolation.

Warren Allmon's chapter addresses terminological issues that have important implications for both evolutionary theory and empirical research. The aim of the chapter is to propose a precise usage of the expression "tempo" and "mode" of evolution as were first discussed by G. G. Simpson (1944) and further developed in punctuated equilibria (Eldredge and Gould 1972). "Tempo" and "mode" acquired a plethora of meanings (e.g., describing patterns of anagenetic change) that deviated widely from the original meaning. This shift was due, for Allmon, both to an emphasis on random processes and to the difficulty of studying speciation in paleontology. The author attempts to revive the initial meaning of the phrase and proposes to regard the directional change, random walk, and stasis as "models" of anagenesis rather than distinct evolutionary "modes." He proposes to use "tempo" and "mode" in reference to the relationship between the rate of change of different parameters and phylogenetic topology and emerging patterns. "Tempo" should be used in all its richness of meaning, because different measures of frequency and duration of speciation have been elaborated, albeit sometimes more theoretically than operationally. The suggested terminology indicates a wider scope of evolutionary theory, where evolutionary trends result not only from within-lineage transformation but also from lineage splitting and where rates are relative to very different processes.

Carlton E. Brett, Andrew Zaffos, and Arnold I. Miller focus on the processes of niche conservatism and niche evolution using the paleontology and stratigraphy of the Hamilton Group of the Middle Devonian as a case study. The history of the Hamilton Group provides a compelling example of evolutionary and ecological stasis. The chapter provides clear operational definitions of terms such as niche (i.e., environmental preferences) and ecological stasis (i.e., stability of interactions) and tries to establish a connection among niche conservatism, evolution, and ecological stasis. The authors address a methodological and epistemological problem that is very relevant to the hierarchy theory of evolution: the problem of the appropriate level of expression of niche conservatism and evolution. To study niche conservatism, the authors operationalize the concept of niche, summarizing preferred environment, environmental tolerance (not very reliable due to statistical error), and peak abundance. The chapter explains in plain terms the statistical techniques (and corrections) used to compare niches at the same level (member-member, formation-formation), as well as to discern niche conservatism at different levels. The latter distinction is particularly important for the results presented show more conservatism at higher level (whole assemblages), revealing the non-aggregative nature of niche conservatism/evolution.

Peter D. Roopnarine and Kenneth D. Angielczyk provide a case study of the relationship between the ecological dynamics, more specifically the guild structure and composition and interguild relations, using the Permian-Triassic terrestrial paleocommunities from the Karoo Basin of South Africa. After examining the data and theoretical models of how a paleocommunity would have responded, in ecological time, after particular types and magnitudes of disturbance, the authors claim that community stability could be an agent of selection, mediated by species dynamics. The hierarchical framework, able to include higher-level community dynamics as an active mechanism (in ecosystem evolution, mass extinction, and system restructuring), improves the understanding of Phanerozoic evolution but may also serve, for the authors, to forecasting the potential futures of current ecosystems.

Finally, *Michael L. McKinney* applies the hierarchy theory of evolution to describe the multidimensional homogenization (genetic, taxonomic, functional, geographical, physical, social, cultural, etc.) of the biosphere characterizing the "Anthropocene," also known—we learn from the author—as "Homogeocene." After laying down some useful theoretical concepts (e.g., biotic homogenization measure, invasion debt, genetic pollution, novel ecosystems, ecosystem services, and functional homogenization) and empirical data, the chapter concentrates on cities as primary engines of genealogical and ecological change (again, in the direction of homogenization). Urban dynamics are analyzed under various respects: physical uniformity, urban ecology, technological support, stages of rapid evolution. The author also applies Eldredge and Grene's view of social systems as reintegrating the genealogical and the ecological domains at a level above the individual, sketching out the position of cities in human evolutionary history. The author's thesis, supported by hierarchy theory, is that urbanization and urban scaling are universal and inevitable, leaving the question open about how the biosphere will evolve in presence of such strong engines of diversity reduction. This discussion contributes a unique theme to the book, extending beyond just biological evolution and shedding light on the significance of hierarchical approaches to approaching current ecological crisis.

CHAPTER 10

Unification of Macroevolutionary Theory

Biologic Hierarchies, Consonance, and the Possibility of Connecting the Dots

William Miller III

"Knowledge has often the character of expectations." (Popper 1999)

Introduction

The purpose of this volume is to explore the varied connections between evolutionary processes and patterns and the modern development of hierarchy theory as it applies to biologic systems. Both empirical and theoretical approaches are followed here. Some of the chapters are more conservative in the approach taken; others are more ambitious. Overall, the promise of this enterprise is an expansion of evolutionary theory to include much more than the focal level of local populations, involving genetic variation and natural selection and the gradual acquisition of adaptations— key ingredients in the original formulation of the modern synthesis of the mid-twentieth century (Fisher 1930; Dobzhansky 1937; Mayr 1942; Simpson 1944—see the contemporary overview by Huxley 1942 and the excellent modern evaluation by Eldredge 1985).

Expansion of the synthesis has been under way for several decades, to include new models of species formation, the possibility of evolution at levels higher than populations or systems of populations, the significance of development, the appearance of major innovations and eruption of new clades, the evolutionary impact of major catastrophes shaping the history of life, and to a limited extent the incorporation of ecologic thinking (to get a sense of the different directions of theoretical expansion, mainly from a paleontologic point of view, see Eldredge and Gould 1972; Gould 1977, 2002; Stanley 1979; Vrba 1980, 1985; Gould and Vrba 1982; Vrba and Eldredge 1984; Raup 1986; Vrba and Gould 1986; Donovan 1989; Eldredge 1985, 1989, 1995; Erwin and Anstey 1995; Raff 1996; Bennett 1997; Hallam and Wignall 1997; Jablonski 2005, 2007; Lieberman and Vrba 2005; Lieberman et al. 2007). These novel ideas about the nature of life on this planet would never have been proposed and evaluated if microevolutionary theory had been the exclusive perspective and if the fossil record had never been consulted seriously as a document of evolutionary history.

This is not merely my biased perspective: "rereading" the fossil record has played a significant role in these advances (Sepkoski and Ruse 2009; Sepkoski 2012). One of the main consequences of these efforts has been the realization (or *expectation*, since I started with a quote from Karl Popper) that macroevolutionary patterns detected in the fossil record could result from both scale independent and scale dependent processes (Stanley 1975, 1979; Gould 1980, 2002; Vrba and Eldredge 1984; Salthe 1985; Eldredge and Salthe 1985; Jablonski 1987; Lieberman et al. 1993; Lieberman 1995; Lieberman and Vrba 1995; Miller 1996; Vrba 2005; Eldredge et al. 2005; Lieberman et al. 2007; Tëmkin and Eldredge 2015). In the language of hierarchy theory, not only do larger, more inclusive systems consist of smaller systems and form working parts of still larger systems, but process rates and products, integration, boundaries, spatiotemporal continuity, and other features can be different at different levels of organization (my entry points to this kind of thinking were the books by Eldredge [1985] and Salthe [1985]—good starting places to explore this perspective).

This has produced a flock of new (or in some cases revived or revamped) theories and problems viewed now from a hierarchical perspective, not deriving exclusively from the traditional idea that macroevolutionary patterns result from gradual microevolutionary processes summed over long intervals of time. The purpose of this chapter is to see if these macroevolutionary theories and problems can be viewed as the components of a grander model of how life works, above the level of individual organisms packaged in local populations. In simple language, can the "dots" be connected? I will propose that the dots actually can be connected with addition of a *macroecologic* component, often absent in evolutionary theory (an earlier attempt to do this is found in Lieberman et al. 2007). This will necessitate an excursion in the land of metatheory, the gathering together

of related theories by discovering (or expecting) logical or required connections—the connections that reinforce, support, or extend more narrowly focused explanations. That places this chapter squarely in the ambitious category.

Hierarchical Underpinning of Macroevolutionary Theory

Evolutionary transformations must include more than transgenerational
phenotypic changes involving individual organisms within local populations before macroevolutionary explanations can be defended as useful/
realistic representations of the development of life on Earth (see discussions by Eldredge 1985, 1989, 1995; Grantham 2007; Jablonski 2007). Empirical evidence from studies of the origin of complex life forms, properties
of species-lineages, speciation patterns including radiations and convergence, trends in clade history and differences in evolutionary rates, the
possibility of species selection, and regional and global mass extinctions
and recoveries paced by climate change, geologic processes, and comet/
asteroid impacts indicates clearly that there is more to evolution than patterns resulting from scaled-up, gradual changes within demes. This takes
nothing away from mid-twentieth-century thinking about evolution but
does shift attention upward to more inclusive levels of pattern and process.
Microevolutionary patterns are embedded in larger patterns—which are
possibly "decoupled" or unique in some ways (Stanley 1975, 1979; Eldredge
1989; Lieberman 1995)—with the expectation that the larger patterns in
evolutionary history are not summations over time of population-level
processes and that the synthesis (as a comprehensive account) can in fact
be expanded to include processes and products of evolution at grander
scales of resolution. Macroevolutionary theory does not displace the modern synthesis as originally developed. Instead, at spatiotemporal scales
familiar to paleontologists, this empirical grounding supports the expectation that patterns in the fossil record will reveal processes unique to higher
levels of organization.

The statement is a bit different with a theoretical orientation. For macroevolution to be more than the summation of microevolutionary patterns and processes, the entities (biologic systems) at levels more inclusive
than individual organisms and local populations have to be real and have
to have their own "rules of operation" (emergent properties—see Grantham 2007 for discussion of emergence in this context). Larger (genealogic

or ecologic) systems should consist of component systems, feature slower rate constants and unique "predicates" (*sensu* Salthe 1985: systems at different hierarchical levels can "do" different things in terms of organization, function, interactions, and development) compared to component systems, and in turn be the working parts of still more inclusive systems (see the recent review by Tëmkin and Eldredge 2015). Discovery of causation or control involves more than dissecting a system and the characterization of mechanisms of component parts: each level of organization is a stage on which possibilities arrive from components and are sorted/modified/filtered by both constraints and forcing arriving from higher levels and interactions at that particular focal level. From an evolutionary point of view, individuals are packaged into demes, the demes are the working parts of species-lineages, and species derived from a common ancestor make up clades of greater or lesser size. From an ecologic perspective, individual organisms make up local populations, which are the working parts of local ecosystems, and these are the components of larger regional systems (Eldredge 1985; Eldredge and Salthe 1985; Salthe 1985; Miller 1996, 2002, 2004, 2008).

Development of Ecologic Systems and Macroevolution

The framing of the modern synthesis and its subsequent elaboration in the late twentieth century occurred mostly in isolation from the growth of population and community ecology. Evolutionary theorists emphasized relative reproductive success of organisms packaged in local populations: natural selection was a struggle for persistence on an ecologic stage, with some individuals dominating resources and solving problems more efficiently than others, assuring that more of their genes survived in subsequent generations. Ecologic processes and patterns were viewed as backdrop or incidental product, and most of what ecologists observed and measured, from an evolutionary perspective, consisted of short-term population fluctuations, mostly weak interactions among the local representatives of different species or the actualization of presumed adaptations to physical and chemical factors in the surrounding environment. From an ecologic viewpoint, description and modeling of population dynamics, interspecific interactions, and the structure and composition of communities rested on accepted microevolutionary assumptions. Organisms (in populations with properties that change over time) and assemblies of organisms (belonging to different species having unique adaptations) were the result of gradual evolutionary changes; natural selection produced all the key adaptations, and the species are supplied to a particular environment "ready-made" for the job. One can point to only a few key publications during this time that signaled attempts to bridge the gap between evolutionary and eco-logic theory—movements toward a common ground that would meld evo-lutionary and ecologic ideas in a more expansive form of the synthesis (e.g., Mayr 1947; Huxley 1942; Huxley et al. 1954; Grant 1963; Hutchinson 1965).

More recently, increased interest in evolutionary and behavioral ecol-ogy, and especially in ecologic speciation, is a promising sign that evolu-tionary theory and ecology are coming together. These movements, how-ever, are still deeply rooted in microevolutionary thinking: organisms are mosaics of adaptive traits, adaptations arise within local populations (with or without spatial separation of demes), and natural selection operating on individuals is the main mechanism of evolutionary transitions.

The truly innovative ideas about the role of ecology in evolutionary his-tory have mostly come from new interpretations of the fossil record. From a paleontologic perspective, the history of life appears to consist of inter-vals of apparent stability punctuated by intervals of relatively rapid reor-ganization, replacements, or turnovers, involving both species and clades and large ecologic systems. The patterns and processes do not appear al-ways to be scaled-up versions of microevolutionary changes played out on local ecologic stages, but rather interwoven or coordinated evolutionary and ecologic dynamics taking place on a grander scale—involving genea-logic and economic entities having both scale independent (e.g., second law of thermodynamics operating in all kinds of ecologic systems) and scale dependent properties (e.g., secondary succession is a community or local ecosystem process, not a population-level process). Examples of mod-els requiring entities or systems more inclusive than individual organisms, evanescent populations, and local communities include Vrba's (1985, 1993, 2005) turnover pulse hypothesis, Eldredge's (2003, 2008) sloshing bucket model, and macroevolutionary consonance theory (Miller 2004). Coordi-nated stasis is a related concept stressing intervals of apparent regional ecologic stability recorded in stratigraphic sequences (Brett and Baird 1995; Morris et al. 1995; Ivany and Schopf 1996; Ivany et al. 2009).

In these models, evolutionary processes and ecologic dynamics have equal importance in the interpretation of evolutionary stasis and change resolved in the fossil record. All these theories and models are driving at essentially the same central ideas: adaptive speciation is paced largely by environmental changes, the magnitude of changes is related to the magni-tude of evolutionary transitions, and both genealogic and economic systems

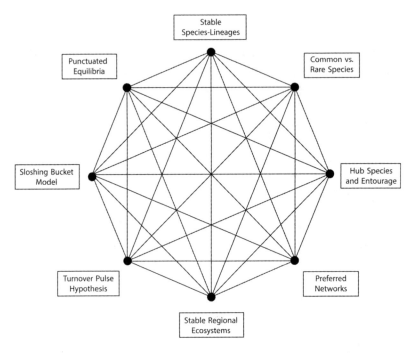

FIGURE 10.1 Representation of the possible expansion of macroevolutionary theory as concatenation of related evolutionary theories, macroecologic models, and some unresolved—but probably related—problems.

undergo transitions together. As environments change, reactions include migration, extinction, and ultimately speciation as ecologic systems undergo reorganization or replacement (Eldredge 2003; Miller 2004; Lieberman et al. 2007). Embedded in these large-scale interactions of ecologic systems and evolving lineages are a host of other phenomena that have resisted explication from a microevolutionary perspective (e.g., persistence of rare species [occurring always as a few individuals in local ecosystems] as stable lineages, development and persistence of networks within regional ecosystems, regional system stability in general, and recurrent system structure involving "hub species" entraining an entourage of dependent species; figure 10.1).

Following out the possible relationships or connections between these ideas/models/problems—attempting to connect the dots in the overall macroevolutionary pattern—leads us ineluctably to the realm of metatheories. An appreciation of hierarchy theory and interest in the possibility of unification are the obvious prerequisites here.

A Note on the Structure and Function of Metatheories

One can think of a metatheory as a kind of averaging of conceptually related generalizations about what nature contains and how it works, or as a con-catenation of related theories that covers a broader territory and explains (or at least highlights) more related phenomena. It will help here to recall Popper's (2004, 37–38) definition: "Theories are nets cast to catch what we call 'the world': to rationalize, to explain, and to master it. We endeavor to make the mesh ever finer and finer." Metatheories are the largest nets con-ceivable, but I think of them not as summations of more narrowly focused, specialized theories but rather as grander generalizations with many con-nected, complementary, and possibly coordinated parts—some being well investigated and supported by empirical experience, some the focus of new interest or approaches but still enigmatic, and some vaguely resolved gen-eralizations (or problems). Connecting the dots would produce an expan-sion of macroevolutionary theory and bring attention to new questions and possible research programs.

And with all kinds of generalizations concerning the entities that make up the world, their deployment in space and time, and processes involved in initiation, development, and reaction to disturbance and eventual extinc-tion/collapse, another distinction needs to be considered. We are mostly familiar (and more readily tolerate) theories that gather observations, in-terpretations, and results into general pictures—casting the net backward, in a sense, into the field of previous work to "catch" truthful generalizations. Theories may also be the kind of net that is thrown *forward*, into places we have only begun to envision and explore.

An Attempt to Connect the Dots

Punctuated Equilibria and Stable Species-Lineages

Punctuated speciation occupies a central position in macroevolutionary theory. Many of the related concepts (species selection, coordinated sta-sis, turnover pulses) simply cannot be explained without this foundational idea of how speciation and lineage stability work. Eldredge and Gould (1972) got it right because their model explains patterns of speciation and stability preserved in the fossil record *and* because the model works as a foundation piece in theories that explain other macroevolutionary and macroecologic patterns (Gould and Eldredge 1977, 1993; Lieberman and Eldredge 2014; Tëmkin and Eldredge 2015). As Princehouse (2009, 171) concluded, "Of all the innovations contributed by paleobiology, re-interpretation of the expectations for empirical evidence of speciation in the fossil record met the most resistance. . . . After the initially stormy reception—both within and outside paleontology, punctuated equilibria gained wide acceptance in the scientific community, even to the point of routine inclusion in biology textbooks by the late 1980s."

The use of the term *expectations* is interesting here. Many species with good fossil records (i.e., skeleton bearing, numerous, and distributed through stratigraphic successions) appear suddenly, persist without much directional phenotypic change, and then vanish. Most adaptive morpho-logic change appears to occur during speciation, and after deployment in multiple populations in often varied local ecologic contexts, species-lineages may remain stable for hundreds of thousands to millions of years (Eldredge et al. 2005). And during recoveries from regional and global dis-turbances, adaptive speciation and reorganization/replacement of large ecologic systems appear to go hand-in-hand (Eldredge 2003; Miller 2004; Lieberman et al. 2007). Empirical evidence is well established for these patterns of diversification, and the general acceptance that punctuated equilibria is the cornerstone of macroevolution prepares us for explora-tion of some rather unfamiliar territory. Microevolutionary processes are definitely embedded in all this, but we now expect that the rules change as one ascends hierarchical levels of organization—many of the dots simply cannot be explained by summing up long intervals of gradual transforma-tion of individuals within local populations. In short, punctuated equilibria turns out to be the primary "dot" in the constellation of macroevolution-ary processes and patterns—the starting place for proposing connections of interrelated large-scale evolutionary and ecologic dynamics.

Development, Stability, and Replacement of Large Ecologic Systems

Ecologic communities or local ecosystems can be composed of organisms with strong interactions (predator-prey, parasite-host, mutualistic connec-tions) or could consist of organisms that simply share the same environ-mental tolerances or even groups of organisms that are thrown together temporarily by chance. Some exhibit stability and appear to recover (more or less) from disturbances, while others are evanescent assemblies that ex-hibit little in the way of recurrent composition and structure. But all these entities could be embedded in larger, more inclusive systems that appear to emerge after regional or global climatic or geologic disturbances, persist for hundreds of thousands to millions of years, and undergo reorganization or replacement during subsequent disturbances. This may not always occur, and regional entities may not always have the chance or raw material to form stable systems.

In many cases dominant players in such systems show up early and are enthroned as centers of organization ("hub species") attracting a retinue of other organisms that depend on them directly or indirectly for food, habitat, or local environmental modulation. Networks of interactors form during intervals of reorganization/replacement (in fact, selection may be working here on the network, not solely on the components; see Ulanowicz 1997), with some species taking on significant roles while others seem to be marginalized or excluded. These organizational elements can then be deployed in different combinations as local systems, which are short-lived and continually reacting to disturbance. It seems extremely unlikely that the establishment, developmental history, and collapse of such regional ecosystems and the bouts of adaptive speciation and concurrent lineage stability apparent in the fossil record are merely coincidences. It is more likely they are two sides of the same coin.

Several models have been proposed to account for these patterns as resolved in the fossil record. In all of them, the environment is the master control, intervals of turnover or replacement are much shorter than intervals of apparent system stability, and evolutionary and ecologic processes are working together to produce the patterns. Vrba's turnover pulse hypothesis is an early attempt to interpret patterns of concurrent stability and replacement at a regional scale. As she summarized it (1993, 449), the pattern is a "concentration of turnover events against the time scale. For example, if a high number of first and last records of species in different lineages occur together within a time interval of one hundred thousand years or less, preceded and postdated by millions of years of predominant stasis in the same monophyletic groups, I would regard this as evidence of a turnover-pulse. I expect that turnover-pulses also occur in more refined time intervals, and that at least some stratigraphic sequences will be of sufficient quality to show this."

Eldredge's sloshing bucket model (2003) is a similar (but broader and unifying) interpretation, linking the magnitude of evolutionary reaction to the intensity and scope of environmental disturbance. Small-scale disturbances may touch off episodes of succession or replacement of local ecologic systems but would not cause much in the way of major migration/invasion, extinction, or adaptive speciation. At the opposite end of the spectrum, rare mass extinctions cause the most dramatic intervals of extinction and recovery involving wholesale removal of incumbents, radiations in surviving clades, and global-scale replacement of large ecologic systems. During the more frequent episodes of intermediate levels of disturbance, where regional systems are collapsing and being replaced by new systems, most of the adaptive speciation is probably taking place. As I have characterized it (Miller 2004, 635), "In this appealing model, Eldredge forces us to look for unique reactions at different scales of environmental disturbance, to acknowledge the overarching control of the environment, and to visualize the probable source of most adaptive evolution: the intermediate levels of the sloshing bucket marking the turnover of large ecologic systems—or in Eldredge's words (2003, 17) '. . . the primary level where adaptive change normally and typically occurs in evolutionary history.' "

In macroevolutionary consonance theory, I attempted to expand these ideas by adding a significant macroecologic component in order to place *both* coincident patterns of adaptive speciation and development of large ecologic systems at center stage. As in the models presented by Vrba and Eldredge, adaptive speciation and regional ecosystem turnover occur when environmental disturbances reach critical thresholds of intensity and scope. Ecologic and evolutionary dynamics are involved in these intervals, which can be brief compared to previous and subsequent intervals of relative stability. Consonance recognizes the reality of hierarchical structure, including both genealogic and economic entities; acknowledges the central role of environmental forcing; and invokes ecologic dynamics not evident at the spatiotemporal scale of neoecologic observations. These include any of the following combinations of large-scale processes, but probably most often occurs in the third form (see the discussion in Miller 2002, 2004):

1. Regional turnovers occur when resident regimes are disrupted to the point where significant extinction/speciation and abandonment/invasion take place. The remnants of the previous system, together with invaders and some new species, form the working parts of the new system. In this form, ecology dominates.
2. Regional extinctions followed by bouts of adaptive evolution are largely responsible for turnovers, with the new ecologic regime emerging gradually from new combinations of the new species. In this view, evolution is primary and ecology is incidental product.
3. Most likely, adaptive speciation and reorganization/replacement of ecologic structure are going on together, with new networks of interactors and new organizational hubs and retinues forming early in the transitions and controlling subsequent speciation and ecologic development simultaneously. In this way, evolutionary and ecologic processes are inseparable.

The participants in these regional turnovers have different roles and statuses. Using the terminology from the original proposal of macroevolutionary consonance (2004, 632–34), the possibilities include the following:

1. *Diehards and durables*: These are the remnants of previous ecologic regimes, which are species that do not suffer tremendous losses and may actually be promoted to key positions in the subsequent system and species that are demoted and remain rare.
2. *Invaders great and small*: These are species invading from adjacent or distant systems taking advantage of the release of resources (food, habitat) as incumbents abandon a region or go extinct; some invaders, however, never rise to prominence in the new regime.
3. *Upstarts and innovators*: These are the results of adaptive speciation producing new species that either replace incumbents or produce new organizational structure owing to novel innovations; regional ecologic transitions are probably both the causes and results of speciation and the appearance of novelties.
4. *The more or less vanquished*: These are the species that go extinct regionally but hang on elsewhere; the removal of certain key components from networks and hub-entourage systems in the previous regime is necessary for reorganization/replacement to take place—their removal makes it appear that the regional system was the entity that went extinct, not merely the local populations.

This does not say much about the role of opportunistic species, which can be the dominant components of local ecosystems. These organisms may have one of the roles listed above, but perhaps some opportunists emerge or persist as ecologic marauders, taking advantage of locally disturbed and degraded systems, never functioning as essential components of durable, recurrent networks. From a macroevolutionary point of view, these species could exhibit phenotypic stasis because, in a sense, they are able to "transcend" regional system structure by having short generation times, explosive population growth, and/or broad ranges and tolerances. In other words, they might persist because they both are broadly deployed *and* stay clear of the game.

As in the turnover pulse hypothesis and the sloshing bucket model, adaptive speciation and ecologic transitions are not merely synchronized, do not seem to occur at a similar spatiotemporal scale, or do not appear coordinated owing to the way fossils are preserved in stratigraphic successions, but they are causally related. Ecologic and genealogic entities interact to produce these macroevolutionary patterns. As I concluded in the same paper (Miller 2004, 638), "The most important idea has to do with recognition that magnitude of ecologic and evolutionary reactions is directly related to scale of associated environmental disruptions. If little adaptive evolution is possible without reaching certain thresholds of disturbance intensity, rate and scope, and if mass extinctions are viewed as being too rare to produce much of the adaptive speciation and innovations in the Phanerozoic, then most of the action simply has to be taking place at intermediate (regional) scales."

The Problem of Species Rarity and Other Loose Ends

Some perplexing problems now may be approached that have resisted resolution by adhering exclusively to the conviction that adaptive traits only arise gradually and continuously from natural selection operating on organisms within local populations. For example, we now have a hypothesis for origin and persistence of rare species. These are kinds of organisms never enthroned in regional systems as dominant components, possibly because as networks were formed, they were excluded or marginalized as relatively inefficient processors of resources or could not preempt other interactors as new networks formed. Rarity then could be part of the fallout from an interval of regional system reorganization/replacement. Rare species are not "kept down" by selection in local populations and communities necessarily; they are leftovers from organizational intervals of stringent natural selection when new regional ecosystems were assembled (Miller 2005).

We might be able to explain the persistence and replication of "preferred networks" of interacting organisms and the recurrence of hub species and their retinue of connected species in a similar way. Certain kinds of networks are more common than others and recur in different areas, and certain species are ecologic dominants, not because natural selection is continuously maintaining or fine-tuning their status, but because they were standardized or enthroned in these positions during relatively short intervals of stringent selection when regional systems were initially organized— involving both coincident and coordinated adaptive evolution and restructuring/replacement of large ecologic systems.

Conclusion

Punctuated equilibria is well established on empirical grounds as a dominant mode of species formation and subsequent stasis (see the review by

Benton and Pearson 2001 and the recent exchange among Pennell et al. 2014, Venditti and Pagel 2014, and Lieberman and Eldredge 2014). As an interpretation of patterns observed in the fossil record, it has become a foundation piece of macroevolutionary theory. Models that attempt to explain processes of coincident or coordinated adaptive speciation linked to turnover of large-scale ecologic systems have stimulated much interest (and controversy) and are areas that need more exploration and more attention to the kinds of ecologic dynamics characteristic of regional systems. If bouts of adaptive evolution are really both the cause and result of turnovers, as I think they are, and if this kind of evolutionary transition occurs when regional ecosystems are reorganized or replaced, much more attention needs to be focused here. And there are other possibilities that we have only dimly glimpsed. The difference between abundant and rare species and the dominance and recurrence of key ecologic networks may be the result of the last episode of large-scale ecologic change paced by climatic and geologic processes.

It is prudent to be cautious about grand generalizations and dreams of a "final theory" (see Weinberg's 1992 account of developments in physics), and it is certainly true that all theories must have empirical foundations. But the phrase often heard in introductory biology and paleontology courses, that evolution is played out on an ecologic stage (a corruption of the title of Hutchinson's 1965 collection of essays), simply is not right. Expansion of macroevolutionary theory requires a macroecologic component (Miller 2002, 2004, 2008; Lieberman et al. 2007). By ignoring the hierarchical organization of evolutionary and ecologic systems, placing evolution exclusively at center and thinking of ecologic processes and structures as by-products or backgrounds, and adhering exclusively to microevolutionary concepts and scaling, interpretation of the patterns and problems outlined here will continue to elude us. Trying to connect the dots will help solve problems—and create new ones.

I began with a quote from Karl Popper and I will conclude with another one (Popper 1999, 14): "The problems are themselves products of theories, and of the difficulties that critical discussion uncovers in theories." Most of us would want to substitute "empirical evidence" or "logical consistency" for "critical discussion" here.

Note on terminology: Readers familiar with Julian Huxley's (1942, 158–59) overview of the modern synthesis will recall that he used terminology similar to mine but with a different and more methodologic meaning. He used the term *consonant* to mean that criteria employed to delineate

different species should be in accord. Morphologic resemblance, separate
reproductive communities, lack of intergradation of defining properties,
and geographic distribution ought to be consonant before organisms are
considered to be separate species. I use the term *consonance* to emphasize the likelihood that macroevolutionary patterns are the result of large-
scale, coordinated, and causally related evolutionary and ecologic dynamics played out at regional scales (Miller 2004).

Acknowledgments

I am grateful to the editors for inviting me to participate in this exploration of hierarchy theory as it applies to biologic systems and for their constructive advice and direction—and not a little tolerance for this kind of
proposal. My special thanks goes to Niles Eldredge for providing the theoretical inspiration for much of my work in hierarchy theory, macroevolutionary and macroecologic dynamics, and species formation. A. Schmidt
prepared the figure.

References

Bennett, Keith D. 1997. *Evolution and Ecology: The Pace of Life*. Cambridge:
    Cambridge University Press.
Benton, Michael J., and Paul N. Pearson. 2001. "Speciation in the Fossil Record."
    Trends in Ecology and Evolution 16: 405–11.
Brett, Carlton E., and Gordon C. Baird. 1995. "Coordinated Stasis and Evolution-
    ary Ecology of Silurian to Middle Devonian Faunas in the Appalachian Basin."
    In *New Approaches to Speciation in the Fossil Record*, edited by Douglas H.
    Erwin and Robert L. Anstey, 285–315. New York: Columbia University Press.
Dobzhansky, Theodosius. 1937. *Genetics and the Origin of Species*. New York: Co-
    lumbia University Press.
Donovan, Stephen K., ed. 1989. *Mass Extinctions: Processes and Evidence*. New
    York: Columbia University Press.
Eldredge, Niles. 1985. *Unfinished Synthesis: Biological Hierarchies and Modern Evo-
    lutionary Thought*. New York: Oxford University Press.
———. 1989. *Macroevolutionary Dynamics: Species, Niches and Adaptive Peaks*.
    New York: McGraw-Hill.
———. 1995. *Reinventing Darwin: The Great Debate at the High Table of Evolu-
    tionary Theory*. New York: Wiley.
———. 2003. "The Sloshing Bucket: How the Physical Realm Controls Evolution."
    In *Evolutionary Dynamics: Exploring the Interplay of Selection, Accident, Neu-

trality, and Function, edited by James P. Crutchfield and Peter Schuster, 3–32. Oxford: Oxford University Press.

———. 2008. "Hierarchies and the Sloshing Bucket: Toward the Unification of Evolutionary Biology." *Evolution: Education and Outreach* 1: 10–15.

Eldredge, Niles, and Stephen J. Gould. 1972. "Punctuated Equilibria: An Alternative to Phyletic Gradualism." In *Models in Paleobiology,* edited by Thomas J. M. Schopf, 82–115. San Francisco: Freeman, Cooper.

Eldredge, Niles, and Stanley N. Salthe. 1985. "Hierarchy and Evolution." *Oxford Surveys in Evolutionary Biology* 1: 184–208.

Eldredge, Niles, John N. Thompson, Paul M. Brakefield, Sergej Gavrilets, David Jablonski, Jeremy B. C. Jackson, Richard E. Lenski, Bruce S. Lieberman, Mark A. McPeek, and William Miller III. 2005. "The Dynamics of Evolutionary Stasis." *Paleobiology* 31: 133–45.

Erwin, Douglas H., and Robert L. Anstey, eds. 1995. *New Approaches to Speciation in the Fossil Record.* New York: Columbia University Press.

Fisher, Ronald A. 1930. *The Genetical Theory of Natural Selection.* Oxford: Clarendon.

Gould, Stephen J. 1977. *Ontogeny and Phylogeny.* Cambridge, MA: Harvard University Press.

———. 1980. "Is a New and General Theory of Evolution Emerging?" *Paleobiology* 6: 119–30.

———. 2002. *The Structure of Evolutionary Theory.* Cambridge, MA: Harvard University Press.

Gould, Stephen J., and Niles Eldredge. 1977. "Punctuated Equilibria: The Tempo and Mode of Evolution Reconsidered." *Paleobiology* 3: 115–51.

———. 1993. "Punctuated Equilibria Comes of Age." *Nature* 366: 223–27.

Gould, Stephen J., and Elisabeth S. Vrba. 1982. "Exaptation: A Missing Term in the Science of Form." *Paleobiology* 8: 4–15.

Grant, Verne. 1963. *The Origins of Adaptations.* New York: Columbia University Press.

Grantham, Todd. 2007. "Is Macroevolution More than Successive Rounds of Microevolution?" *Palaeontology* 50: 75–85.

Hallam, Antony, and Paul B. Wignall. 1997. *Mass Extinctions and Their Aftermath.* Oxford: Oxford University Press.

Hutchinson, G. Evelyn. 1965. *The Ecological Theater and the Evolutionary Play.* New Haven: Yale University Press.

Huxley, Julian. 1942. *Evolution: The Modern Synthesis.* London: Allen and Unwin.

Huxley, Julian, A. C. Hardy, and E. B. Ford, eds. 1954. *Evolution as a Process.* London: Allen and Unwin.

Ivany, Linda C., Carlton E. Brett, Heather L. B. Wall, Patrick D. Wall, and John C. Handley. 2009. "Relative Taxonomic and Ecologic Stability in Devonian Marine Faunas of New York State: A Test of Coordinated Stasis." *Paleobiology* 35: 499–524.

Ivany, Linda C., and Kenneth M. Schopf, eds. 1996. "New Perspectives on Faunal Stability in the Fossil Record." *Palaeogeography, Palaeoclimatology, Palaeoecology* 127: 1–359.

Jablonski, David. 1987. "Heritability at the Species Level: Analysis of Geographic Ranges of Cretaceous Mollusks." *Science* 238: 360–63.

———. 2005. "Mass Extinctions and Macroevolution." *Paleobiology* 31: 192–210.

———. 2007. "Scale and Hierarchy in Macroevolution." *Palaeontology* 50: 87–109.

Lieberman, Bruce S. 1995. "Phylogenetic Trends and Speciation: Analyzing Macroevolutionary Processes and Levels of Selection." In *New Approaches to Speciation in the Fossil Record*, edited by Douglas H. Erwin and Robert L. Anstey, 316–37. New York: Columbia University Press.

Lieberman, Bruce S., Warren D. Allmon, and Niles Eldredge. 1993. "Levels of Selection and Macroevolutionary Patterns in the Turritellid Gastropods." *Paleobiology* 19: 205–15.

Lieberman, Bruce S., and Niles Eldredge. 2014. "What Is Punctuated Equilibrium? What Is Macroevolution? A Response to Pennell et al." *Trends in Ecology and Evolution* 29: 185–86.

Lieberman, Bruce S., William Miller III, and Niles Eldredge. 2007. "Paleontological Patterns, Macroecological Dynamics and the Evolutionary Process." *Evolutionary Biology* 34: 28–48.

Lieberman, Bruce S., and Elisabeth Vrba. 1995. "Hierarchy Theory, Selection, and Sorting." *BioScience* 45: 394–99.

———. 2005. "Stephen Jay Gould on Species Selection: 30 Years of Insight." *Paleobiology* 31: 113–21.

Mayr, Ernst. 1942. *Systematics and the Origin of Species.* New York: Columbia University Press.

———. 1947. "Ecological Factors in Speciation." *Evolution* 1: 263–88.

Miller, William, III. 1996. "Ecology of Coordinated Stasis." *Palaeogeography, Palaeoclimatology, Palaeoecology* 127: 177–90.

———. 2002. "Regional Ecosystems and the Origin of Species." *Neues Jahrbuch für Geologie und Paläontologie Abh.* 225: 137–56.

———. 2004. "Assembly of Large Ecologic Systems: Macroevolutionary Connections." *Neues Jahrbuch für Geologie und Paläontologie Mh.* 2004: 629–40.

———. 2005. "The Paleobiology of Rarity: Some New Ideas." *Neues Jahrbuch für Geologie und Paläontologie Mh.* 2005: 683–93.

———. 2008. "The Hierarchical Structure of Ecosystems: Connections to Evolution." *Evolution—Education and Outreach* 1: 16–24.

Morris, Paul J., Linda C. Ivany, Kenneth M. Schopf, and Carlton E. Brett. 1995. "The Challenge of Paleoecological Stasis: Reassessing Sources of Evolutionary Stability." *Proceedings of the National Academy of Sciences of the United States of America* 92: 11269–73.

Pennell, Matthew W., Luke J. Harmon, and Josef C. Uyeda. 2014. "Is There Room for Punctuated Equilibrium in Macroevolution?" *Trends in Ecology and Evolution* 29: 23–32.

Popper, Karl. 1999. *All Life Is Problem Solving.* London: Routledge.

———. 2004. *The Logic of Scientific Discovery.* London: Routledge.

Princehouse, Patricia. 2009. "Punctuated Equilibria and Speciation: What Does It Mean to Be a Darwinian?" In *The Paleobiological Revolution: Essays on the Growth of Modern Paleontology*, edited by David Sepkoski and Michael Ruse, 149–75. Chicago: University of Chicago Press.

Raff, Rudolf A. 1996. *The Shape of Life: Genes, Development, and the Evolution of Animal Form.* Chicago: University of Chicago Press.

Raup, David M. 1986. "Biological Extinction in Earth History." *Science* 231: 1528–33.

Salthe, Stanley N. 1985. *Evolving Hierarchical Systems: Their Structure and Representation.* New York: Columbia University Press.

Sepkoski, David. 2012. *Rereading the Fossil Record: The Growth of Paleobiology as an Evolutionary Discipline.* Chicago: University of Chicago Press.

Sepkoski, David, and Michael Ruse, eds. 2009. *The Paleobiological Revolution: Essays on the Growth of Modern Paleontology.* Chicago: University of Chicago Press.

Simpson, George G. 1944. *Tempo and Mode in Evolution.* New York: Columbia University Press.

Stanley, Steven M. 1975. "A Theory of Evolution above the Species Level." *Proceedings of the National Academy of Sciences of the United States of America* 72: 646–50.

———. 1979. *Macroevolution: Pattern and Process.* San Francisco: W. H. Freeman.

Tëmkin, Ilya, and Niles Eldredge. 2015. "Networks and Hierarchies: Approaching Complexity in Evolutionary Theory." In *Macroevolution: Explanation Interpretation and Evidence,* edited by Emanuele Serrelli and Nathalie Gontier, 183–226. Berlin: Springer.

Ulanowicz, Robert E. 1997. *Ecology, the Ascendent Perspective.* New York: Columbia University Press.

Venditti, Chris, and Mark Pagel. 2014. "Plenty of Room for Punctuational Change." *Trends in Ecology and Evolution* 29: 71–72.

Vrba, Elisabeth S. 1980. "Evolution, Species and Fossils: How Does Life Evolve?" *South African Journal of Science* 76: 61–84.

———. 1985. "Environment and Evolution: Alternative Causes of the Temporal Distribution of Evolutionary Events." *South African Journal of Science* 81: 229–36.

———. 1993. "Turnover-Pulses, the Red Queen, and Related Topics." *American Journal of Science* 293-A: 418–52.

———. 2005. "Mass Turnover and Heterochrony Events in Response to Physical Change." *Paleobiology* 31: 157–74.

Vrba, Elisabeth S., and Niles Eldredge. 1984. "Individuals, Hierarchies and Processes: Toward a More Complete Evolutionary Theory." *Paleobiology* 10: 146–71.

Vrba, Elisabeth S., and Stephen J. Gould. 1986. "The Hierarchical Expansion of Sorting and Selection: Sorting and Selection Cannot Be Equated." *Paleobiology* 12: 217–28.

Weinberg, Steven. 1992. *Dreams of a Final Theory: The Scientist's Search for the Ultimate Laws of Nature.* New York: Vintage.

CHAPTER 11

Coming to Terms with *Tempo and Mode*

Speciation, Anagenesis, and Assessing Relative Frequencies in Macroevolution

Warren D. Allmon

"The intent of scientific vocabulary or terminology is to create a precision of definition such that any topic may be discussed with unambiguous meaning." (Jackson 1999)

"Most controversies would soon be ended, if those engaged in them would first accurately define their terms, and then adhere to their definitions." (Edwards 1908, 88)

"Terms in themselves are trivial, but taxonomies revised for a different ordering of thought are not without interest. Taxonomies are not neutral or arbitrary hat-racks for a set of unvarying concepts; they reflect (or even create) different theories about the structure of the world. . . . when you know why people classify in a certain way, you understand how they think." (Gould and Vrba 1982, 351)

Introduction

At least since the Renaissance, when radical new thinkers tried to distance themselves from their intellectual predecessors by characterizing (not altogether accurately) medieval scholastic thought as little more than endless debate about the meanings of words, intense attention to definitions has frequently been thought of as not really relevant for our understanding of the physical world. As Stephen J. Gould once characterized this view, "What could be more tedious, more dull, more devoid of meaning and importance for understanding the real world out there than debates about terminology?" (Gould 1988, 24). At the same time, however, the use of at least some specialized vocabulary ("good jargon" in the sense of Hirst [2003]) is essential to all science. Without it, scientific communication would require a much larger number of vague and imprecise words and would be far less able to identify clearly what was being observed, tested, examined, or debated. More generally, technical terms represent the things and ideas that we study. Their meaning and usage reflect what we think (or don't think) those things and ideas are and why they are important.

Like all language, scientific terminology is not static but evolves by a process resembling natural selection. As noted by Koonin (2001), new terms are proposed; those that are useful survive and grow in usage, "while the rest will die out or will lead a marginal existence." Preexisting words can be co-opted for new functions (and so, as Koonin observes, may be thought of as "exapted"; *sensu* Gould and Vrba 1982, itself a technical neologism with a mixed history of acceptance; Pievani and Serrelli 2011). Definitions and usage of terms can change. For example, the original meaning of the term *adaptive radiation* (Osborn 1902, 1910) was long ago discarded, and it has a very different and widely used definition today (e.g., Losos and Mahler 2010). New knowledge can also render previous meanings and usage obsolete or confusing. In the early days of molecular biology, for example, similarity between sequences of amino acids or nucleic acid bases was widely referred to as "sequence homology," until evolutionarily minded biologists succeeded in convincing their colleagues that not all similarity was "homology" (the latter word long defined as only those similarities resulting from inheritance from a common ancestor). The result was the introduction of the term *orthologous* to refer to a molecular sequence found in two organisms as a result of evolutionary descent (Reek et al. 1987; Gould 1988; Fitch 2000; Koonin 2001).

In other words, the terms by which we refer to different patterns in nature are not just labels; they both reflect and affect the way we think about the reality and causes of those patterns. For new technical terms, we should ask something to the effect of, "Is the world a better place because of them?" (Koonin 2001, 2). For changes in existing terms, we should ask "if that change is a refinement that increases clarity of present-day thought and exposition." (Fitch 2000, 227).

When modern paleobiology began in the 1970s (Sepkoski 2012), it introduced a number of new terms to explore evolution as revealed in the fossil record as well as modified or repurposed existing terms. Considerable attention has been devoted to explicating some of these terms, such as "species selection" (see, e.g., Vrba 1984; Gould 2002; Jablonski 2008) and "heterochrony" and related terms (e.g., Gould 1977, 2000; Alberch et al. 1979).

Much less attention has been devoted to the term *tempo and mode*, which was introduced in the title of one of the founding documents of the modern synthesis and then a half century later became almost synonymous with many of the ideas associated with the founding of modern paleobiology. Yet its usage in recent decades has become so varied that it has become almost meaningless as a technical description of evolution.

The meaning of *tempo and mode* is central to investigation of hierarchy in macroevolution (i.e., evolution above the species level) because much of this investigation depends on the relative frequency of various evolutionary patterns within and between species, such as stasis, speciation, or anagenesis (e.g., Gould 2002, 772ff.; Eldredge 2015; Lidgard and Hopkins 2015). We cannot tabulate relative frequency unless we are clear on the question "frequency of what?"

In this chapter, I argue that the decline in precision in the use of the term *tempo and mode* in technical evolutionary writing threatens not only progress in the hierarchical approach to understanding evolution but also our recognition of two of the major messages of the paleobiological revolution: that evolutionary trends are not just the products of phyletic change (anagenetic transformation) within lineages but also the splitting of lineages by speciation and that "rates" of evolution can refer to more than a single process. I propose more precise definitions for both *tempo* and *mode* and the variety of evolutionary patterns produced by their interaction and recommend that other usages be avoided.

Tempo and Mode: A Brief History

In introducing the term *tempo and mode*, Simpson (1944) wrote,

> On two topics in particular, the paleontologist enjoys special advantages. . . . The first of these . . . has to do with evolutionary rates under natural conditions, the measurement and interpretation of rates, their acceleration and deceleration, the conditions of exceptionally slow or rapid evolution. . . . In the present study all these problems are meant to be suggested by the word "tempo." The group of related problems implied by the word "mode" involves the study of the way, manner, or pattern of evolution, a study in which tempo is a basic factor, but which embraces considerably more than

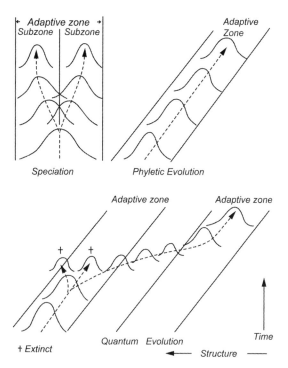

FIGURE 11.1 Simpson's (1944) illustration of his three evolutionary *modes*.

tempo. The purpose is to determine how populations became genetically and morphologically differentiated, to see how they passed from one way of living to another or failed to do so, to examine the figurative outline of the stream of life and the circumstances surrounding each characteristic element in that pattern. (1944, xxix–xxx)

Then toward the end of the book, he specified "three major styles, the basic modes of evolution" as "speciation, phyletic evolution, and quantum evolution" (1944, 197–98; figure 11.1).

In 1977, Gould and Eldredge recalled the term but limited it to two of Simpson's three modes: speciation and phyletic evolution (which had been labeled cladogenesis and anagenesis, respectively, by Rensch 1959, 281). They noted that in paleontological studies of evolution, "tempos can be observed and measured: modes must be inferred. . . . Our model of punctuated equilibria is a hypothesis about mode. We claim that speciation is orders of magnitude more important than phyletic evolution as a mode of evolutionary change" (Gould and Eldredge 1977, 115–16).

Importantly, here and later, Gould and Eldredge made the argument that data about evolutionary tempo (rate) could be used to infer evolutionary mode (process): "Tempos are a unique paleontological domain; modes may be inferred from them, and status as a source for theory thus conferred upon paleontology" (Gould 1994, 6767).

It is interesting that Eldredge (2013, 2015) has recently recalled that his "major contribution" to this paper "was to insist that wherever Steve had written 'tempo', I said he should write 'and mode'. . . . [because] Simpson had made a fundamental distinction between tempos (rates) of genetically based morphological evolution and the modes of such change—by which he meant phyletic evolution, speciation, and his own concept of 'quantum evolution'" (Eldredge 2015, 265).

Since 1977, however, authors have frequently taken a looser approach to these terms. While most have continued to use the term *tempo and mode* in the Simpson/Gould-Eldredge (SGE) sense (e.g., Cheetham 1986; Geary 1987, 1990; Geyssant 1988; Jackson and Cheetham 1994, 1999; McHenry 1994; Bleiweiss 1998; Kinnison and Hendry 2004; Baker et al. 2005; Esselstyn et al. 2009; Nagorsen and Cardini 2009; Pachut and Anstey 2009; Goldberg and Igić 2012), the term has also been used in a wide variety of other meanings, including the following:

- very generally to refer to evolution on a broad scale within most or all life, especially in the Precambrian (e.g., Woese et al. 1985; Schopf 1994; Peterson et al. 2005; see also Blomberg and Garland 2002, who, ironically, discuss similarly imprecise use of the term *phylogenetic inertia*)
- overall rates and/or patterns of change in particular characters and/or taxonomic diversity within particular clades (e.g., Kohn 1990; Padian 1994; Bernardi and Lape 2005; Pérez-Losada et al. 2008; Cooper and Purvis 2010; Meloro and Raia 2010; Olsen et al. 2011; Pettengill and Moeller 2012)
- rates of speciation (e.g., Nee et al. 1992; Day et al. 2008)
- geographic pattern of speciation (e.g., allopatric by vicariance vs. allopatric by dispersal; Rosenzweig 1997; Bernardi and Lape 2005; Gante et al. 2009; Allmon and Smith 2011)
- rate and mechanism of molecular or genetic evolutionary change (e.g., Brown et al. 1982; Doolittle and Brown 1994; Adachi and Hasegawa 1996; Delmotte et al. 2006; Romano et al. 2006; Dacks and Field 2007; Oliver et al. 2007; Schürch et al. 2010; Averbeck and Eickbush 2005; Feyereisen 2011; Gordon and Ruvinsky 2012; Coolon et al. 2014)
- patterns of adaptive radiation (e.g., Harmon et al 2003; Van Tuinen 2006)

The term has even been used to describe purely ecological processes (such as gopher mound production; Klaas et al. 2000). Adding "tempo and mode" to the title of a paper about evolution frequently seems more of a literary decoration than a thoughtful choice of technical words.

In parallel with the development of the SGE use of *tempo and mode* was the introduction of a variety of statistical methods for analysis of evolutionary change at a variety of levels, combined with a general interest in random patterns in evolution (Raup et al. 1973; Raup 1977a,b; Bookstein 1987, 1988). For microevolution, authors emphasized that it was statistically difficult to unambiguously recognize a particular pattern of anagenetic change and suggested that varying patterns (from static to directional) lay on a continuum exhibited by unbiased (symmetric) and biased random walks. They therefore used the random walk, in which trait increases and decreases are equally likely, as a null model (Bookstein 1987, 1988; Gingerich 1993; Roopnarine et al. 1999; Roopnarine 2001, 2003). Roopnarine (2001) referred to three patterns of anagenetic change—stasis, random walk (or "oscillatory variation"), and directional change (at a variety of rates)—as "microevolutionary modes."

It has been suggested that this consideration of stochastic processes "revolutionized the study of tempo and mode in evolution, as framed by Simpson (1944)" and "the resulting shift in perspective transformed paleontological studies" (Goldberg and Igić 2012, 3701). Whether this is an accurate characterization or not, it does seem that the focus on random processes contributed to a focus of paleontological attention toward anagenetic change and away from speciation.

For example, in his influential analyses of phenotypic change within fossil lineages, Hunt (2006, 2007, 2008a,b, 2012; Hunt and Carano 2010)

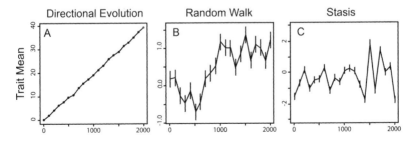

FIGURE 11.2 The three "models" or "modes" of evolution identified by Hunt, herein referred to as "models" of anagenesis (from Hunt 2012).

notes that "the term 'modes of evolution' descends from Simpson (1944), and it has become standard when referring to qualitatively distinct patterns of change in fossil lineages" (2010, S67). He then identifies the "three canonical modes of evolution" as "directional change, random walk, and stasis" (2012, 353; figure 11.2). Hunt also refers to these as "models" (e.g., 2012, 352): "[evolutionary] rates are usefully construed as parameters in models that predict the magnitude of change over specified intervals of time. Rates are not data but rather follow from data when combined with a model of how phenotypes evolve."

What Is Evolutionary Tempo?

Tempo or rate of evolution has been expressed in several ways. It can refer to the rate of anagenetic phyletic change—that is, transformation within a lineage without branching (e.g., Haldane 1949; Gingerich 2009; Hunt 2012). This is probably close to what Simpson (1944) meant when he discussed his various categories of evolutionary rate.

It can also refer to "rate of speciation" (Stanley 1979, 1985), which itself can have at least two meanings, not necessarily closely related. The first is the time interval between speciation events within a clade ("species per million years" or the "biological speciation interval," BSI; Raup 1978; Coyne and Orr 2004; Futuyma 2009, 493). The second is the time it takes for a daughter population to diverge sufficiently from the parent population to achieve reproductive isolation (the "transition time," "waiting time to speciation," "duration of speciation," or "time for speciation," TFS; Coyne and Orr 2004; McCune 2004; Curnoe et al. 2006; Futuyma 2009; figure 11.3). BSIs can be longer than TFSs (sometimes much longer), because lineages may not begin to branch immediately after they arise and because at least some transition times can be fairly short (Coyne and Orr 2004). BSIs are estimated in a variety of ways, usually based on the fossil record or (increasingly) on molecular phylogenies (e.g., Mendelson and Shaw 2005; Weir and Schluter 2007). TFSs are generally more difficult to estimate, because this requires knowledge of when speciation begins and ends, and so most estimates of "speciation rates" in the literature are actually estimates of BSI. In rare instances, such as in deep-sea cores and lakes, which preserve extremely high-resolution fossil records, TFS can be estimated with high precision (e.g., McCune 2004). Estimates of "divergence times" between geminate species, which are commonly used to calibrate molecular clocks (e.g., Marko 2002), are useful metrics of rates of

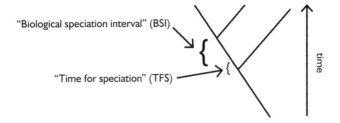

FIGURE 11.3 Two different measures of "rate of speciation."

genetic divergence between populations—equivalent to estimates of "speciation duration" in phylogeography (Avise 2000). These are not, however, necessarily equal to TFS; differences between geminates could have been generated more quickly than suggested by divergence times based on the fossil record, and so divergence rates based on germinates may represent minimum values. Other estimates of TFS have been made from the fossil record (summarized by Gould 2002, 852), ranging from 5,000 years for the Quaternary origin of dwarfed woolly mammoths on Wrangel Island, Alaska, to 73,000–275,000 years for the marginellid gastropod genus *Prunum*, to 100,000–200,000 years in a genus of Cretaceous marine ostracodes.

TFSs in the fossil record have also been discussed in the context of the theory of punctuated equilibria (or "punctuated equilibrium" as it came to be frequently called; Eldredge and Gould 1972; Eldredge 2013). For example, Eldredge (e.g., 1985, 189; 2015, 273–74) suggested—without much direct empirical support—that speciation usually takes "five to fifty thousand years." Similarly, in attempting to define when a speciation event is "punctuated," as opposed to gradual, Gould repeatedly argued for a relative metric, such as 1–2 percent of the total subsequent duration of the species produced (Gould 1982, 2002, 768). Such a definition, he said, would allow "up to 100,000 years for the origin of a species with a subsequent life span of 10 million years" (the estimate for the average species of marine bivalves; Stanley 1979). Gould added, however, that he believed "that most events of speciation occur much more rapidly" (Gould 1982, 84).

Compilations of BSIs (Stanley 1979; Coyne and Orr 2004, 419) show that, although values range from very short (4,000 years for cichlid fishes in Lake Nabugabo, Uganda) to very long (several hundred million years for notostracan crustaceans), most cluster between 1 and 20 million years. The compilation of Coyne and Orr (2004) yields a mean BSI for all taxa (protists to mammals) of 6.5 million years, with marine invertebrates showing higher values (most groups between 6 and 16 million years) than terrestrial plants, which in turn show longer BSIs than terrestrial and freshwater insects and vertebrates. Marine gastropods, for example, show a mean BSI of 14.9 million years (Stanley 1979). Several more recent studies of divergence times of geminate pairs of marine gastropods across the Central American Isthmus indicate somewhat lower values (5–10 million years, which, as mentioned above, probably represent minimum rate values; Collins 1996; Marko 2002). Based on all these data, the average lineage of marine gastropod appears to produce a new species roughly every 5–15 million years. Each of these new species, however, may actually arise in less than 100,000 years.

What Is a *Mode* of Evolution?

In common English, *mode* refers to, among other meanings, "a particular form or variety of something; a form . . . style . . . a possible, customary, or preferred way of doing something . . . a manifestation, form, or arrangement of being . . . [especially] a particular form or manifestation of an underlying substance . . . a particular functioning arrangement or condition . . . the most frequent value of a set of data" (*Webster's New Collegiate Dictionary* 1974, 739). The statistical meaning of *mode*—the most abundant value in a distribution—suggests both discreteness of such "classes" and description of parts of a continuum. Thus the word itself might or might not most suitably refer to a qualitatively discrete thing, as opposed to a restricted piece of a continuous series.

In coining the term *mode* for referring to evolution, however, Simpson appeared to be emphasizing not just quantitative differences but qualitatively different categories of phenomena. In doing so, he was trying to encompass both the legacy of Darwin and the emerging modern synthesis, both of which recognized that evolution must include both change within lineages and the origin and multiplication of new lineages. Change within lineages (what Simpson called phyletic change) can happen at a variety of rates (tempos), from zero (a.k.a. "stasis") to slow, moderate, and very rapid (which Simpson called "bradytely," "horotely," and "tachytely," respectively). In the SGE sense, then, *mode* describes the topology of the tree of life, whereas *tempo* describes the slope of a line on a time-phenotype graph (see Wilkinson 2015 for further discussion of this graphical conception).

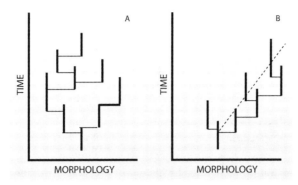

FIGURE 11.4 Two of the "models" of anagenesis (figure 11.2) applied to evolutionary trends in branching clades, in which all the individual species are themselves showing morphological stasis. Modified from Rasskin-Gutman and Esteve-Altava (2008). A. No net change in morphology (stasis) during the history of the clade. B. A net directional change ("trend") in morphology in the history of the clade.

In labeling three different patterns of anagenesis (stasis, random walk, and directional change) as *modes*, more recent discussions (e.g., Roopnarine 2001; Hunt 2010) have focused on what emerged as one of the central points of contention in the debates over punctuated equilibrium: different authors could look at the same pattern of change in the fossil record and label it as stasis or "gradualism" (directional phyletic change) depending on their preferred point of view (reviewed in Erwin and Anstey 1995 and Hunt 2008b). That is, one person's stasis was another's gradualism. Rigorous quantitative techniques were introduced to address this confusion, with the empirical result that directional change is rare, while stasis and random walks are more common (Hunt 2007).

Yet (as noted, e.g., by Rasskin-Gutman and Esteve-Altava 2008), such a taxonomy of *mode* largely ignores speciation (cladogenesis), which was also central to the punctuated equilibrium debates. (Unfortunately, Rasskin-Gutman and Esteve-Altava confuse tempo and mode in their attempt at a more comprehensive taxonomy of evolutionary patterns; see below and figure 11.4.)

Tempo and Mode Reconsidered, Again

As Simpson cogently observed, when we investigate large-scale evolution, we instinctively want to know both "how" (mode) and "how fast" (tempo).

According to Simpson, paleontology could contribute insights about both. Both he and Gould and Eldredge noted, however, that tempos of change within lineages can (in principle at least) be measured, whereas modes must be inferred. In this context, an emphasis on tempo via quantitative approaches to distinguishing different patterns of anagenesis therefore makes sense in terms of what is most available for paleontologists to study and also reflects the widely recognized analytical difficulty of recognizing and studying speciation in the fossil record (Hunt 2010, 2013).

Yet just because it is more difficult to study lineage splitting does not mean it was not important in the history of life, and excluding speciation from the taxonomy of evolutionary mode just because it is harder to quantify than anagenesis carries risks to our gaining a more adequate assessment of the overall pattern (and process) of the history of life. First, as Darwin recognized (1859), albeit through a glass darkly (Allmon 2013; Eldredge 2013, 2015), a complete theory of evolution must include not just the transformation of lineages but their origin and multiplication. Second, and more specifically, the modern discussion of macroevolution in paleontology was stimulated in large part by punctuated equilibrium, which is explicitly not just a theory of rates of evolutionary change but also a theory about lineage branching and how that indisputable (although still poorly understood) phenomenon is connected to phenotypic change (Gould 1982, 2002).

Several empirical tests have been proposed for punctuated equilibrium (table 11.1). Estimating the relative frequency of stasis—which is one of the most noted results of application of recent quantitative techniques (Hunt 2007)—was one that was emphasized by supporters of punctuated equilibrium (e.g., Gould 2002, 854–74). Yet stasis is a pattern of tempo, which

TABLE 11.1 **Methods for testing tempo and mode generally (and punctuated equilibrium sometimes)**

Living taxa	Testing whether variance in phenotypes increases as the number of speciation events (inferring a "punctuational mode") or of total branch length—that is, time (inferring a "gradualist mode"; see, e.g., Hopkins and Lidgard 2012)
Fossils	Persistence or survival of ancestor (Gould 2002, 840)
Fossils	Prevalence of stasis—implying punctuation (but not necessarily punctuated equilibrium; Gould 2002, 854ff.)
Fossils	"Magnitude of morphological difference between sister species compared with the changes within species" (Hunt 2013, 716)

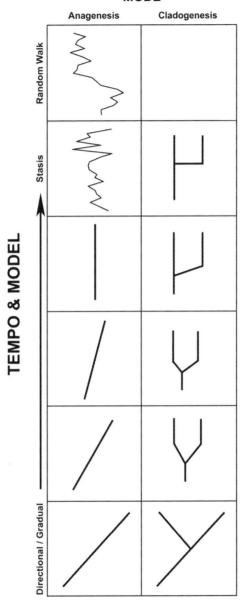

FIGURE 11.5 Visual taxonomy of the various possible patterns of macroevolutionary dynamics. Based in part on Barton et al. (2009, 280). Stasis is an end-member (zero rate, no direction) for both tempo and model of anagenesis. The "stasis" and "random walk" graphs are modified slightly from Hunt (2012; see figure 11.2). See text for further discussion.

is only one element of punctuated equilibrium. One of the other crucial tests for mode—the recognition of the persistence or survival of ancestors (e.g., Gould 2002, 840)—requires consideration of species and speciation in the fossil record. Paleontologists have long struggled with recognition of species in fossils (Allmon and Yacobucci 2016). Reservations of some paleontologists notwithstanding, however, there is in reality a robust common working conception of species in fossils and a strong argument that the morphospecies recognized by paleontologists approximate many of the most common evolutionary characteristics of species as recognized by neontologists, such as evolutionary separateness (Allmon 2016).

Whether something is labeled as a *mode* of evolution or not matters because it can affect what we think is important—to know and to do research on. Lineage splitting is clearly one of the major patterns characterizing the history of life, caused by the apparently multitudinous processes of speciation (e.g., Coyne and Orr 2004). Questions about the connection between lineage splitting and transformation within lineages (which is also caused by a wide variety of processes, including random change) cannot be answered if we demote speciation to less than a major mode of evolution or if we use the term *mode* as merely a synonym for *kind*.

With these distinctions in mind, when we look most generally at the large-scale history of life, three major phenomena emerge as major (not to say "the" major) characteristics of that history: evolutionary change occurs within lineages, new lineages arise (and old ones go extinct), and these changes happen at a variety of rates. The "inseparability of tempo and mode" (Hunt 2012) is therefore not just about anagenesis; it is at the core of understanding macroevolution. As many authors have noted (e.g., Erwin and Anstey 1995; Geary 1987, 1995; Gould 2002, 772ff., 854ff.; Barton et al. 2009; Futuyma 2009; Pievani 2015), the goal of macroevolutionary research is not to determine whether "punctuated equilibrium" is true or not but to quantify the relative frequency of the wide array of evolutionary patterns produced by the interaction among tempos and modes (e.g., figure 11.5; table 11.2).

Conclusions and Recommendations

1. The word *mode* in discussion of evolution should be used to refer to the topology of the tree of life. Using this definition and putting aside episodes of anastomosis or reticulate evolution (e.g., Lawrence and Ochman 2002; Vriesendorp

TABLE 11.2 **A classification of evolutionary patterns**

Mode	Model
Anagenesis	Stasis
	Random walk
	Directional
Cladogenesis	Allpataric-Vicariance
	Allopatric-Dispersal
	Sympatric

and Bakker 2005; Mallet 2007), only two modes of evolutionary change are currently recognized: anagenesis (a.k.a. phyletic evolution) and cladogenesis (a.k.a. speciation).

2. The three anagenetic modes of Hunt (2010, etc.; "microevolutionary modes" of Roopnarine 2001)—stasis, random walk, and directional change (figure 11.2)—are more usefully referred to as "models" of anagenesis (*sensu* Hunt 2012, 352). Importantly for the theme of this volume, at a higher hierarchical level, these three models can also be applied to patterns of change (or lack of change)—"trends"—across a branched clade (cf. Rasskin-Gutman and Esteve-Altava 2008; figure 11.4).

3. In the interest of simplification of taxonomy and terminology, what have previously been referred to as different "modes" of speciation might also be better referred to as "models" (contra, e.g., Allmon and Smith 2011; table 11.2). Patterns consistent with these models are, in turn, caused by a wide variety of "mechanisms" of speciation (e.g., Coyne and Orr 2004).

4. Evolutionary tempo can refer to rates at different hierarchical levels: rate of phenotypic or genotypic change within a single unbranching lineage (i.e., anagenetic trends; as noted by Hunt 2012, such rates cannot be specified independent of the model of anagenetic change; see point number 2, above); rate of speciation (in either of its two meanings, BSI and TFS); or rate of phenotypic or genotypic change across (at the level of) the clade, as a macroevolutionary trend (figures 11.4 and 11.5).

5. The term *tempo and mode* in evolutionary literature should be reinstated as a technical expression for the varied relationship between evolutionary rate and topology, and other uses should be discouraged. The "relative frequency" of tempos and modes of evolution (*sensu* Gould 2002, 772ff., 854ff.) can be assessed accurately only by considering the full possible range of combinations of tempo, mode, and model of change (figure 11.5; table 11.2).

Acknowledgments

I am grateful to Emanuele Serrelli and Ilya Tëmkin for the invitation to participate in this volume, to Andrielle Swaby and Alana McGillis for assistance with figures, and to Brendan Anderson, Dana Friend, Dana Geary, Amy McCune, Rob Ross, Bruce Wilkinson, the editors, and especially Gene Hunt, for discussion and/or comments on previous drafts of the manuscript.

References

Adachi, Jun, and Masami Hasegawa. 1996. "Tempo and Mode of Synonymous Substitutions in Mitochondrial DNA of Primates." *Molecular Biology and Evolution* 13 (1): 200–208.

Alberch, Pere, Stephen J. Gould, George F. Oster, and David B. Wake. 1979. "Size and Shape in Ontogeny and Phylogeny." *Paleobiology* 5 (3): 296–317.

Allmon, Warren D. 2013. "Species, Speciation, and Paleontology up to the Modern Synthesis: Persistent Themes and Unanswered Questions." *Palaeontology* 56 (6): 1199–1223.

———. 2016. "Studying Species in the Fossil Record: A Review and Recommendations for a More Unified Approach." In *Species and Speciation in the Fossil Record*, edited by Warren D. Allmon and Margaret M. Yacobucci. Chicago: University of Chicago Press, in press.

Allmon, Warren D., and Ursula E. Smith. 2011. "What, If Anything, Can We Learn from the Fossil Record about Speciation in Marine Gastropods? Biological and Geological Considerations." *American Malacological Bulletin* 29 (1): 247–76.

Allmon, Warren D., and Margaret M. Yacobucci, eds. 2016. *Species and Speciation in the Fossil Record*. Chicago: University of Chicago Press, in press.

Averbeck, Karin T., and Thomas H. Eickbush. 2005. "Monitoring the Mode and Tempo of Concerted Evolution in the *Drosophila* Melanogaster rDNA Locus." *Genetics* 171: 1837–46.

Avise, John C. 2000. *Phylogeography: The History and Formation of Species*. Cambridge, MA: Harvard University Press.

Baker, Allan J., Leon J. Huynen, Oliver Haddrath, Craig D. Millar, and David M. Lambert. 2005. "Reconstructing the Tempo and Mode of Evolution in an Extinct Clade of Birds with Ancient DNA: The Giant Moas of New Zealand." *Proceedings of the National Academy of Sciences of the United States of America* 102 (23): 8257–62.

Barton, Nicholas H., Derek E. G. Briggs, Jonathan A. Eisen, David B. Goldstein, and Nipam H. Patel. 2009. *Evolution*. Cold Spring Harbor, NY: Cold Spring Harbor Laboratory Press.

Bernardi, Giacomo, and Jennifer Lape. 2005. "Tempo and Mode of Speciation in the Baja California Disjunct Fish Species *Anistremus davidsonii*." *Molecular Ecology* 14: 4085–96.

Bleiweiss, Robert. 1998. "Tempo and Mode of Hummingbird Evolution." *Biological Journal of the Linnean Society* 65 (1): 63–76.

Blomberg, Simon P., and Theodore Garland Jr. 2002. "Tempo and Mode in Evolution: Phylogenetic Inertia, Adaptation and Comparative Methods." *Journal of Evolutionary Biology* 15: 899–910.

Bookstein, Fred L. 1987. "Random Walk and the Existence of Evolutionary Rates." *Paleobiology* 13: 446–64.

———. 1988. "Random-Walk and the Biometrics of Morphological Characters." *Evolutionary Biology* 23: 369–98.

Brown, Wesley M., Ellen M. Prager, Alice Wang, and Allan C. Wilson. 1982. "Mitochondrial DNA Sequences of Primates: Tempo and Mode of Evolution." *Journal of Molecular Ecology* 18: 225–39.

Cheetham, Alan H. 1986. "Tempo of Evolution in a Neogene Bryozoan: Rates of Morphological Change within and across Species Boundaries." *Paleobiology* 12: 190–202.

Collins, Timothy M. 1996. "Molecular Comparisons of Transisthmian Species Pairs: Rates and Patterns of Evolution." In *Evolution and Environment in Tropical America*, edited by Jeremy B. C. Jackson, Ann F. Budd, and Anthony G. Coates, 303–34. Chicago: University of Chicago Press.

Coolon, Joseph D., C. Joel McManus, Kraig R. Stevenson, Brenton R. Graveley, and Patricia J. Wittkopp. 2014. "Tempo and Mode of Regulatory Evolution in *Drosophila*." *Genome Research* 24: 797–808.

Cooper, Natalie, and Andy Purvis. 2010. "Body Size Evolution in Mammals: Complexity in Tempo and Mode." *American Naturalist* 175 (6): 727–38.

Coyne, Jerry A., and H. Allen Orr. 2004. *Speciation*. Sunderland, MA: Sinauer Associates.

Curnoe, Darren, Alan Thorne, and J. A. Coate. 2006. "Timing and Tempo of Primate Speciation." *Journal of Evolutionary Biology* 19: 59–65.

Dacks, Joel B., and Mark C. Field. 2007. "Evolution of the Eukaryotic Membrane-Trafficking System: Origin, Tempo and Mode." *Journal of Cell Science* 120: 2977–85.

Darwin, Charles R. 1859. *On the Origin of Species by Means of Natural Selection, or the Preservation of Favoured Races in the Struggle for Life*. London: John Murray.

Day, Julia J., James A. Cotton, and Timothy G. Barraclough. 2008. "Tempo and Mode of Diversification of Lake Tanganyika Cichlid Fishes." *PLoS ONE* 3 (3): e1730.

Delmotte, François, Claude Rispe, Jörg Schaber, Francisco J. Silva, and Andrés Moya. 2006. "Tempo and Mode of Early Gene Loss in Endosymbiotic Bacteria from Insects." *BMC Evolutionary Biology* 6: 56. doi:10.1186/1471-2148-6-56.

Doolittle, W. Ford, and James R. Brown. 1994. "Tempo, Mode, the Progenote, and the Universal Root." *Proceedings of the National Academy of Sciences of the United States of America* 91 (15): 6721–28.

Edwards, Tyron, ed. 1908. *A Dictionary of Thoughts: Being a Cyclopedia of La-conic Quotations from the Best Authors of the World, Both Ancient and Mod-ern.* Detroit: F. B. Dickerson.

Eldredge, Niles. 1985. *Time Frames: The Re-thinking of Darwinian Evolution and the Theory of Punctuated Equilibria.* New York: Simon and Schuster.

———. 2013. "Stephen J. Gould in the 1960s and 1970s, and the Origin of 'Punc-tuated Equilibria.'" In *Stephen J. Gould: The Scientific Legacy,* edited by Gian Antonio Danieli, Alessandro Minelli, and Telmo Pievani, 3–19. Milan: Springer-Verlag Italia.

———. 2015. *Eternal Ephemera. Adaptation and the Origin of Species from the Nineteenth Century through Punctuated Equilibrium and Beyond.* New York: Columbia University Press.

Eldredge, Niles, and Stephen J. Gould. 1972. "Punctuated Equilibria: An Alterna-tive to Phyletic Gradualism." In *Models in Paleobiology,* edited by Thomas J. M. Schopf, 82–115. San Francisco: Freeman, Cooper.

Erwin, Douglas H., and Robert L. Anstey. 1995. "Speciation in the Fossil Record." In *New Approaches to Speciation in the Fossil Record,* edited by Douglas H. Erwin and Robert L. Anstey, 11–38. New York: Columbia University Press.

Esselstyn, Jacob A., Robert M. Timm, and Rafe M. Brown. 2009. "Do Geological or Climatic Processes Drive Speciation in Dynamic Archipelagos? The Tempo and Mode of Diversification in Southeast Asian Shrews." *Evolution* 63 (10): 2595–2610.

Feyereisen, René. 2011. "Arthropod CYPomes Illustrate the Tempo and Mode in P450 Evolution." *Biochimica et Biophysica Acta (BBA)-Proteins and Pro-teomics* 1814 (1): 19–28.

Fitch, Walter M. 2000. "Homology: A Personal View on Some of the Problems." *Trends in Genetics* 16 (5): 227–31.

Futuyma, Douglas J. 2009. *Evolution.* 2nd ed. Sunderland, MA: Sinauer Associates.

Gante, Hugo F., Joana Micael, Francisco J. Oliva-Paterna, Ignatio Doadrio, Thomas E. Dowling, and Maria J. Alves. 2009. "Diversification within Glacial Refugia: Tempo and Mode of Evolution of the Polytypic Fish *Barbus sclateri.*" *Molecular Ecology* 18 (15): 3240–55.

Geary, Dana H. 1987. "Evolutionary Tempo and Mode in a Sequence of the Up-per Cretaceous Bivalve *Pleuriocardia.*" *Paleobiology* 13: 140–51.

———. 1990. "Patterns of Evolutionary Tempo and Mode in the Radiation of Melanopsid Gastropods." *Paleobiology* 16: 492–511.

———. 1995. "Investigating Species-Level Transitions in the Fossil Record: The Importance of Geologically Gradual Change." In *New Approaches to Specia-tion in the Fossil Record,* edited by Douglas H. Erwin and Robert A. Anstey, 67–86. New York: Columbia University Press.

Geyssant, J. R. 1988. "Diversity in Mode and Tempo of Evolution within One Tithonian Ammonite Family, the Simoceratids." In *Cephalopods Present and Past: Proceedings of the 2nd International Cephalopod Symposium,* edited by Jost Wiedmann and Jurgen Kullmann, 79–88. Stuttgart: Schweizerbart'sche Verlagsbuchhandlung.

Gingerich, Philip D. 1993. "Quantification and Comparison of Evolutionary Rates." *American Journal of Science* 293-A: 453–78.

———. 2009. "Rates of Evolution." *Annual Review of Ecology, Evolution, and Systematics* 40: 657–75.

Goldberg, Emma E., and Boris Igić. 2012. "Tempo and Mode in Plant Breeding System Evolution." *Evolution* 66 (12): 3701–9.

Gordon, Kacy L., and Ilya Ruvinsky. 2012. "Tempo and Mode in Evolution of Transcriptional Regulation." *PLoS Genetics* 8 (1): e1002432.

Gould, Stephen J. 1977. *Ontogeny and Phylogeny*. Cambridge, MA: Harvard University Press.

———. 1982. "The Meaning of Punctuated Equilibrium and Its Role in Validating a Hierarchical Approach to Macroevolution." In *Perspectives on Evolution*, edited by R. Milkman, 83–104. Sunderland, MA: Sinauer Associates.

———. 1988. "The Heart of Terminology." *Natural History* 97 (2): 24–31.

———. 1994. "Tempo and Mode in the Macroevolutionary Reconstruction of Darwinism." *Proceedings of the National Academy of Sciences of the United States of America* 91 (15): 6764–71.

———. 2000. "Of Coiled Oysters and Big Brains: How to Rescue the Terminology of Heterochrony, Now Gone Astray." *Evolution and Development* 2: 241–48.

———. 2002. *The Structure of Evolutionary Theory*. Cambridge, MA: Harvard University Press.

Gould, Stephen J., and Niles Eldredge. 1977. "Punctuated Equilibria: The Tempo and Mode of Evolution Reconsidered." *Paleobiology* 3: 115–51.

Gould, Stephen J., and Elisabeth S. Vrba. 1982. "Exaptation—A Missing Term in the Science of Form." *Paleobiology* 8 (1): 4–15.

Haldane, John B. S. 1949. "Suggestions as to the Quantitative Measurement of Rates of Evolution." *Evolution* 3: 51–56.

Harmon, Luke J., James A. Schulte II, Allan Larson, and Jonathan B. Losos. 2003. "Tempo and Mode of Evolutionary Radiation in Iguanian Lizards." *Science* 301: 961–64.

Hirst, Russel. 2003. "Scientific Jargon, Good and Bad." *Journal of Technical Writing and Communication* 33 (3): 201–29.

Hopkins, Melanie J., and Scott Lidgard. 2012. "Evolutionary Mode Routinely Varies among Morphological Traits within Fossil Species Lineages." *Proceedings of the National Academy of Sciences of the United States of America* 109: 20520–25.

Hunt, Gene. 2006. "Fitting and Comparing Models of Phyletic Evolution: Random Walks and Beyond." *Paleobiology* 32: 578–601.

———. 2007. "The Relative Importance of Directional Change, Random Walks, and Stasis in the Evolution of Fossil Lineages." *Proceedings of the National Academy of Sciences of the United States of America* 104: 18404–8.

———. 2008a. "Evolutionary Patterns within Fossil Lineages: Model-Based Assessment of Modes, Rates, Punctuations and Process." *From Evolution to Geobiology: Research Questions Driving Paleontology at the Start of a New Century. The Paleontological Society Papers*, edited by Patricia H. Kelley and Richard K. Bambach, 14: 117–31.

———. 2008b. "Gradual or Pulsed Evolution: When Should Punctuational Explanations Be Preferred?" *Paleobiology* 34: 360–77.

———. 2010. "Evolution in Fossil Lineages: Paleontology and the Origin of Species." *American Naturalist* 176: S61–S76.

———. 2012. "Measuring Rates of Phenotypic Evolution and the Inseparability of Tempo and Mode." *Paleobiology* 38: 351–73.

———. 2013. "Testing the Link between Phenotypic Evolution and Speciation: An Integrated Palaeontological and Phylogenetic Analysis." *Methods in Ecology and Evolution* 4: 714–23.

Hunt, Gene, and Matthew T. Carano. 2010. "Models and Methods for Analyzing Phenotypic Evolution in Lineages and Clades." *Quantitative Methods in Paleobiology: The Paleontological Society Papers*, edited by John Alroy and Gene Hunt, 16: 245–69.

Hunt, Gene, and Daniel L. Rabosky. 2014. "Phenotypic Evolution in Fossil Species: Pattern and Process." *Annual Review of Earth and Planetary Sciences* 42: 421–41.

Jackson, Jeremy B. C., and Alan H. Cheetham. 1994. "Phylogeny Reconstruction and the Tempo of Speciation in Cheilostome Bryozoa." *Paleobiology* 20: 407–23.

———. 1999. "Tempo and Mode of Speciation in the Sea." *Trends in Ecology and Evolution* 14: 72–77.

Jackson, Julius H. 1999. "Terminologies for Gene and Protein Similarity." In *Technical Reports & Reviews*, Michigan State University, Department of Microbiology, Report no. TR 99-01. Accessed November 10, 2015. https://www.msu.edu /~jhjacksn/Reports/similarity.htm.

Kinnison, Michael T., and Andrew P. Hendry. 2004. "From Macro- to Micro-Evolution: Tempo and Mode in Salmonid Evolution." In *Evolution Illuminated: Salmon and Their Relatives*, edited by Andrew P. Hendry and Stephen C. Stearns, 209–31. Oxford: Oxford University Press.

Klaas, Benjamin A., Kirk A. Maloney, and Brent J. Danielson. 2000. "The Tempo and Mode of Gopher Mound Production in a Tallgrass Prairie Remnant." *Ecography* 23: 246–56.

Kohn, Alan J. 1990. "Tempo and Mode of Evolution in Conidae." *Malacologia* 32 (1): 55–67.

Koonin, Eugene V. 2001. "An Apology for Orthologs—or Brave New Memes." *Genome Biology* 2 (4): 1–2.

Lawrence, Jeffrey G., and Howard Ochman. 2002. "Reconciling the Many Faces of Lateral Gene Transfer." *Trends in Microbiology* 10: 1–4.

Lidgard, Scott, and Melanie Hopkins. 2015. "Stasis." In *Oxford Bibliographies in Evolutionary Biology*, edited by Jonathan B. Losos. New York: Oxford University Press. Last modified August 31, 2015. http://www.oxfordbibliographies .com/view/document/obo-9780199941728/obo-9780199941728-0067.xml.

Losos, Jonathan B., and D. Luke Mahler. 2010. "Adaptive Radiation: The Interaction of Ecological Opportunity, Adaptation, and Speciation." In *Evolution since Darwin: The First 150 Years*, edited by Michael A. Bell, Douglas J. Futuyma, Walter F. Eanes, and Jeffrey S. Levinton, 381–420. Sunderland, MA: Sinauer Associates.

Mallet, James. 2007. "Hybrid Speciation." *Nature* 446: 279–83.

Marko, Peter B. 2002. "Fossil Calibration of Molecular Clocks and the Divergence Times of Geminate Species Pairs Separated by the Isthmus of Panama." *Molecular Biology and Evolution* 19: 2005–21.

McCune, Amy R. 2004. "Diversity and Speciation of Semionotid Fishes in Meso-zoic Rift Lakes." In *Adaptive Speciation*, edited by Ulf Dieckmann, Michael Doebeli, Johan A. J. Metz, and Diethard Tautz, 362–79. Cambridge: Cambridge University Press.

McHenry, Henry M. 1994. "Tempo and Mode in Human Evolution." *Proceedings of the National Academy of Sciences of the United States of America* 91: 6780–86.

Meloro, Carlo, and Pasquale Raia. 2010. "Cats and Dogs Down the Tree: The Tempo and Mode of Evolution in the Lower Carnassial of Fossil and Living Carnivore." *Evolutionary Biology* 37: 177–86.

Mendelson, Tamra C., and Kerry L. Shaw. 2005. "Sexual Behaviour: Rapid Speciation in an Arthropod." *Nature* 433: 375–76.

Nagorsen, David W., and Andrea Cardini. 2009. "Tempo and Mode of Evolutionary Divergence in Modern and Holocene Vancouver Island Marmots (*Marmota vancouverensis*) (Mammalia, Rodentia)." *Journal of Zoological Systematics and Evolutionary Research* 47 (3): 258–67.

Nee, Sean, Arne Ø. Mooers, and Paul H. Harvey. 1992. "Tempo and Mode of Evolution Revealed by Molecular Phylogenies." *Proceedings of the National Academy of Sciences of the United States of America* 89: 8322–26.

Oliver, Matthew J., Dmitri Petrov, David Ackerly, Paul Falkowski, and Oscar M. Schofield. 2007. "The Mode and Tempo of Genome Size Evolution in Eukaryotes." *Genome Research* 17: 594–601.

Olsen, Paul E., Dennis V. Kent, and Jessica H. Whiteside. 2011. "Implications of the Newark Supergroup-Based Astrochronology and Geomagnetic Polarity Time Scale (Newark-APTS) for the Tempo and Mode of Early Diversification of the Dinosauria." *Earth and Environmental Science Transactions of the Royal Society of Edinburgh* 101: 201–29.

Osborn, Henry F. 1902. "The Law of Adaptive Radiation." *American Naturalist* 36: 353–63.

———. 1910. "Paleontologic Evidences of Adaptive Radiation." *Popular Science Monthly* 77: 77–81.

Pachut, Joseph F., and Robert L. Anstey. 2009. "Inferring Evolutionary Modes in a Fossil Lineage (Bryozoa: *Peronopora*) from the Middle and Late Ordovician." *Paleobiology* 35 (2): 209–30.

Padian, Kevin. 1994. "What Were the Tempo and Mode of Evolutionary Change in the Late Triassic to Middle Jurassic?" In *In the Shadow of the Dinosaurs: Early Mesozoic Tetrapods*, edited by Nicholas C. Fraser and Hans-Dieter Sues, 401–7. Cambridge: Cambridge University Press.

Pérez-Losada, Marcos, Margaret Harp, Jens T. Høeg, Yair Achituv, Diana Jones, Hiromi Watanabe, and Keith A. Crandall. 2008. "The Tempo and Mode of Barnacle Evolution." *Molecular Phylogenetics and Evolution* 46: 328–46.

Peterson, Kevin J., Mark A. McPeek, and David A. D. Evans. 2005. "Tempo and Mode in Early Animal Evolution: Inferences from Rocks, Hox, and Molecular Clocks." *Paleobiology* 31 (2, Suppl.): 36–55.

Pettengill, James B., and David A. Moeller. 2012. "Tempo and Mode of Mating System Evolution between Incipient *Clarkia* Species." *Evolution* 66 (4): 1210–25.

Pievani, Telmo. 2015. "How to Rethink Evolutionary Theory: A Plurality of Evolutionary Patterns." *Evolutionary Biology.* doi:10.1007/s11692-015-9338-3.

Pievani, Telmo, and Emanuele Serrelli. 2011. "Exaptation in Human Evolution: How to Test Adaptive vs Exaptive Evolutionary Hypotheses." *Journal of Anthropological Sciences* 89: 9–23.

Rasskin-Gutman, Diego, and Borja Esteve-Altava. 2008. "The Multiple Directions of Evolutionary Change." *BioEssays* 30: 521–25.

Raup, David M. 1977a. "Probabilisitc Models in Evolutionary Paleobiology." *American Scientist* 65: 50–57.

———. 1977b. "Stochastic Models in Evolutionary Paleobiology." In *Patterns of Evolution as Illustrated by the Fossil Record*, edited by Anthony Hallam, 59–78. Amsterdam: Elsevier.

———. 1978. "Cohort Analysis of Genetic Survivorship." *Paleobiology* 4: 1–16.

Raup, David M., Stephen J. Gould, Thomas J. M. Schopf, and David S. Simberloff. 1973. "Stochastic Models of Phylogeny and the Evolution of Diversity." *Journal of Geology* 81 (5): 525–42.

Reeck, Gerald R., Christoph de Haën, David C. Teller, Russell F. Doolittle, Walter F. Fitch, Richard E. Dickerson, Pierre Chambon, Andrew O. McLachlan, Emanuel Margoliash, Thomas H. Jukes, and Emile Zuckerkandl. 1987. " 'Homology' in Proteins and Nucleic Acids: A Terminology Muddle and a Way Out of It." *Cell* 50: 667.

Rensch, Bernhard. 1959. *Evolution above the Species Level.* New York: Columbia University Press.

Romano, Camila M., Rodrigo F. Ramalho, and Paolo M. de A. Zanotto. 2006. "Tempo and Mode of ERV-K Evolution in Human and Chimpanzee Genomes." *Archives of Virology* 151 (11): 2215–28.

Roopnarine, Peter D. 2001. "The Description and Classification of Evolutionary Mode: A Computational Approach." *Paleobiology* 27 (3): 446–65.

———. 2003. "Analysis of Rates of Morphologic Evolution." *Annual Review of Ecology, Evolution, and Systematics* 34: 605–32.

Roopnarine, Peter D., Gabe Byars, and Paul Fitzgerald. 1999. "Anagenetic Evolution, Stratophenetic Patterns, and Random Walk Models." *Paleobiology* 25: 41–57.

Rosenzweig, Michael L. 1997. "Tempo and Mode of Speciation." *Science* 277: 1622–23.

Schopf, J. William. 1994. "Disparate Rates, Differing Fates: Tempo and Mode of Evolution Changed from the Precambrian to the Phanerozoic." *Proceedings of the National Academy of Sciences of the United States of America* 91: 6735–42.

Schürch, Anita C., Kristin Kremer, Albert Kiers, Olaf Daviena, Martin J. Boeree, Roland J. Siezen, Noel H. Smith, and Dick van Soolingen. 2010. "The Tempo and Mode of Molecular Evolution of *Mycobacterium* Tuberculosis at Patient-to-Patient Scale." *Infection, Genetics and Evolution* 10 (1): 108–14.

Sepkoski, David. 2012. *Rereading the Fossil Record: The Growth of Paleobiology as an Evolutionary Discipline.* Chicago: University of Chicago Press.

Simpson, George G. 1944. *Tempo and Mode in Evolution.* New York: Columbia University Press.

Smith, Craig R. 1994. "Tempo and Mode in Deep-Sea Benthic Ecology: Punctuated Equilibrium Revisited." *Palaios* 9 (1): 3–13.

Stanley, Steven M. 1979. *Macroevolution:. Pattern and Process*. San Francisco: W. H. Freeman.

———. 1985. "Rates of Evolution." *Paleobiology* 11 (1): 13–26.

Tuinen, Marcel van, Thomas A. Stidham, and Elizabeth A. Hadly. 2006. "Tempo and Mode of Modern Bird Evolution Observed with Large-Scale Taxonomic Sampling." *Historical Biology* 18 (2): 205–21.

Vanderpoorten, Alain, and Anne-Laure Jacquemart. 2004. "Evolutionary Mode, Tempo, and Phylogenetic Association of Continuous Morphological Traits in the Aquatic Moss Genus *Amblystegium*." *Journal of Evolutionary Biology* 17: 279–87.

Vriesendorp, Bastiaantje, and Freek T. Bakker. 2005. "Reconstructing Patterns of Reticulate Evolution in Angiosperms: What Can We Do?" *Taxon* 54 (3): 593–604.

Webster's New Collegiate Dictionary. 1974. Springfield, MA: G. C. Merriam.

Weir, Jason T., and Dolph Schluter. 2007. "The Latitudinal Gradient in Recent Speciation and Extinction Rates of Birds and Mammals." *Science* 315: 1574–76.

Wilkinson, Bruce H. 2015. "Precipitation as Meteoric Sediment and Scaling Laws of Bedrock Incision: Assessing the Sadler Effect." *Journal of Geology* 123. (Electronically published April 16, 2015.)

Woese, Carl R., Erko Stackebrandt, and W. Ludwig. 1985. "What Are Mycoplasmas: The Relationship of Tempo and Mode in Bacterial Evolution." *Journal of Molecular Evolution* 21 (4): 305–16.

CHAPTER 12

Niche Conservatism, Tracking, and Ecological Stasis

A Hierarchical Perspective

Carlton E. Brett, Andrew Zaffos, and Arnold I. Miller

Introduction

Ecological stasis occurs when interactions between taxa and environments are unchanged for some period of geologic time. One way to achieve stasis is if a preponderance of taxa in a community maintain stable environmental preferences throughout their evolutionary histories, a phenomenon commonly labeled *niche conservatism* (Wiens and Graham 2005; Pearman et al. 2008; Wiens et al. 2010). In contrast, if taxa regularly alter their environmental preferences, consequently changing their distributions among environments, then it seems unlikely that ecological stasis could be sustained. Genuine change in environmental and ecological preferences has been termed "niche evolution" (e.g., Stigall 2012, 2015); however, ecological entities, such as niches of species, are not part of the genealogical hierarchy (see Eldredge 2003, 2008) and technically do not evolve. Therefore, we use the term *niche modification* in place of "niche evolution" throughout this chapter to indicate temporal changes in niche parameters and environmental tolerances of species through time.

Despite the importance of niche conservatism to the issue of ecological stasis, its generality in the fossil record remains unclear. This is partly a consequence of multiple, competing definitions of the concept of ecological niche (see Soberón 2007) and different ways to measure and test for conservatism (cf. Holland and Zaffos 2011; Stigall 2012, 2015; Hopkins et al.

2014). For example, at what level of the taxonomic hierarchy is niche conservatism most likely to be expressed (Hadly et al. 2009)? Ecological evolutionary (EE) subunits or "faunas" were recognized as blocks of relative similarity or "dynasties" in which up to 80 percent of species persist throughout intervals of up to several million years, wherein biocoenoses ("communities" in a broad sense) and faunal gradients appear to persist with similar structure (Brett and Baird 1995; Brett et al. 1996, 2007, 2009). These are subdivisions of more broadly defined EE units, even more prolonged intervals (tens of millions of years) in which onshore-offshore gradients and benthic assemblages of genera and/or families of organisms remain relatively consistent through time (Boucot 1975, 1983; Sheehan 1985). The existence of EE units and subunits (Boucot 1975, 1983; Sheehan 1985; Brett and Baird 1995) implies niche conservatism at the level of genera, and possibly families, for millions to tens of millions of years. In contrast, the pattern of onshore-to-offshore drift of marine orders and families over tens of millions of years suggests that niches are not fixed over the long term at higher taxonomic levels (Jablonski et al. 1983; Sepkoski and Miller 1985; Bottjer and Jablonski 1988; Jablonski and Bottjer 1990). This is perhaps not surprising given the variability of life habits exhibited by the species- and genus-level constituents of families and, especially, orders. Similarly, at what spatiotemporal scales is niche conservatism expected? Holland and Zaffos (2011) found sustained niche conservatism throughout eight depositional sequences (approximately ten million years total duration) of Late Ordovician age in the Cincinnati Arch region, whereas others have reported a more complex pattern of intrasequence (approximately one million years duration) conservatism as well as niche modification *within* those same strata (Dudei and Stigall 2010; Walls and Stigall 2011; Stigall 2015).

Here, using the data-rich Middle Devonian Hamilton Group as a "natural laboratory," we investigate the extent to which niche conservatism is maintained stratigraphically in a hierarchical context. Importantly, we define explicitly our conception of ecological niche, niche conservatism, and a stratigraphic hierarchy. The hierarchy of Hamilton stratigraphic units is a temporal one with smaller scale sedimentary cycles nested in larger ones; thus, it does not specifically relate to a static ecological hierarchy (organisms, avatars [populations], biocoenoses, ecological gradients, regional biotas) but rather permits examination of faunal patterns at multiple time scales from thousands to millions of years. Following the precedent set by Holland and Zaffos (2011), we characterize the niches of taxa from the upper Hamilton as unimodal response curves along an inferred water depth

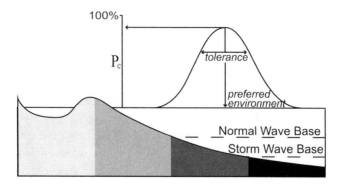

FIGURE 12.1 A schematic representation of a response curve along an onshore-offshore gradient. Gradient theory (Whittaker 1967) predicts that the abundance (probability of collection, pc) of a genus will rise and fall symmetrically along an environmental gradient. These Gaussian curves are parameterized with the three parameters of peak abundance, environmental tolerance, and preferred environment, which, respectively, describe the height, width, and central position of the curve. Modified from Holland and Zaffos (2011).

gradient. These curves describe niches with three parameters: preferred environment, environmental tolerance, and peak abundance (figure 12.1). We define conservatism as the strength of the correlation coefficient of the preferred environment and peak abundance parameters between different temporal intervals of the Hamilton. We do not assess niche conservatism within individual lineages but rather examine its prevalence for assemblages as a whole. Lastly, and most central to our theme of hierarchy and stasis, we make comparisons among patterns exhibited at different scales of stratigraphic resolution to understand the importance, if any, of spatiotemporal scale.

The Hamilton Group of western and central New York State is subdivided into formations of several decameter scale that correspond roughly to intervals of several hundred thousand years. In turn, formations are subdivided into thinner, lithologically distinctive intervals termed "members" that probably represent approximately one hundred thousand to four hundred thousand years (see more detailed discussion below). We compared niche parameters from member to member and from formation to formation, as well as individual members versus the Hamilton Group, as a whole, and individual formations versus the entire Hamilton.

We find that niche conservatism is ubiquitous in the Hamilton Group, as evidenced by strong correlation of niche parameters from formation to formation and formation to group. Niche conservatism is less pronounced at a finer stratigraphic scale, among stratigraphic members, and in some cases, patterns appear more consistent with niche modification. At face value, these results therefore suggest a scale-dependent pattern of niche conservatism. We discuss potential biological, geological, and methodological drivers of this member-versus-formation-level dichotomy and conclude with a discussion of the evolutionary consequences of niche conservatism.

Geologic Setting

The Middle Devonian (latest Eifelian-Givetian) Hamilton Group of western and central New York figures prominently in studies of evolutionary-ecological stability. The Hamilton Group and overlying Tully, which are commonly classified as an EE subunit (e.g., Brett and Baird 1995), represent the bulk of the lower and middle Givetian, with the much shorter upper Givetian recorded in the overlying Geneseo Formation. Although there is some disagreement on the duration of the Givetian, most recent estimates place it around 5 to 6 million years. Based on cyclostratigraphy, Ellwood et al. (2011) suggest a duration of 5.6 million years, although some recent work would make it markedly shorter, perhaps 4.3 million years (DeVleeshouer et al. 2014). Thus the duration of the Hamilton Group is probably somewhere between 3 and 5 million years.

The Hamilton Group provides an excellent framework for studies in a hierarchical perspective for several reasons. First, it has a very well developed stratigraphic framework. Many individual cycles and component fossil beds have been traced for more than three hundred kilometers along the outcrop belt throughout western and central New York State and show substantial internal facies change such that they provide time-averaged samples of facies (environmental) transects from more calcareous mudrock facies in the west into dark, organic-rich clay shales near the basin center of the Finger Lakes region, and into coarser siltstone and sandstone-rich facies of central New York (Brett et al. 1990, 1996, 2011).

Second, these beds have been placed in a sequence-stratigraphic framework, with a hierarchy of nested cycles that have been correlated more than three hundred kilometers across New York State and into Pennsylvania and the midcontinent (Brett and Bartholomew 2007; Brett et al. 2011). These include five third-order depositional sequences, with approximately one-million-year time spans, which correspond approximately to formations as traditionally defined in the Hamilton Group of New York (Cooper

1930, 1933); each ranges from tens of meters to more than one hundred meters thick. Hamilton third-order sequences are correlated more than three hundred kilometers across New York State and into Pennsylvania and the midcontinent (Brett and Baird 1996; Brett and Bartholomew 2007; Brett et al. 2011). Moreover, these larger sequences are now tentatively linked with global sea level cycles (Brett et al. 2011). Each sequence commences at the base of a thin, compact carbonate, which records sea level lowstand to transgressive conditions. This package is, in turn, overlain by a thick shale/mudstone interval representing sea level highstand to falling phases of cycles. The relatively complex internal structure of Hamilton formations is composed of three or four member-scale subsequences ("fourth-order" cycles) that each possesses architectures similar to, but thinner than, the larger sequences; these are estimated to record timespans of approximately four hundred thousand years. Finally, most members have been subdivided into 0.5–3.0 meter-scale submembers or "beds" representing fifth- or sixth-order cycles. Assuming a Milankovitch cycle-driving mechanism, these probably reflect approximately twenty to forty thousand year precession or obliquity driven oscillations in climate and/or sea level (Ellwood et al. 2011). As with Hamilton formations, the member-scale and bed-scale units have all been traced very widely (typically more than two hundred kilometers) in the New York outcrop belt (Brett et al. 2011). Units at all scales display lateral facies variation, thickening and becoming increasingly siliciclastic dominated to the east.

Third, the Hamilton strata yield highly diverse (greater than three hundred species) and well-documented faunas in a variety of biofacies that have been characterized in terms of sediment-turbidity, oxygenation, and water depth (Brett and Baird 1995; Brett et al. 1996; Brett et al. 2007b). These faunas have been described as exhibiting long-term biological stasis in terms of low extinction rates, low origination rates, static morphology, abundance structure, and habitat associations (Brett and Baird 1995; Ivany et al. 2009), making them an excellent candidate for niche conservatism studies.

The third-order sequences of the Hamilton Group correspond to the Union Springs (lower Marcellus), Oatka Creek (upper Marcellus), Skaneateles, Ludlowville, and Moscow formations (Brett et al. 2011). The Moscow Formations have been subdivided into two sequences, but whether these should be interpreted as groups of fourth-order sequences composing a single third-order sequence or as two separate third-order sequences is ambiguous (Brett et al. 2011); here, we use the former interpretation. Each

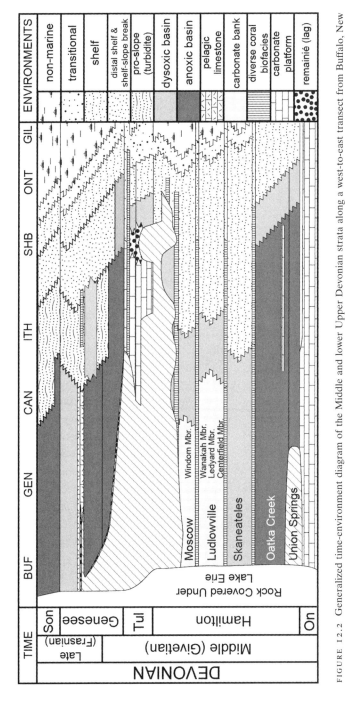

FIGURE 12.2 Generalized time-environment diagram of the Middle and lower Upper Devonian strata along a west-to-east transect from Buffalo, New York, to Gilboa, New York. Two environmental gradients are spatially collinear with longitude in these strata. First, there is a carbonate-siliciclastic gradient typified by siliciclastic shelf facies in the east and carbonate bank facies in the west. Second, there is a nonlinear water depth gradient characterized by deeper-water dysoxic and anoxic basin deposits in central New York. Key to environment patterns on right. Modified from Zambito et al. (2012).

Hamilton formation is a relatively conformable stratigraphic unit, consisting of a sharp erosive base (sequence boundary); a transgressive or deepening portion, commonly with thin, widespread limestones; and a shale-rich highstand to regressive unit, and is laterally divisible into three broad zones: an eastern shelf dominated by sandstones, siltstones, and other terrigenous influx from the second tectophase of the Acadian Orogeny (Ettensohn 2008), a western dominated by thin concretionary limestones and calcareous ramp shale mudstones, and an intervening basin center comprised of dysoxic-anoxic shale/mudstone facies (Brett et al. 2011; Zambito et al. 2012; figure 12.2).

Data

Regional occurrences of upper Hamilton genera through the study interval were compiled in a synoptic data set that includes previously published data (Cooper 1930; Baird and Brett 1983; Bezusko 2001; Bonelli et al. 2006; Ivany et al. 2009; Boyer and Droser 2009), unpublished data collected by C. E. Brett and utilized in the study of Ivany et al. (2009), and a new set of samples collected for this study along a transect in New York from Buffalo to Syracuse. The lower Hamilton Group (i.e., the Oatka Creek Formation) is not included in the present data set because of its low diversity and poor preservational quality in western New York. The newly collected samples span fifteen localities with approximately two to five samples per locality. In aggregate, the final synoptic data set contains 183 taxa, mostly at the genus level, represented by a total of 449 samples across 110 localities with 7,215 fossil occurrences. After removing depauperate samples (minimum richness = 5) and rare taxa (minimum occurrences = 5), the final data set was reduced to 134 taxonomic groups and 367 samples. Identification of taxa was to genus level; however, it should be noted that a majority of Hamilton Group genera are known only from single species, so that this is a reasonable proxy for species.

Because the final matrix includes data from a variety of studies, each with different sampling regimens, all abundance data are converted to presence-absence, and the high-resolution Boyer and Droser (2009) centimeter-scale samples are combined into larger outcrop level samples to make them more similar in size and scope to other data, though some variation in the size of samples undoubtedly remains even after these adjustments. An additional adjustment is made for beds formerly assigned to the lowest portions of the Centerfield Member of the Ludlowville Formation, which lie below the unconformity separating the Skaneateles and Ludlowville sequences (Brett et al. 2011). Those beds are used in the Skaneateles Formation data set when making comparisons among formations but are placed in the Centerfield Member data set when making comparisons among members. Because members are never compared against formations in this analysis, these samples are never used twice in the same comparison.

Analytical Methods

Detrended correspondence analysis (DCA) is used to reconstruct the environmental gradient of each Hamilton unit (e.g., Hill and Gauch 1980; Holland et al. 2001; Miller et al. 2001). DCA is a variant of the multivariate ordination technique reciprocal averaging (RA), which averages the (initially arbitrary) scores of samples bearing a taxon to create its taxon score, then averages the taxon scores of all taxa in a sample to create new sample scores, and continues this process recursively until converging on a unique solution (Hill 1973). This methodology is both theoretically and operationally linked to weighted averaging (ter Braak and Looman 1986), a technique used to estimate unimodal response curves along such environmental gradients (figure 12.1). Just as in RA, the ordination scores of all samples bearing a taxon are averaged to find the central position of the response curve, which is the *preferred environment*. The standard deviation of sample scores is used to calculate the *environmental tolerance*. The proportion of samples bearing a taxon, relative to samples not bearing the taxon within a standard deviation of the preferred environment, is the *peak abundance* (Holland 2000). We did not use environmental tolerance to test for niche conservatism because previous simulations indicate high statistical error associated with this measurement (Holland and Zaffos 2011).

Parameters are reconstructed in this manner for each unit in the stratigraphic hierarchy of the Hamilton (figure 12.2)—that is, at the group, formation, and member levels. Not all members are sampled sufficiently for an extractable gradient, and the ordination analyses are restricted to units with twenty-five or more samples. Sufficiently sampled members are the Centerfield, Ledyard-Otisco, Wanakah-Ivy Point, and Windom (figure 12.2). Once parameters are estimated, they are correlated both within and between spatiotemporal scales to measure whether niche parameters are conserved. As noted earlier, correlation does not assess whether any individual taxon's niche is conserved (see Dudei and Stigall 2010; Walls and Stigall 2011) but rather assesses niche conservatism in aggregate across entire faunal assemblages (Holland and Zaffos 2011). This is because the scaling of ordination axes is unique to each ordination, which makes it impossible to meaningfully interpret changes in the ordination scores of a single taxon from interval to interval.

We focus on two different levels of organization. In the first, parameters at the formation and member level are compared against parameters calculated at the group level. These two variables are not statistically independent because formation and member data are incorporated into the group data, meaning that formations and members will be at least partially correlated with the overall group. We correct for this by removing any shared data between the comparisons. For example, if comparing the Centerfield Member to the entire Hamilton Group, samples from the Centerfield are removed prior to the ordination of the entire Hamilton Group. In the second, units of the same scale are compared with each other (i.e., formation to formation, member to member) in chronological succession. This latter set of comparisons is similar to previous studies of niche conservatism that compare the conservation of niche parameters between successive third-order depositional sequences (Holland and Zaffos 2011) or within individual third-order depositional sequences (Dudei and Stigall 2010; Walls and Stigall 2011).

Environmental Interpretation

The niche parameterization process is only valid if DCA extracts the same environmental gradient in all compared intervals; otherwise, the niche parameters will not be commensurate. For example, preferred temperature should not be compared against preferred salinity as a measure of niche conservatism. Therefore, ordination axes need to be placed into an environmental context before comparison. This is traditionally accomplished by correlating ordination axes with an independent environmental proxy, such as sedimentary rock type (facies; e.g., Holland and Patzkowsky 2007), biology (e.g., Scarponi and Kowalewski 2004), or geography (Zambito et al. 2012). For example, if an onshore-offshore gradient runs from north to south in a set of strata, and ordination separates samples in a north to south fashion, then the ordination can be interpreted as an onshore-offshore gra-

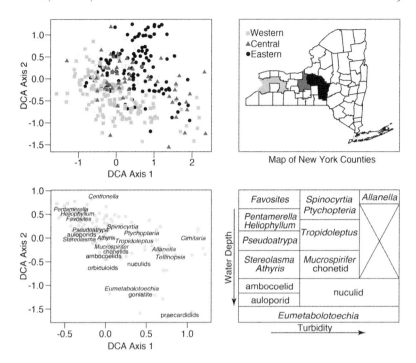

FIGURE 12.3 Hamilton ordinations in a geographic, environmental, and biological context. A = An unrotated plot of Hamilton Group DCA sample scores, with symbol shape and coded to correspond to those in panel B. B = Coded map of western (light gray), central (dark gray), eastern (black), and other (white) New York Counties; note key for gray scale tones with symbols used in figure 12.3a. There is a clear separation by county in the unrotated ordination, but this geographic clustering is oblique to DCA axis 1 and axis 2. C = A plot of Hamilton preferred environment scores for common taxa, rotated to maximize the geographic separation of samples along DCA axis 1. Only a select subset of characteristic taxa is represented with text for legibility; other taxa are represented as gray points. D = The qualitative depth-turbidity diagram of Brett et al. (1986, 1990), which arrays named biofacies in terms of water depth and turbidity. The orientation of the rotated preferred environment scores closely matches the qualitative depth-turbidity diagram.

dient. In the case of the Hamilton, there is a geographically nonlinear water depth gradient that runs roughly perpendicular to longitude across the state, wherein deeper-water dark shales representing the basin depocenter are bracketed to the east and west by shallower-water carbonate bank and siliciclastic shelf facies (figure 12.2). A corresponding west-to-east separation of samples is observed in an ordination of the entire Hamilton Group, implying a water depth signal (figure 12.3a).

However, a closer examination reveals that the geographic clustering of samples in ordination space is oblique to both DCA axis 1 and DCA axis 2 (figure 12.3a). This is because a sedimentation-turbidity gradient also runs from west to east across the state. Clean-water carbonates in the west give way to siliciclastic deposits in the east (figure 12.2). The ordination responds to the spatial collinearity of depth and turbidity by expressing both gradients on DCA axis 1, which leads to the oblique orientation. We can correct for this by performing an affine rotation of the samples (see Holland and Zaffos 2011), which maximizes the correlation of west-to-east sample geography and DCA axis 1 (figure 12.3c).

Once this correction is applied, the biological clustering of taxa in ordination space closely follows the qualitative depth and turbidity diagram of Brett et al. (1986; also see Brett et al. 2007a; figure 12.3d). The turbidity axis (carbonate-siliciclastic; west-east) is expressed along DCA axis 1. Lower DCA 1 scores are dominated by taxa characteristic of "clean-water" carbonates, particularly large corals (e.g., *Favosites*, *Cystiphylloides*, and *Heliophyllum*) and brachiopods (e.g., *Pentamerella*, *Centronella*, *Pseudoatrypa*; see discussions in Brett et al. 1990; Brett et al. 2007a). Siliciclastic tolerant faunas dominate higher DCA 1 scores, particularly large bivalves (e.g., *Grammysoidea*, *Cimitaria*, *Tellinopsis*) characteristic of shelf settings. The water depth gradient is expressed along DCA axis 2. Shallow-water faunas from calcareous facies dominate the top of the ordination, while deeper-water diminutive brachiopods (e.g., orbiculoids, leiorhynchids, chonetids, and ambocoeliids), bivalves (e.g., nuculids and praecardids), and pelagic taxa (e.g., orthocone nautiloids and goniatites) populate the bottom (figure 12.3c).

We therefore use DCA axis 2 to calculate Hamilton Group niche parameters because our focus here is on the inferred water depth gradient. Other, more confined stratigraphic units within the Hamilton exhibit their own unique responses to the competing turbidity and depth gradients, expressing water depth along either DCA axis 1 or DCA axis 2. In some cases, applying the geographic rotation worsens expression of the environmental gradient, possibly because of sampling or because the geographic orientation of environments is changing between intervals. As documented previously, the depocenter of the Appalachian foreland basin, which ran obliquely northeast to southwest relative to the modern east-west outcrop belt, shifted progressively westward throughout deposition of the upper Hamilton Group (Brett and Baird 1994, 1996). To compensate for this, niche parameters are only calculated using whichever of the two *unrotated* axes most strongly correlates with DCA axis 2 in the *rotated* Hamilton Group ordination (figure 12.3)—because the water depth signal is already established in the Hamilton Group ordination. What this means for the overall statistical power of the study is that there is a stronger tendency toward type II error (i.e., a false negative). If niche parameters are not correlated between two intervals, this could be because of genuine niche modification or because no DCA axes accurately represent water depth in a particular stratigraphic unit. In contrast, the possibility of a false positive is reduced, as it is unlikely that niche parameters will be strongly correlated between two unrelated environmental axes.

Results

We assess the strength of niche conservatism with the Pearson's correlation coefficient of each niche parameter among the different stratigraphic scales of analysis (table 12.1); following Holland and Zaffos (2011), we consider a correlation coefficient >0.5 to be indicative of niche conservatism

TABLE 12.1 **Pearson's correlation coefficients for each set of comparisons**

	Preferred environment	Peak abundance
Formation-to-group		
Skaneateles-to-group	**0.76**	**0.57**
Ludlowville-to-group	**0.77**	**0.63**
Moscow-to-group	**0.76**	**0.71**
Formation-to-formation		
Skaneateles-to-Ludlowville	**0.75**	**0.54**
Ludlowville-to-Moscow	**0.86**	**0.72**
Member-to-group		
Centerfield-to-group	0.23	**0.71**
Ledyard-to-group	**0.89**	0.38
Wanakah-to-group	**0.63**	**0.65**
Windom-to-group	**0.73**	**0.70**
Member-to-member		
Centerfield-to-Ledyard	0.26	0.22
Ledyard-to-Wanakah	**0.74**	**0.55**
Wanakah-to-Windom	**0.63**	**0.58**

Correlation coefficients >0.5 are set in boldface. Formation-to-group comparisons exhibit the highest correlations. Formation-to-formation comparisons are strongly correlated in preferred environment and peak abundance. Member-to-group and member-to-member comparisons are more variable.

and lower values to be indicative of niche modification. Other thresholds could, of course, be chosen, but this would not change the overall tenor of our results.

The correlation of niche parameters between different stratigraphic units supports near-ubiquitous niche conservatism throughout Hamilton strata (table 12.1). All formation-to-group and formation-to-formation comparisons meet our criterion for niche conservatism. Only member-level comparisons—Centerfield-to-group, Centerfield-to-Ledyard, and Ledyard-to-group—exhibit low correlation coefficients, suggestive of niche modification. Of these, only the Centerfield-to-Ledyard comparison falls below the threshold for both preferred environment and peak abundance (table 12.1). Even if the cutoff is deemphasized, the average value of formation-level correlation coefficients (0.71) is larger than the member-level average (0.57).

Discussion

Niche Conservatism and Hamilton Hierarchy

A straightforward reading of the correlation coefficients suggests that niche parameters are more labile between members than formations, suggesting that the intensity of niche conservatism scales with the stratigraphic hierarchy. We propose, however, that there are three possible explanations. First, the pattern might be accepted at face value from a biological standpoint, suggesting increased niche modification between members, but overall niche conservatism between formations. Alternatively, methodological bias may falsely indicate niche modification at comparatively fine stratigraphic scales. In such a scenario, members would be just as conservative as formations if the study were perhaps conducted with increased sampling and more sophisticated methods of analysis. Finally, there may be a systematic geological bias toward apparent niche modification at the member level such that formations will inherently appear more conservative than members regardless of sampling and analytical methodology, even if there is no underlying biological difference in conservatism patterns at different scales.

Although it might seem paradoxical from a biological standpoint for niches to be modified at a finer scale (members) but appear to be stable at a coarser scale (formations), it is nevertheless plausible. Geologically brief fluctuations in the abundance and geographic extent of taxa, so-called epiboles, are common in the Hamilton group (e.g., Cooper 1930, 1933; see review in Brett and Baird 1996, 1997). Epiboles are bed-scale phenomena that can be traced over hundreds of kilometers in some cases, wherein a particular taxon or set of taxa suddenly becomes extraordinarily common across a range of environments. Importantly, epiboles can be viewed as ephemeral changes in the realized niches of taxa (i.e., the observed environmental distributions; Soberón 2007), and it is perhaps to be expected that these brief explosions in abundance and environmental distribution translate into weakly correlated peak abundance and preferred environment at shorter time scales of analysis. That said, epiboles can be found throughout the Hamilton, not just within the Centerfield and Ledyard members, which are the primary intervals exhibiting weaker correlations in niche parameters (table 12.1). If epiboles drive weaker correlations, then this should also be visible in the other member-level comparisons, if not also at the formation level.

The second scenario, a methodological failure, is also plausible. As we noted earlier, our analysis is biased toward type II error—that is, the false appearance of niche modification—because there is some uncertainty as to which, if any, ordination axis best captures water depth. It is, therefore, possible that water depth is poorly expressed in our Centerfield and Ledyard ordinations, and this is why the niche parameters do not match. Although possible, we do not favor this explanation because a closer examination of the correlation coefficients shows that in the Centerfield-to-Hamilton Group and Ledyard-to-Hamilton Group comparisons, at least one of the two niche parameters is strongly correlated (table 12.1). It seems unlikely that *any* niche parameters would be correlated if the ordination axis did not at least partially characterize water depth.

Perhaps the most likely explanation is that members, as subsets of formations, will necessarily record a smaller portion of the full environmental gradient. This is true for any formation and its constituent members, but the problem is exacerbated in the Hamilton where formations are chronostratigraphic units that laterally record large portions of the environmental gradient from east to west. However, Hamilton members are closer to traditional lithostratigraphic units characterized by a limited range of lithologies (Brett et al. 2011). This could cause the correlation of gradients between members to be more volatile. If so, the response curves of eurytopic (broadly adapted) taxa are vulnerable to truncation in members because other parts of their environmental range are not recorded. If a taxon's range is truncated in one member but not in another, then there might be the *appearance* of niche modification because the response curve will transition from larger in the untruncated member to smaller in the truncated member. In other words, if members do not record the full breadth of depositional environments, then they cannot record the full breadth of niches that span those (missing) environments. Importantly, this bias, if real, would persist regardless of data collection methods, and observations of apparent niche modification would be more common at shorter time scales (e.g., Olszewski and Erwin 2009; Dudei and Stigall 2010; Walls and Stigall 2011) than at longer time scales (e.g., Olszewski and Erwin 2009; Holland and Zaffos 2011).

Niche Conservatism, Tracking, and Ecological Stasis

Niche conservatism is the ecological counterpart of morphological stasis. Of critical importance is whether as new species arise, they rapidly develop a series of environmental tolerances, which control their distribution, and whether this range of tolerances is "hardwired" into the species, such that environmental tolerances are maintained and that all individuals within a lineage must find suitable environments if they are to survive. In short, this view suggests that niche modification, like anagenesis in lineages, is rare and acts only to fine-tune organisms to their environments. Indeed, niche modification and anagenesis may well be linked as changes in fundamental niche imply evolving physiological tolerances.

Conversely, under the concept of niche conservatism, members of a given species must occupy environments that fall within the range of physiological tolerance built into the species. A species pool must inhabit areas of favorable environment as they shift geographically, become restricted or expanded—the concept of habitat tracking (see Brett et al. 2007b), or become extinct. In turn, habitat tracking may facilitate a lack of directional evolution and, thus, promote morphological as well as ecological stasis.

Darwinian evolutionary models suggest that local populations are constantly evolving and adapting to a changing ecological landscape and adjusting their physiological tolerances. However, as long ago argued by Eldredge and Gould (1972) and Gould and Eldredge (1977), it is improbable that niche modification could operate to produce consistent species-level changes, as adaptive pressures to environmental change are unlikely to be the same in different populations. Hence some minor physiological changes might occur in one population but opposite or at least different changes might occur in another. With gene flow among populations, it is unlikely that this could lead to any consistent directional change (Lieberman et al. 1996; Eldredge et al. 2005).

Niche Conservatism as a General Phenomenon

Instances of niche conservatism and niche modification have been claimed in a number of different studies. Only some studies, however, are concerned with the similarity of ecological gradients (Olszewski and Patzkowsky 2001; Scarponi and Kowalewski 2004; Holland and Patzkowsky 2007; Olszewski and Erwin 2009; Holland and Zaffos 2011). Those that are, particularly from deep-time analyses of Paleozoic marine faunas, generally report patterns comparable to what we observe in the Hamilton Group, broadly enduring similarity of gradients between larger depositional sequences (e.g., formations and groups), with more nuanced change at smaller scales. In this light, we suggest there is emerging evidence for niche conservatism and ecological stasis as regular features of the Paleozoic fossil record, although niche conservatism itself has not been evaluated directly in several of the aforementioned studies.

In recent detailed studies of the upper Ordovician Cincinnatian faunas, Holland and Zaffos (2011) observed a strong degree of niche conservatism based on DCA scores that permitted identification of the distribution of species along hypothetical environmental gradients. Stigall (2012) and Dudei and Stigall (2010) studied hypothetical niche breadth of Cincinnatian brachiopod and trilobite species before and during a major ecological restructuring event known as the "Richmondian invasion" (Holland 1997; Holland and Patzkowsky 2007) and characterized by the influx of some sixty new taxa from multiple biogeographic source areas. They observed that a majority of species maintained their preferred niches for a prolonged period (two to three million years) prior to the Richmondian invasion. However, during that event, there were both some local extirpation of taxa and a contraction of niche space exhibited by incumbents that survived, as evidenced by geographic restriction of previously more widespread populations. Thus, while the fundamental niche may have been unchanged throughout this crisis interval, the realized niche contracted in response to increased ecological pressure associated with the influx of new taxa with similar environmental preferences and tolerances.

It is true that in some cases a degree of actual niche modification appears possible, but these tend to be somewhat ambiguous. Holland and

Patzkowsky (2007), for example, note that among Cincinnatian brachiopods, the genus *Sowerbyella* failed to occupy a consistent position as defined by DCA, occurring in relatively deep, offshore areas during the early Cincinnatian but reappearing and occupying shallow water settings after the Richmondian invasion. However, it is now acknowledged that the latter forms belong to not only a different species but a different genus, *Eochonetes* (Bauer and Stigall in review) than the older sowerbyellids. Thus this may indicate evolution of different tolerances within members of a single family but not necessarily within a genus or species.

Niche conservatism was implied previously for the classic Middle Devonian fossil assemblages of the Appalachian basin, the subject of the present study. Brett, Bartholomew, and Baird (2007) tested for similar position of fossil associations, identified independently using cluster analysis, along environmental gradient related to water depth identified using DCA. They found closely overlapping clusters of points arrayed from shallow to deep for assemblages ranging in age by some three to four million years. This strongly indicated that a majority of species maintained similar positions in an onshore-offshore gradient through time, at least with respect to one another, and Brett et al. (2007b) suggested that the recurring species were tracking their preferred environments. It was acknowledged, however, that species associations do not recur with perfect fidelity (see also Bonelli et al. 2005) and must have different suites of environmental tolerances, such that they independently track their preferred environmental range.

Implications of Niche Conservatism to Hierarchical Evolutionary Models

To bring this study into the realm of hierarchy theory, we need to consider the interaction of both ecological and evolutionary processes. Eldredge (2003, 2008) proposed a dual hierarchy of genealogical and economic systems involved in the history of life. Herein, we are primarily concerned with the economic (ecological) entities that participate in transfer of energy in Earth's ecosystems. In his "sloshing bucket" model of evolution, Eldredge argues that evolution is shaped by environmental perturbations of various magnitudes acting upon the ecological-economic entities. Minor local disturbances may impact individual organisms or avatars with little impact on the overall species/genera or biocoenoses. Local disruptions may even eradicate entire populations but these are rapidly restocked from undisturbed areas. In contrast, major regional crises may completely eliminate entire taxa and indirectly facilitate relatively rapid evolution of new species.

The corroboration of relative niche conservatism over time scales of millions of years during the Givetian (Middle Devonian) reconciles well with the sloshing bucket model. During the time represented by the Hamilton Group, the northern part of the Appalachian Basin in New York and adjacent regions no doubt experienced hundreds of thousands of disruptive events and the occurrence of regional obrution (mass mortality and burial) horizons bears witness to widespread disruptions of the seafloor. These perturbations were probably followed by recolonization of organisms and perhaps a set of successional seres in disturbed areas, with essentially no net effect (e.g., Parsons et al. 1988; Miller et al. 1988). The stratigraphic units of the Hamilton Group record disturbances of larger scale; the members represent oscillations of sea level at the scale of tens of meters in tens to hundreds of thousands of years, whereas the large-scale formations reflect rather larger changes in sea level over hundreds of thousands of years and wholesale migration of environments changing in some cases from deep, dysoxic mud bottoms (perhaps fifty to one hundred meters of water) to essentially exposed land (Brett et al. 2007a). The geographic displacement of environments was on the order of several hundred kilometers. Yet the ecological-evolutionary response to this type of regional disturbance was minimal.

"Ecological niche" is a complex concept, and, of course, various species of organisms and their local populations were responding to more than a simple combination of water depth and sediment/substrate type, the two factors that are most evidently recorded in the strata. In some cases, particular species, responding to as yet unknown factors, underwent widespread proliferation (epiboles) or retrenchment (local outages). On one hand, this phenomenon could imply accelerated niche modification in temporally restricted avatars of the species. On the other hand, it is probable that epiboles and outages do not represent niche modification but instead reflect species responding to aspects of their current niche that are not adequately summarized by the proxies of water depth and turbidity used here.

The epibole issue aside, strong evidence for niche conservatism, provided in this study, suggests that populations or avatars of most taxa studied were able to persist and maintain similar modes of life, with similar ranges of environmental tolerance, over hundreds of thousands to a few million years. Only a very few taxa are thought to have become extinct or regionally exterminated during this three- to four-million-year span. This apparent stability in the face of evident environmental shift implies that the basic environmental parameters required for a vast majority of species or at least genera persisted throughout this time even though environments were locally or even regionally shifted through large distances. Populations (avatars) of species appear to have been able to track or at least repopulate the shifting environments (see Brett et al. 2007a,b).

At the next higher level—that of the biocoenoses—the issue of conservatism versus modification becomes less clear. Certainly, evidence for niche conservancy suggests that genera (most of which were represented by a single species) maintained similar positions in environmental gradients for the duration of the Givetian Stage. But does this imply true stability of biocoenoses or persistence of communities per se? Some authors have implied a kind of "ecological locking" via networks of interactions in complex ecosystems (see e.g., Morris et al. 1995; Eldredge 2003; Tëmkin and Eldredge 2015). There is some evidence that a high degree of similarity of more complex (higher diversity) assemblages is maintained throughout the Hamilton Group (Ivany et al. 2009) and to a greater extent than that in low diversity associations (Brett et al. 2007a). However, other research has shown evidence for substantial change at least in the relative abundance of species in analogous communities (e.g. coral thickets; see Bonelli et al. 2005) through time. Thus apparent similarities of community structure may be more of an epiphenomenon of the niche conservancy of component species—that is, groups of taxa individually tracking similar portions of environmental gradients—than an emergent property of the biocoenosis level.

Conclusions

1. The Hamilton faunas are geographically distributed in response to two underlying environmental gradients: a carbonate-siliciclastic gradient and a water depth gradient.

2. Correlations of response curve parameters among different spatiotemporal scales in the Middle Devonian upper Hamilton Group collectively support a pattern of niche conservatism, wherein the distribution of marine genera along an inferred water depth gradient remains similar over millions of years.

3. Patterns of niche conservatism are most apparent at the scale of Hamilton for-

mations (third-order depositional sequences) and least apparent at the scale of Hamilton members (fourth-order sequences).

4. Apparent niche modification within some members is most likely an artefact of scale rather than a genuine ecological pattern, although a case could be made that the biological phenomenon of epiboles represent instances of short-term (realized) niche modification.

5. Overall, the implication of niche conservatism in the Hamilton Group for evolution is that organisms, populations, and species may persist for periods hundreds of thousands of years and through substantial regional environmental disturbance and shifting. In the context of the sloshing bucket model, the degree of sloshing (almost literally in the case of sea level oscillations) was not strong enough during an interval of three to four million years to produce any lasting disruption of the ecosystem. However, the existence of major changes that bracket periods of relative stability ("ecological-evolutionary subunits" of Brett and Baird 1995), including the Hamilton, imply, as in Eldredge's (2003) model, that there were thresholds in terms of intensity and/or timing of disruption, beyond which cascading effects on the ecosystem included loss species, immigration/emigration, and evolution of new species (see Brett et al. 1996, 2009; Zambito et al. 2012). At these times the likelihood of niche modification probably increases.

Acknowledgments

We appreciate the perceptive comments of the editors and outside reviewers, which have helped to integrate the ideas of this study with the broader issues of this book. Doctoral research by Andrew Zaffos was supported by grants or logistical aid from American Museum of Natural History, the Paleontological Society, the Geological Society of America, the National Aeronautics and Space Administration, the Yale Peabody Museum, the Genesee Valley Conservancy, the Florida Museum of Natural History, the Florida Geological Survey, the Paleobiology Database, and the University of Cincinnati.

References

Baird, Gordon C., and Carlton E. Brett. 1983. "Regional Variation and Paleontology of Two Coral Beds in the Middle Devonian Hamilton Group of Western New York." *Journal of Paleontology* 57: 417–46.

Bauer, Jennifer E., and Alycia L. Stigall. In review. "Systematics and Evolution of the Late Ordovician Laurentian Sowerbyellid Brachiopod Genera *Eochonetes* and *Thaerodonta*." *Journal of Paleontology.*

Bezusko, Karen M. 2001. "Biotic Interaction versus Abiotic Response as Mediators of Biodiversity in the Middle Devonian (Givetian) Upper Hamilton Group of New York State." Master's thesis, University of Cincinnati.

Bonelli, James R., Carlton E. Brett, Arnold I. Miller, and J. Bret Bennington. 2006. "Testing for Faunal Stability across a Regional Biotic Transition: Quantifying Stasis and Variation among Recurring Coral-Rich Biofacies in the Middle Devonian Appalachian Basin." *Paleobiology* 32: 20–37.

Bottjer, David J., and David Jablonski. 1988. "Paleoenvironmental Patterns in the Evolution of Post-Paleozoic Benthic Marine Invertebrates." *Palaios* 3: 540–60.

Boucot, Arthur J. 1975. *Evolution and Extinction Rate Controls.* Amsterdam: Elsevier.

———. 1983. "Does Evolution Take Place in an Ecological Vacuum? II." *Journal of Paleontology* 57: 1–30.

Boyer, Diana L., and Mary L. Droser. 2009. "Palaeoecological Patterns within the Dysaerobic Biofacies: Examples from Devonian Black Shales of New York State." *Palaeogeography, Palaeoclimatology, Palaeoecology* 276: 206–16.

Brett, Carlton E., and Gordon C. Baird. 1994. "Depositional Sequences, Cycles, and Foreland Basin Dynamics in the Late Middle Devonian (Givetian) of the Genesee Valley and Western Finger Lakes Region." In *Field Trip Guidebook: New York State Geological Association 66th Annual Meeting*, edited by Carlton E. Brett and James Scatterday, 505–85. New York: University of Rochester.

———. 1995. "Coordinated Stasis and Evolutionary Ecology of Silurian to Middle Devonian Faunas in the Appalachian Basin." In *New Approaches to Speciation in the Fossil Record*, edited by Douglas H. Erwin and Robert L. Anstey, 285–315. New York: Columbia University Press.

———. 1996. "Middle Devonian Sedimentary Cycles and Sequences in the Northern Appalachian Basin." *Paleozoic Sequence Stratigraphy: Views from the North American Craton. Geological Society of America Special Papers*, edited by Brian J. Witzke, Greg A. Ludvigson, and Jed Day, 306: 213–41.

———. 1997. "Epiboles, Outages and Ecological Evolutionary Events." In *Paleontologic Events, Stratigraphic, Ecological and Evolutionary Implications*, edited by Carlton E. Brett and Gordon C. Baird, 249–85. New York: Columbia University Press.

Brett, Carlton E., Gordon C. Baird, Alexander J. Bartholomew, Michael K. DeSantis, and Charles A. Ver Straeten. 2011. "Sequence Stratigraphy and a Revised Sea-Level Curve for the Middle Devonian of Eastern North America." *Palaeogeography, Palaeoclimatology, Palaeoecology* 304: 21–53.

Brett, Carlton E., Gordon C. Baird, and K. B. Miller. 1986. "Sedimentary Cycles and Lateral Facies Gradients across a Middle Devonian Shelf to Basin Ramp, Ludlowville Formation, Cayuga Basin." In *Field Trip Guidebook. New York State Geological Association 58th Annual Meeting*, 81–127. New York: Cornell University.

Brett, Carlton E., and A. J. Bartholomew. 2007. "Revised Correlations and Sequence Stratigraphy of the Middle Devonian (Givetian) in Ohio, USA and Ontario, Canada: Implications for Paleogeography, Sedimentology and Paleoecology." In *Devonian Events and Correlations*, edited by R. T. Becker and W. T. Kirchgasser, 278: 105–31. London: Geological Society, Special Publications.

Brett, Carlton E., Alexander J. Bartholomew, and Gordon C. Baird. 2007a. "Biofacies Recurrence in the Middle Devonian of New York State: An Example with Implications for Habitat Tracking." *Palaios* 22: 306–24.

Brett, Carlton E., Austin J. W. Hendy, Alexander J. Bartholomew, James R. Bonelli, and Patrick I. McLaughlin. 2007b. "Response of Shallow Marine Biotas to Sea-Level Fluctuations: A Review of Faunal Replacement and the Process of Habitat Tracking." *Palaios* 22: 228–44.

Brett, Carlton E., Linda C. Ivany, Alexander J. Bartholomew, Michael K. DeSantis, and Gordon C. Baird. 2009. "Devonian Ecological-Evolutionary Subunits in the Appalachian Basin: A Revision and a Test of Persistence and Discreteness." *Geological Society London Special Publications* 314: 7–36.

Brett, Carlton E., Linda Ivany, and Kenneth Schopf. 1996. "Coordinated Stasis: An Overview." *Palaeogeography, Palaeoclimatology, Palaeoecology* 125: 1–20.

Brett, Carlton E., Keith B. Miller, and Gordon C. Baird. 1990. "A Temporal Hierarchy of Paleoecological Processes in a Middle Devonian Epeiric Sea." In *Paleocomunity Temporal Dynamics: The Long-Term Development of Multispecies Assemblies*, edited by William Miller III, *Paleontological Society Special Paper* 5, 178–209.

Cooper, G. Arthur. 1930. "Stratigraphy of the Hamilton Group of New York." *American Journal of Science* 19: 116–34, 214–36.

———. 1933. "Stratigraphy of the Hamilton Group of Eastern New York." *American Journal of Science* 26: 537–51; 27: 1–12.

De Vleeschouwer, David, Frédéric Boulvain, Anne-Christine DaSilva, Damien Pas, Corentin LaBaye, and Philippe Claeys. 2014. "The Astronomical Calibration of the Givetian (Middle Devonian) Timescale (Dinant Synclinorium, Belgium)." *Geological Society of London* 414. First published online November 21. doi:10.1144/SP414.3.

Dudei, Nicole L., and Alycia L. Stigall. 2010. "Using Ecological Niche Modeling to Assess Biogeographic and Niche Response of Brachiopod Species to the Richmondian Invasion (Late Ordovician) in the Cincinnati Arch." *Palaeogeography, Palaeoclimatology, Palaeoecology* 296: 28–43.

Eldredge, Niles. 2003. "The Sloshing Bucket: How the Physical Realm Controls Evolution." In *Evolutionary Dynamics. Exploring the Interplay of Selection, Accident, Neutrality, and Function*, edited by James P. Crutchfield and Peter Schuster, 3–32. Oxford: Oxford University Press.

———. 2008. "Hierarchies and the Sloshing Bucket: Toward Unification of Evolutionary Biology." *Evolution: Education and Outreach* 1 (1): 10–15.

Eldredge, Niles, and Stephen J. Gould. 1972. "Punctuated Equilibria: An Alternative to Phyletic Gradualism." In *Models in Paleobiology*, edited by Thomas J. Schopf, 82–115. San Francisco: Freeman, Cooper.

Eldredge, Niles, John N. Thompson, Paul M. Brakefield, Sergey Gavrilets, David Jablonski, Jeremy B. C. Jackson, Richard E. Lenski, Bruce S. Lieberman,

Mark A. McPeek, and William Miller III. 2005. "The Dynamics of Evolutionary Stasis." *Paleobiology* 31 (2): 133–45.

Ellwood, Brooks B., Jonathan H. Tomkin, Ahmed El-Hassani, Pierre Bultynk, Carlton E. Brett, Eberhard Schindler, Raimund Feist, and Alexander J. Bartholomew. 2011. "A Climate-Driven Model and Development of a Floating Point Time Scale for the Entire Middle Devonian Givetian Stage: A Test Using Magnetostratigraphy Susceptibility as a Climate Proxy." *Palaeogeography, Palaeoclimatology, Palaeoecology* 304: 85–95.

Ettensohn, Frank R. 2008. "The Appalachian Foreland Basin in Eastern United States." In *Sedimentary Basins of the World, Vol. 5. The Sedimentary Basins of the United States and Canada*, edited by Andrew D. Miall, 105–79. Amsterdam: Elsevier.

Gould, Stephen J., and Niles Eldredge. 1977. "Punctuated Equilibria: The Tempo and Mode of Evolution Reconsidered." *Paleobiology* 3: 115–51.

Hadly, Elisabeth A., Paula A. Spaeth, and Cheng Li. 2009. "Niche Conservatism above the Species Level." *Proceedings of the National Academy of Sciences of the United States of America* 106: 19707–14.

Hill, Mark O. 1973. "Reciprocal Averaging: An Eigenvector Method of Ordination." *Journal of Ecology* 61: 237–49.

Hill, Mark O., and Hugh G. Gauch. 1980. "Detrended Correspondence Analysis: An Improved Ordination Technique." *Vegetatio* 42: 47–58.

Holland, Steven M. 1997. "Using Time/Environment Analysis to Recognize Faunal Events in the Upper Ordovician of the Cincinnati Arch." In *Paleontological Events: Stratigraphic, Ecological, and Evolutionary Implications*, edited by Carlton E. Brett and Gordon C. Baird, 309–34. New York: Columbia University Press.

———. 2000. "The Quality of the Fossil Record: A Sequence Stratigraphic Perspective." *Paleobiology* 26: 148–68.

Holland, Steven M., Arnold I. Miller, David L. Meyer, and Benjamin F. Dattilo. 2001. "The Detection and Importance of Subtle Biofacies within a Single Lithofacies: The Upper Ordovician Kope Formation of the Cincinnati, Ohio Region." *Palaios* 16: 205–17.

Holland, Steven M., and Mark E. Patzkowsky. 2007. "Gradient Ecology of a Biotic Invasion: Biofacies of the Type Cincinnatian Series (Upper Ordovician), Cincinnati, Ohio Region, USA." *Palaios* 22: 392–407.

Holland, Steven M., and Andrew Zaffos. 2011. "Niche Conservatism along an Onshore-Offshore Gradient." *Paleobiology* 37: 270–86.

Hopkins, Melanie J., Carl Simpson, and Wolfgang Kiessling. 2014. "Differential Niche Dynamics among Major Marine Invertebrate Clades." *Ecology Letters* 17: 314–23.

Ivany, Linda C., Carlton E. Brett, Heather L. B. Wall, Patrick D. Wall, and John C. Handley. 2009. "Relative Taxonomic and Ecologic Stability in Devonian Marine Faunas of New York State: A Test of Coordinated Stasis." *Paleobiology* 35: 499–524.

Jablonski, David, and David J. Bottjer. 1990. "Onshore-Offshore Trends in Marine Invertebrate Evolution." In *Causes of Evolution: A Paleontologic Perspective*, edited by Robert M. Ross and Warren D. Allmon, 21–75. Chicago: University of Chicago Press.

Jablonski, David, J. John Sepkoski Jr., David J. Bottjer, and Peter M. Sheehan. 1983. "Onshore-Offshore Patterns in the Evolution of Phanerozoic Shelf Communities." *Science* 222: 1123–25.

Lieberman, Bruce S., and Steven Dudgeon. 1996. "An Evaluation of Stabilizing Selection as a Mechanism for Stasis." *Palaeogeography, Palaeoclimatology, Palaeoecology* 127: 229–38.

Miller, Arnold I., Steven M. Holland, David L. Meyer, and Benjamin F. Dattilo. 2001. "The Use of Faunal Gradient Analysis for Intraregional Correlation and Assessment of Changes in Sea-Floor Topography in the Type-Cincinnatian." *Journal of Geology* 109: 603–13.

Miller, Keith B., Carlton E. Brett, and Karla M. Parsons. 1988. "The Paleoecological Significance of Storm-Generated Disturbance within a Middle Devonian Muddy Epeiric Sea." *Palaios* 3: 35–52.

Morris, Paul, Linda Ivany, Kenneth Schopf, and Carlton E. Brett. 1995. "The Challenge of Paleoecological Stasis: Reassessing Sources of Evolutionary Stability." *Proceedings of the National Academy of Sciences of the United States of America* 92: 11269–73.

Olszewski, Thomas D., and Douglas H. Erwin. 2009. "Change and Stability in Permian Brachiopod Communities from Western Texas." *Palaios* 24: 27–40.

Olszewski, Thomas D., and Mark E. Patzkowsky. 2001. "Measuring Recurrence of Marine Biotic Gradients: A Case Study from the Pennsylvanian-Permian Midcontinent." *Palaios* 16: 444–60.

Parsons, Karla M., Carlton E. Brett, and Keith B. Miller. 1988. "Taphonomy and Depositional Dynamics of Devonian Shell-Rich Mudstones." *Palaeogeography, Palaeoclimatology, Palaeoecology* 63: 109–39.

Pearman, Peter B., Antoine Guisan, Olivier Broennimann, and Christophe F. Randin. 2008. "Niche Dynamics in Space and Time." *Trends in Ecology and Evolution* 23: 149–58.

Scarponi, Daniele, and Michal Kowalewski. 2004. "Stratigraphic Paleoecology: Bathymetric Signature and Sequence Overprint of Mollusk Associations from Upper Quaternary Sequences of the Po Plain Italy." *Geology* 32: 989–92.

Sepkoski, John J., Jr., and Arnold I. Miller. 1985. "Evolutionary Faunas and the Distribution of Paleozoic Benthic Communities in Space and Time." In *Phanerozoic Diversity Patterns*, edited by James W. Valentine, 153–90. Princeton: Princeton University Press.

Sheehan, Peter R. 1985. "Reefs Are Not So Different—They Follow the Evolutionary Pattern of Level-Bottom Communities." *Geology* 13: 46–49.

Soberón, Jorge. 2007. "Grinellian and Eltonian Niches and Geographic Distributions of Species." *Ecology Letters* 10: 1115–23.

Stigall, Alycia L. 2012. "Using Ecological Niche Modeling to Analyze Niche Stability in Deeper Time." *Journal of Biogeography* 39: 772–81.

———. 2015. "Speciation: Expanding the Role of Biogeography and Niche Evolution Macroevolutionary Theory." In *Macroevolution: Explanation, Interpretation and Evidence*, edited by Emanuele Serrelli and Nathalie Gontier, 301–27. Cham, Switzerland: Springer International.

Tëmkin, Ilya, and Niles Eldredge. 2015. "Networks and Hierarchies: Approaching Complexity in Evolutionary Theory." In *Macroevolution: Explanation,*

Interpretation, Evidence, edited by Emanuele Serrelli and Nathalie Gontier, 183–226. Cham, Switzerland: Springer International.

ter Braak, Cajo J. F., and Caspar W. N. Looman. 1986. "Weighted Averaging, Logistic Regression and the Gaussian Response Model." *Vegetatio* 65: 3–11.

Walls, Bradley J., and Alycia L. Stigall. 2011. "Analyzing Niche Stability and Biogeography of Late Ordovician Brachiopod Species Using Ecological Niche Modeling." *Palaeogeography, Palaeoclimatology, Palaeoecology* 299: 15–29.

Whittaker, Robert H. 1967. "Gradient Analysis of Vegetation." *Biological Reviews* 49: 207–64.

Wiens, John J., David D. Ackerly, Andrew P. Allen, Brian L. Anacker, Lauren B. Buckley, Howard V. Cornell, Ellen I. Damschen, T. Jonathan Davies, John A. Grytnes, Susan P. Harrison, Bradford A. Hawkins, Robert D. Holt, Christy M. McCain, and Patrick R. Stephens. 2010. "Niche Conservatism as an Emerging Principle in Ecology and Conservation Biology." *Ecology Letters* 13: 1310–24.

Wiens, John J., and Catherine H. Graham. 2005. "Niche Conservatism: Integrating Evolution, Ecology, and Conservation Biology." *Annual Review of Ecology, Evolution, and Systematics* 36: 519–39.

Zambito, James J., Carlton E. Brett, Gordon C. Baird, Sarah E. Kolbe, and Arnold I. Miller. 2012. "New Perspectives on Transitions between Ecological-Evolutionary Subunits in the 'Type Interval' for Coordinated Stasis." *Paleobiology* 38: 664–81.

CHAPTER 13

The Stability of Ecological Communities as an Agent of Evolutionary Selection

Evidence from the Permian-Triassic Mass Extinction

Peter D. Roopnarine and Kenneth D. Angielczyk

Introduction

B iological communities are systems comprising collections of individuals of multiple species who interact with their environments, members of their own species, and those of others. Individuals interact competitively, trophically, and mutualistically, moving energy and matter through the system. The demographic, energetic, and chemical dynamics of the system are determined directly by the actions and properties of individual organisms within it and by the manner in which they are organized hierarchically into subsets of the community. The most obvious subset, or compartment, is the species population. Other subsets that affect system dynamics include functional groups, such as primary producers and decomposers, and the numbers of other species with which an individual typically interacts, such as specialist and generalist predators. Intraspecific interactions, within populations, are generally distinct in process and impact from interspecific interactions—for example, intraspecific competition versus interspecific trophic interactions.

The properties of aggregation into subgroups or compartments, flux of quantities of energy and matter through the system, and a diversity of interacting entities are identified by Holland (1995) as three of four properties of complex adaptive systems. Levin (1998) argued that Holland's fourth property, nonlinearity, is evident in the highly nonlinear, path-dependent histories of ecosystems and community assembly. To this should be added evolutionary history, because the properties of species during community assembly on short, ecological time scales have already been determined over evolutionary time. It would be incorrect, however, to assume that ecosystem properties are determined solely in a bottom-up fashion by the ecological properties and evolutionary histories of individuals and the subsets to which they belong, for the evolutionary histories of species within an ecosystem are very much influenced by the ecological dynamics of the system.

The hierarchy theory of evolution, the focus of this volume, identifies two hierarchies into which biological diversity can be organized: the genealogical and the ecological. Eldredge's "sloshing bucket" model (Eldredge 2008) proposes that one of the most important interactions between the two hierarchies occurs as a consequence of major perturbations to the ecological hierarchy. The higher the hierarchical level or extent of the perturbation—for example, at the level of regional ecosystems in contrast to local populations or avatars—the greater the effect on the genealogical or evolutionary hierarchy. For example, more severe mass extinction events will result in greater losses of higher taxa and greater differences between taxa belonging to pre- and postextinction ecological systems. Here, we explore the potential for community-level properties to influence species' evolutionary histories and present a hypothesis where the differential fate of communities based on their stabilities acts as an agent of selection on the evolution and extinction of species. Importantly, we show that ecological control of species evolution and adaptation extends beyond intervals of major extinction or species turnover and persists even between major extinction intervals. Severe extinction events may indeed lead to a relaxation of that control, but although the succeeding genealogical hierarchy might differ compositionally from its predecessor, the ecological dynamics of the postextinction system converges toward those of the pre-extinction systems.

We use a model system of Permian-Triassic terrestrial paleocommunities from the Karoo Basin of South Africa that span approximately fifteen million years, dated approximately 261–245 million years ago, and include the Permian-Triassic mass extinction (PTME). In a series of previous studies (Angielczyk et al. 2005; Roopnarine et al. 2007; Roopnarine and Angielczyk 2012, 2015), we characterized two aspects of community stability in this system: resistance to secondary extinction and transient instability in response to minor perturbations. Secondary extinctions occur when the loss of one or more species cascades through an ecological network with negative demographic consequences. The two aspects are described by models of how a paleocommunity would have responded, in ecological time, after particular types and magnitudes of disturbance. The models suggest that guild structure, or the manner in which species were categorized according to habitat and feeding ecologies, was the foundation of paleocommunity food web stability during the Permian. We will argue that the paleocommunities exhibited top-down regulation of species by the guild level of organization, recognizing that populations of interacting species may be organized within food webs into nonnested hierarchical levels (Cuming and Norberg 2008), such as guilds or trophic levels, where the types or impacts of interactions between levels often differ from those within levels. We present evidence that although community stability was determined by the ecological dynamics of species and interspecific interactions, there was feedback from the community level whereby stability (1) influenced probabilities of species extinction and (2) constrained the evolutionary pathways of species via selection acting on community stability. Our claims are based on three observations. First, Karoo paleocommunities during times of low extinction (background extinction rates) were more stable than hypothetical alternatives of equal species richness and functional diversity (Roopnarine and Angielczyk 2015) as well as resistant to secondary extinction (Roopnarine et al. 2007). Second, the PTME was selectively biased: levels of extinction were significantly lower for guilds whose presence maximized system stability during the extinction. Third, the rapid recovery of species richness in the Early Triassic resulted in an unstable community with low resistance that exhibited significant turnover of taxa in spite of their initial evolutionary success. This community was replaced, by the Middle Triassic, by one as stable as the Permian communities but with a new and fundamentally different guild structure.

The Origin of Guilds

The organization of community richness into groups of functionally similar species, or guilds, is central to our dynamic models of the Karoo paleocommunities. Our work has shown that the Permian and Middle Triassic guild structures were more stable than would be expected were species partitioned among alternative guild arrangements. Understanding how guilds and such patterns might arise in the first place and how they are maintained is crucial to explaining the history of the Karoo paleocommunities. There are multiple perspectives from which to view the compartmentalization of a community or its food web, and guilds reflect the interface between evolutionary processes and fundamental energetic constraints on community development and assembly.

Trophic guilds are groups of species sharing similar prey, predators, and habitat. The existence of a guild in a community depends on several features of the community, foremost of which is that sufficient energy must be fixed by producers to support the community's biomass. Furthermore, the manner in which the biomass is partitioned among various types of taxa determines the amount of energy required because of the diversity of metabolic strategies employed. This is manifested in the constraint whereby sufficient biomass must be available at a trophic level to support the next higher one. In other words, the existence of a trophic level and the guilds that occupy it is not possible until a biomass threshold has been crossed by the lower trophic level upon which it depends, permitting stable persistence of species within the higher trophic level (Moore and de Ruiter 2012). This must also be the case evolutionarily as an ecosystem takes form: successful trophic innovations are feasible only if energetically supportable by the existing community.

The other requirement for guilds is that sufficient similarity exists between two or more species for them to share the same set of potential predators and prey or other resources. One might expect species that are phylogenetically close to occupy the same or similar guilds or that more distantly related species will do so as a result of convergent evolution toward common solutions or opportunities present in the food web. Conversely, one could expect exactly the opposite as a result of competition driving divergent evolution. One potential solution to this paradox is that a balance is struck whereby species that are similar enough to approach neutrality (Hubbell 2001) will converge toward overlap in resource use while at the same time diverging from less similar competitors (Scheffer and van Nes 2006; Scheffer 2009). Occupation of an intermediate state would be the least favorable condition, so the result is a compartmentalization of ecologically similar species into competing guilds that are separated by unoccupied ecological "space." Thus as ecosystems develop over evolutionary time, one would expect the formation of guilds as species originate, immigrate, or evolve and self-organize into groups of trophically similar species.

The Permian-Triassic Mass Extinction in the Karoo Basin

Our empirical data set consists of six terrestrial paleocommunities from the Permian and Triassic of the Karoo Basin that document the effects of and initial recovery from the PTME (figure 13.1). Specifically, we focused on the rocks of the Beaufort Group, which are highly fossiliferous (Nicolas and Rubidge 2009, 2010; Smith et al. 2012) and provide a nearly continuous record of terrestrial community evolution from the Middle Permian to the Middle Triassic (Rubidge 1995, 2005; Rubidge et al. 2013). The PTME has been studied extensively in the Karoo over the past two decades (e.g., Smith 1995; Ward et al. 2000, 2005; Smith and Ward 2001; Retallack et al. 2003; Schwindt et al. 2003; Steiner et al. 2003; Angielczyk et al. 2005; Gastaldo et al. 2005; Botha and Smith 2006; Coney et al. 2007; Irmis and Whiteside 2012; Irmis et al. 2013; Botha-Brink et al. 2014; Fröbisch 2014; Smith and Botha-Brink 2014; Gastaldo et al. 2015), and the prevailing model of environmental changes and extinction in the Karoo stemming from the PTME posits increased aridity and seasonality proceeding from the latest Permian to the Early Triassic; changes in sedimentary environments from large meandering river systems to braided river systems with higher bedloads; extensive plant die-offs; and rapid, geologically intermittent extinctions of terrestrial vertebrates (e.g. Smith and Botha-Brink 2014). It is clear that a major faunal turnover, with far-reaching biological and ecological implications, took place in the Karoo at about the time of the Permo-Triassic boundary (PTB; 252 million years ago). The decreased richness in the Early Triassic does not appear to be an artifact of factors such as reduced fossiliferous outcrop area in the basin (Fröbisch 2014). Likewise, the cosmopolitan southern Gondwana tetrapod fauna was fragmented by the extinction (Sidor et al. 2013), reduced body sizes and altered growth and life history patterns have been documented among surviving clades (Angielczyk and Walsh 2008; Botha-Brink and Angielczyk 2010; Huttenlocker and Botha-Brink 2013, 2014; Huttenlocker 2014), and anomalous community structures have been proposed for Early Triassic Karoo communities (Roopnarine et al. 2007; Roopnarine and Angielczyk 2012, 2015).

The Karoo paleocommunities also document intriguing patterns of

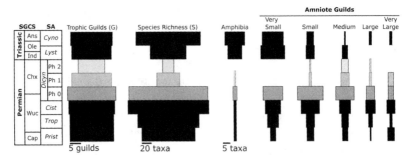

FIGURE 13.1 Functional diversity and taxon richness of the Karoo Permian-Triassic paleocommunities. The left spindle diagram illustrates the number of guilds present in each
paleocommunity, which reaches a maximum at the end of the Permian in the *Dicynodon* Assemblage Zone Phase 0 (Pho). Species richness illustrates the total number of metazoan species in each paleocommunity model, including invertebrate and vertebrate taxa. Subsequent
spindles represent subsets of tetrapod species, divided between temnospondyl amphibians
and amniotes, with the latter subdivided further according to body size. The phases of the
mass extinction are distinguished as gray-colored boxes. Several features can be noted from
the spindles: the decline of both functional diversity (number of guilds) and species richness during the mass extinction, the significant loss of small-bodied amniotes during Phase 1
(Ph1), the more rapid recovery of species richness among small-bodied amniotes in contrast
to larger amniotes during the Early Triassic, and the increase of amphibian diversity in the
Triassic relative to the Permian. The stratigraphic table on the left shows correlations be-
tween Beaufort Group assemblage zones and the geological time scale. The subdivisions
of the *Dicynodon* Assemblage Zone, illustrated as equal length, were of unequal durations
(Smith and Botha-Brink 2014). Stratigraphic units: Ans = Anisian; Cap = Capitanian; Chx =
Changhsingian; Ind = Induan; Ole = Olenekian; Wuc = Wuchiapingian. Assemblage zones:
Cist = *Cistecephalus*; Cyno = *Cynognathus*; Dicyn = *Dicynodon*; Lyst = *Lystrosaurus*; Prist =
Pristerognathus; Trop = *Tropidostoma*.

bias in both the extinction and recovery that hint at community-level feedback to species extinction and diversification. For example, Roopnarine
et al. (2007) documented that the earliest Triassic *Lystrosaurus* Assemblage
Zone (LAZ; assemblage zones are biostratigraphic units consisting of a
series of rocks that are distinguished by the presence of three or more fossil taxa, and we use the assemblages of fossils known from these units as
approximations of paleocommunities; they are generally named for one
of the dominant taxa in the assemblage; figure 13.1) food web differed in
structure from those of preceding and succeeding communities and that
these differences made species in the LAZ community more susceptible
to secondary extinction if primary production was disrupted. Roopnarine
and Angielczyk (2012) extended the results on the LAZ, showing that
its vulnerability was caused by unusually high levels of resource overlap

among carnivores and insectivores of small body size, in part reflecting the rapid rediversification of small faunivore guilds in the PTME's aftermath. As the recovery proceeded, a more typical food web structure was rebuilt in the *Cynognathus* Assemblage Zone through an increase of herbivore richness. Roopnarine and Angielczyk (2012) noted that the herbivore diversification could be driven in part by community-level selection for resistance, favoring the expansion of herbivores relative to carnivores.

Food Web Reconstruction

The general procedure for compiling the data underlying our trophic networks, or food web of interactions between species and the construction of the networks themselves, is described by Roopnarine et al. (2007) and Roopnarine and Angielczyk (2015). Predator-prey interactions of fossil species can be inferred from predation traces, gut contents, functional morphology, body size ratios, and habitat (Roopnarine 2009; Dunne et al. 2014). Using these data, we assigned species with the same sets of potential predators and prey to guilds. Because of uncertainty about the detailed ecological properties of the Karoo species, our guild definitions are intentionally broad, and our assignment of species to particular guilds and the reconstructed interactions between those guilds are based on sets of simple rules. It is noteworthy that despite its simplicity, this approach yields food webs with plausible structures and responses to perturbations (Roopnarine et al. 2007; Mitchell et al. 2012; Roopnarine and Angielczyk 2012, 2015).

A total of twenty-six guilds, or taxa aggregated according to trophic function, interactions, and habitat, are recognized for the Karoo paleocommunities (figure 13.2). Community data include four primary producer guilds and both terrestrial and aquatic invertebrates and vertebrates. Most of the records in the data set represent genera, reflecting the fact that many Karoo vertebrate genera are effectively monotypic. We recognized multiple species within genera where they were supported by strong taxonomic evidence and had distinct stratigraphic ranges (e.g., *Lystrosaurus*). Tetrapod body size categories were based on skull length as follows: very small, 0–100 mm; small, 101–200 mm; medium, 201–300 mm; large, 301–400 mm; very large, 401 mm and above. The insect body fossil record of the Beaufort Group is limited (e.g., Geertsema et al. 2002), so we used the approach of Mitchell et al. (2012) to estimate insect richness in communities with poorly preserved insect fauna (DAZ Phase 0 [Ph0]

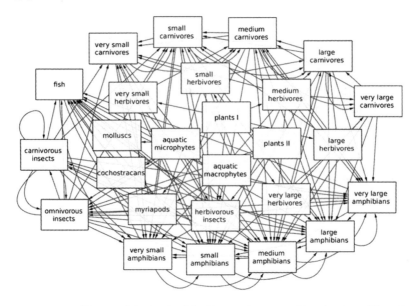

FIGURE 13.2 The overall food web of the Karoo paleocommunities, showing guilds and interguild interactions. Paleocommunities differ in the number of guilds occupied and the number of species per guild. Species composition and richness for each guild in each paleocommunity are given in Roopnarine and Angielczyk (2015).

and the *Cynognathus* Assemblage Zone are exceptions). This approach utilizes a significant linear relationship between insect species richness and the ratio of insectivore richness to the number of insectivore prey guilds, recognizing that most of the insectivores belong to insectivore-carnivore guilds. Species richnesses of each guild vary among the communities and not all guilds are represented in each community, with maximum functional diversity being reached in the *Dicynodon* Assemblage Zone (DAZ; figure 13.1; Roopnarine and Angielczyk 2015).

The current set of biostratigraphic assemblage zones for the Beaufort Group date to Rubidge (1995). In the time since that work, collection of additional data has opened the possibility for further subdivision of at least some of these assemblage zones (e.g., Hancox et al. 1995; Rubidge 2005; Lucas 2010; Neveling 2004; Angielczyk and Rubidge 2013; Day 2014). Of most importance to the present study is Smith and Botha-Brink's (2014) observation of a stepped pattern of disappearance of tetrapods approaching the Permo-Triassic boundary in the upper portion of the latest Permian *Dicynodon* Assemblage Zone. Although they did not suggest a biostratigraphic subdivision of the *Dicynodon* zone based on these data, Smith and Botha-Brink did interpret them as indicating that the PTME in the Karoo basin was manifested as a series of extinction episodes. We, therefore, constructed three subcommunities for the *Dicynodon* Assemblage Zone: (1) a Phase 0 (Ph0) background community comprising all taxa documented as occurring anywhere within the *Dicynodon* zone; (2) a Phase 1 (Ph1) survivor community consisting of the fifteen tetrapods that Smith and Botha-Brink documented as reaching the lower part of their extinction interval; and (3) a Phase 2 (Ph2) survivor community comprising the seven tetrapod taxa that go extinct just before the Permo-Triassic boundary or cross the boundary to persist in the LAZ. With the exception of the *Dicynodon* Assemblage Zone, we considered each assemblage zone to represent a single community.

Paleocommunity Stability

"Stability" has multiple meanings when applied to ecological communities; therefore, in this section we explain both the types of stability that we have applied to paleocommunities and the quantitative models behind each type. In general, stability means that the rate of change of the system is effectively zero or that the system is persistent. The system itself may be described from multiple aspects; therefore, in our work on the Karoo paleocommunities, we have examined two types of stability—namely, transient instability, or the manner in which a locally stable community would respond to minor perturbations of one or more of its species on ecological time scales, and resistance to secondary extinction after some component of the system has been removed. We will use a familiar heuristic device, the stability landscape, to illustrate both models. Mathematical details are given elsewhere (Roopnarine 2006, 2009; Roopnarine and Angielczyk 2015).

We imagine the state of the community to be defined by its position on a mathematical landscape, where the topology of the landscape reflects stability dynamics (figure 13.3). Stable communities, in which rates of change of population properties are zero, occupy the lowest points of basins on the landscape, better termed attractors. A minor perturbation to the community is envisaged as a displacement of the community within the basin. A locally stable community will return to the lowest point of the basin; this is one definition of resilience (Pimm 1982). The landscape heuristic is a useful

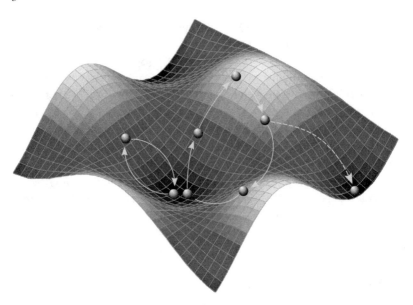

FIGURE 13.3 A theoretical community stability landscape. Lower topography indicates greater stability, with areas of greatest stability, or stable attractors, being represented as basins (darker area). The scenario shown here is of a community (sphere) perturbed from a stable point in a basin, with the magnitude of perturbation equal to the length of the first arrow. The community on the left begins a return to a stable equilibrium immediately after being perturbed, whereas that on the right experiences several amplifications of the perturbation before returning to equilibrium. The amplifications are illustrated by the sequence of arrows. The dashed arrow shows the possibility for transient dynamics to lead to the transition to a new community equilibrium.

one and can be extended to include unstable equilibria—for example, saddle points at which any displacement will result in movement away from the starting point—and multiple alternative states, represented as multiple attractors that can be reached if the initial displacement is large enough or the basin too small (Scheffer 2009). The heuristic does have limitations, however, and cannot usefully reflect more complicated dynamics, such as stable limit cycles and chaotic attractors (Murray 1989). Additionally, the perturbation is "minor" because models of local stability do not describe dynamics if the community is displaced to an extent where alternative states—that is, additional basins—become additional attractors. This simple model also does not describe how the community responds between the time of perturbation and when equilibrium is once again attained. Marten Scheffer (2009) has likened this to a ball on the landscape having no momentum and always moving as if immersed in a viscous, dampening fluid (figure 13.3).

Transient Stability

Models of transient instability add momentum to the ball, because once displaced from equilibrium, the displacement may become amplified, and although a return to equilibrium is inevitable, there may be numerous times during the return when the displacement from equilibrium actually increases (Neubert and Caswell 1997; figure 13.3). This means that the effects of an initial perturbation on a food web do not necessarily begin to decline immediately and that the return to equilibrium is not smooth and monotonic (figure 13.4). The return may in fact be complicated, and the rate at which the system responds to the perturbation and the extent to which the effects are amplified are functions of the structure of the community. Importantly, frequent perturbations can maintain the system in this transient state, and amplification may lead to critical transitions to a new state (Scheffer et al. 2009; Dai et al. 2012).

Roopnarine and Angielczyk (2015) modeled the transient stability dynamics of the Karoo paleocommunities, comparing the communities to model communities with equal species richness and guild number but

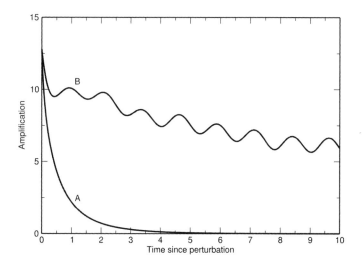

FIGURE 13.4 An alternative illustration comparing a simple model of local stability to transient behavior. The lower curve shows the expected dynamics of a locally stable community after perturbation, with an asymptotically monotonic decline of the displacement from equilibrium. The upper curve is a more realistic expectation, showing both short-term amplifications of the perturbation and the long-term return to equilibrium.

randomized numbers of species in guilds or randomized interguild interactions. Paleocommunity models were also compared to model communities of equal species richness but lacking guild structure and having varying
patterns of interspecific interactions. All Permian communities, including
those during the PTME, exhibited as little or significantly less transient instability relative to the alternative models (Roopnarine and Angielczyk
2015). Of the Triassic communities, the Early Triassic LAZ community
in the immediate aftermath of the PTME was no more stable than models
with alternative guild structures, whereas the Middle Triassic *Cynognathus* Assemblage Zone (CAZ) community exhibited dynamics as stable as
the Permian communities. Furthermore, stability was maintained during
the two phases of the PTME, identified by Smith and Botha-Brink (2014),
by an initially strong selective bias of the extinction against amniotes of
small body size. This bias is counterintuitive if one accepts that extinction
risk generally increases with body size for vertebrates (Purvis et al. 2000).
In contrast, the addition of new species during the Early Triassic resulted
in a LAZ community that was less stable than model communities of alternative guild structure. The model comparisons lead to the conclusion
that the number of guilds and the interactions among them are the major
factors controlling the variability of transient stability/instability in each
paleocommunity.

The CEG Model

The CEG (cascading extinctions on graphs) model was developed to examine a paleocommunity's resistance to the secondary extinction of species in response to perturbation of the food web (Roopnarine 2005). Species in this model are considered to be in energetic equilibrium between
incoming energy from production or consumption and outgoing energy
to herbivory, predation, or parasitism. The effects of a perturbation can
initiate a cascade of changes through the web as the strengths of biotic interactions are modified to maintain equilibrium. Such changes in interaction strengths are not always sufficient to return a species to equilibrium,
though, in which case it becomes extinct in the community. CEG models
of the Karoo paleocommunities have concentrated on perturbations to
primary productivity as tests of the hypothesis that disruptions to primary
productivity were significant drivers of the PTME on land (Angielczyk
et al. 2005; Roopnarine et al. 2007). The paleocommunities in general are
resistant to extinction at low levels of perturbation but then transition

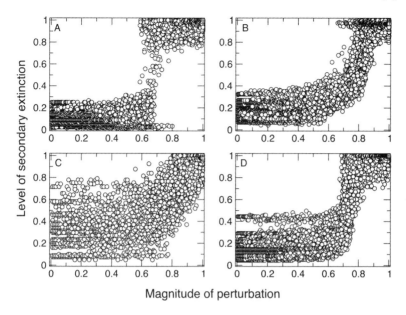

Level of secondary extinction

Magnitude of perturbation

FIGURE 13.5 Paleocommunity resistance to secondary extinction as simulated with the CEG model (Roopnarine 2006). Each plot shows the results of perturbations to one hundred food webs generated stochastically from a paleocommunity model (Roopnarine et al. 2007). The perturbation (x-axis) is an incrementally increasing removal of primary production from the community, with secondary extinction measuring the proportion of species that become extinct as a consequence. A = Tropidostoma Assemblage Zone; B = *Dicynodon* Assemblage Zone Pho (prior to the mass extinction); C = *Lystrosaurus* Assemblage Zone (LAZ); D = *Cynognathus* Assemblage Zone. Communities A, B, and D exhibit resistance to secondary extinction over broad ranges of perturbation magnitude, but rapidly accelerated secondary extinction as a threshold is approached or crossed. The LAZ community, however, exhibits high variance of secondary extinction even at low levels of perturbation, with potentially catastrophic collapse of the community.

within a narrow range of perturbation magnitudes to significantly higher levels of extinction (figure 13.5). The dampened responses at low levels of perturbation demonstrate engineering resilience, wherein communities are able to maintain a level of functioning similar to that prior to perturbation (Holling 1996). The nature of the transition to higher levels of secondary extinction might be incremental over the transitional range of perturbations, very rapid, or even discontinuous. Some communities also show the existence of alternative states of low and high extinction during the transition, "flickering" between those states at a critical level of perturbation (Scheffer et al. 2012). Roopnarine et al. (2007; see also Roopnarine and Angielczyk 2012) reported that the Early Triassic LAZ community is an exception to this generality. LAZ exhibits the potential for very high secondary extinction at low levels of perturbation, suggesting that this community would have had low engineering resilience and occupied a very small basin on the stability landscape.

Biased Extinction and Survival

The biased extinction during the first phase of the PTME (Ph1) and high turnover in the first phase of the recovery LAZ were both associated with small body-sized terrestrial amniotes. Those taxa, herbivores and faunivores, suffered levels of extinction significantly greater than those of larger amniotes during Ph1. The resulting community would have exhibited less transient instability in comparison to communities that could have been produced had the extinctions been distributed randomly among the amniote guilds (Roopnarine and Angielczyk 2015). Furthermore, if the bias had instead been against larger amniotes, the resulting community would have had very low resistance to disruptions of primary productivity of any magnitude (figure 13.6). The differences between the alternatives based on body size are the result of smaller and larger amniotes being involved in different subwebs of the overall food web. Smaller amniotes are inferred to have interacted strongly with insects as insectivores, and a major proportion of energy supplied to smaller faunivores would have been via herbivorous, omnivorous (including detritivorous), and carnivorous insects. The remainder of their energy budget arrived via predation on small, herbivorous amniotes. Larger carnivorous amniotes, on the other hand, would have received all their energy via predation on other amniotes, including large herbivores. Given that insects were most likely not immune to the PTME (Roopnarine and Angielczyk 2015), a reduction of insect richness without a corresponding biased loss of amniote insectivores would have resulted in the unstable, hypothetical food webs. The first, biased phase of the PTME therefore seems to have produced a community as stable as possible given the magnitude of the extinction.

In contrast, the Early Triassic LAZ community exhibited no more transient stability than the hypothetical models that were structurally altered in various ways, and its resistance to the effects of disruptions to primary productivity had a high likelihood of being poor (figure 13.5c). These model results are congruent with high taxon extinction and turnover associated with the transition to the next assemblage zone, the Middle Triassic

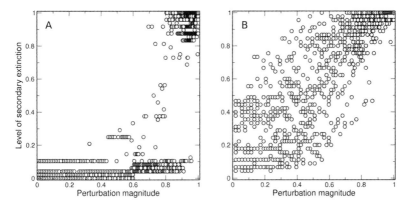

FIGURE 13.6 CEG dynamics during the first phase (Ph1) of the mass extinction. A = Secondary extinction and resistance of the actual Ph1 community, exhibiting resistance at low levels of perturbation, followed by a critical increase. B = An alternative model scenario in which large body-sized amniotes become extinct in the first phase of the extinction in contrast to the observed extinction of smaller amniotes (A). In this case secondary extinction would have increased at lower levels of perturbation and would have been much greater. The alternative model was produced by reducing guild richnesses of the largest body-sized amniote guilds in the premass extinction community by a number of species equal to the number of small-bodied amniotes that actually went extinct. Both models are therefore of equal species richness.

Cynognathus Assemblage Zone, as well as inferences of demographic instability of LAZ species (Irmis and Whiteside 2012). Roopnarine and Angielczyk (2012) demonstrated that the low resistance of the LAZ CEG model is generated by the community's high diversity of small to medium body-sized insectivorous and carnivorous amniotes and a highly reduced amniote herbivore diversity. The small amniote faunivore guilds might have suffered during the PTME, but they were also among the first to rebound in the aftermath. Roopnarine and Angielczyk (2012) argued that the initial evolutionary successes of taxa within those guilds constituted a type of tragedy of the commons, where their own success and the resulting trophic imbalance of the community resulted in high rates of turnover and ultimately the end of the LAZ community. Alternatively, the community could have been significantly more resistant if the taxa were far more trophically specialized than is generally expected within communities. Nevertheless, such extreme specialization would also have increased probabilities of extinction. It seems that the early recovery community was doomed by its own success. Important unanswered questions include the degree to which the unusual properties of the LAZ stemmed directly from the effects the PTME and degraded environmental conditions in its aftermath as well as whether other so-called disaster faunas commonly thought to exist after mass extinctions are similarly anomalous (Dineen et al. 2015; Hautmann et al. 2015).

Adaptive Cycles of Community Transformation

The transient stability and resistance of each paleocommunity can be summarized qualitatively by classifying each as either resistant or not and as having a guild structure that either constrains transient instability or plays no role at all—for example, DAZ versus LAZ (table 13.1). All the Karoo communities belong to a set in which both transient stability and resistance are high, with the exception of the Early Triassic LAZ community. This seemingly nonrandom distribution raises the questions of what mechanism(s) keeps communities within the set of greatest stability and what relationship it bears to species that originate, evolve, and become extinct during the geological duration of a community. In the remaining text, we argue that when ecosystems are subjected to extreme environmental disturbances, they undergo phases of collapse and reorganization, and community dynamics during those intervals exert top-down influence on species extinction. Integral to this argument is the scenario presented earlier, whereby guild structure within a community is built on the bases of energetic requirements and constraints, self-organization among species, and community dynamics. Stability of the guild structure is generally

TABLE 13.1 **Matrix of four pairwise possibilities of transient instability and resistance to secondary extinction**

Secondary extinction	Transient instability	
	Low amplification	High amplification
High resistance	*Pristerognathus, Tropidostoma, Cistecephalus, Dicynodon* Phases 0–2, *Cynognathus*	Empty
Low resistance	Empty	*Lystrosaurus*

The most stable communities under the models described here are those exhibiting both low transient amplification of perturbations and high resistance to secondary extinction. All the Karoo paleocommunities are located in that quadrant, with the exception of the *Lystrosaurus* Assemblage Zone, which has levels of transient amplification indistinguishable from randomized models. Assemblage zones are denoted by their associated taxon names only— for example, *Pristerognathus*. See figure 13.1 for stratigraphic details.

conserved on geological and evolutionary time scales but is potentially destroyed during severe extinction events and replaced by a reorganized structure. Our description of the scenario begins with a consideration of the Early Triassic LAZ community, which existed in the immediate aftermath of the extinction and captures the construction of a new community during the initial recovery from the PTME.

The Karoo terrestrial ecosystem was depauperate by the end of the second phase of the PTME. Only four vertebrate genera crossed the boundary into the Early Triassic: *Notaelurodon* (*Promoschorhynchus*), *Elonichthys*, *Lystrosaurus*, and *Tigrisuchus* (*Moschorhinus*). Nevertheless, the Early Triassic LAZ community is characterized by a significant rebound of taxon richness, eventually comprising more than forty vertebrate species and a presumably equally rebounded though mostly unpreserved invertebrate fauna. Structurally, temnospondyl amphibians were both richer and more functionally diverse than in the Permian paleocommunities, with ten species distributed among four guilds, whereas prior to the PTME (DAZ Pho), there were three species distributed among two guilds. Furthermore, prior to the PTME, there were twenty species of herbivorous amniotes distributed among five guilds but only five predominantly small-bodied species in the Early Triassic. In contrast, faunivorous amniotes rebounded in terms of species richness, with twenty-two species in the LAZ compared to thirty-four in DAZ Pho, though again the rebound was dominated by taxa of small body size. The LAZ community therefore differed substantially from the Permian communities because of the predominance of amphibians and smaller, insectivorous/carnivorous amniotes.

The transition from the LAZ to the Middle Triassic *Cynognathus* Assemblage Zone (CAZ) is as dramatic as the PTME and subsequent rebound. The two communities have only three vertebrate genera in common: *Elonichthys*, *Kestrosaurus*, and *Watsonisuchus* (based on stratigraphic range data presented in Kitching 1995; Damiani et al. 2001; and Damiani 2004). Herbivorous amniotes diversify, with nineteen species distributed among four guilds, but carnivorous amniotes are poorly represented with only four species. Amphibians are still both rich and functionally diverse, with fourteen species distributed among four guilds. CAZ is therefore a continuation of LAZ from the perspective of amphibian richness but differs because of taxonomic discontinuity and the reversal of diversity between amniote herbivores and faunivores. Its ecological dynamics also differ significantly, and CAZ exhibits both the transient stability and resistance characteristic of the structurally and taxonomically different Permian communities (figure 13.5d).

The entire transformational sequence of paleocommunities can thus be summarized as an adaptive cycle (Holling and Gunderson 2002), where a series of ecologically and structurally stable Permian communities came to an end during the PTME and were succeeded by a structurally different, ecologically unstable community in the Early Triassic, which was itself replaced by yet another structurally different but ecologically stable community. A similar transition may have occurred at the end of the Cretaceous Period in North America (Mitchell et al. 2012), but lack of data for Paleocene communities currently makes the comparison incomplete. The adaptive cycle concept stresses the ability of systems to respond to crises because of the gradual accumulation of system resilience. The Ph1 community is an example, whereby the maintenance of stability resulted from elevated extinction within specific guilds, facilitated by the guild structure of the earlier Permian communities. The severity of the PTME eventually overwhelmed the system, however, leading to its replacement by the structurally reorganized Triassic communities.

Guild Structure and Community Stability

Although there are many combinatorial ways in which species can be arranged into guilds, even with the added energetic constraints outlined above, the models of the Karoo paleocommunities inform us that some combinations may be significantly more stable than others. The remarkable finding is that the observed paleocommunities throughout the Permian and that of the Middle Triassic are always the most stable or among the most stable of the models tested. This is true of even the Ph1 and Ph2 communities that were reduced significantly by extinction. We consider it likely that stability was maintained during the Permian by a conservatism of guild structure that, once established, became relatively fixed. This in turn implies that species within the communities were constrained to trophic properties that would not affect stability negatively. The instability that would be generated otherwise would elevate probabilities of local extinction and the community would then either remain in its current state or transition to a similar state with a similar guild structure but an altered species composition. This hypothesis is supported by the Ph1 community, whereby stability was maintained by the significantly biased extinction of the smaller amniotes.

The Early Triassic LAZ community also lends support to the hypothesis. There would have been little constraint following Ph2 of the PTME because of the severe reduction of species richness and the elimination of most guilds. LAZ represents a new community structure, generated by the rapid diversification of some lineages and geographic expansions of others. Its stability, however, was not exceptional, and we further hypothesize that rates of evolution and probabilities of extinction would have been significantly greater relative to the pre-PTME communities. Supporting evidence is provided by the extensive compositional differences between LAZ and the Middle Triassic *Cynognathus* community (CAZ). Furthermore, although CAZ's guild structure is similar to that of LAZ, with the greatest amphibian richness of the entire series, the turnover of species and changes in the relative richnesses of different guilds resulted in a return to community stability.

A Hypothesis of Community Stability and Species Evolution

The relationship between community functional structure, community stability, and species extinction in the Karoo paleocommunities leads us to propose that the effects of community instability on species demographics at ecological time scales affect the geological durations of species and their evolution. The current idea is distinguishable from other theories of biotic interactions and evolution, such as the Red Queen's hypothesis (Van Valen 1973), the theory of escalation (Vermeij 1987), and other coevolutionary concepts (Thompson 1994), because we argue that community stability itself is an agent of selection, mediated by species dynamics.

The addition of a species to a community, whether by immigration or in situ speciation, changes the manner in which the community will respond to further perturbations. Those changes may be insignificant or enhance resilience or resistance, or they could be disruptive—for example, by increasing the transient amplification of perturbations or lowering resistance. Disruptive changes could, under conditions of extreme environmental stress, result in the loss of species because of direct effects of the ecological downturn on their fitnesses. The biased extinction of the PTME Ph1, however, showed that the species that were lost were the ones that would have been thought to have lower probabilities of extinction compared to those that survived, based on our conventional understanding of the relationship between vertebrate body size and extinction. Because their extinction preserved community stability, it would seem that in this case selection acting on community level properties was able to overcome the benefits conferred by their species-level properties, increasing their probabilities of extinction. One might expect a major mass extinction, such as the PTME, to have strong contingent effects simply because of its magnitude, resulting in a more or less random pattern of extinction among species. The biased pattern of extinction of smaller animals during the first phase of the extinction and the comparisons of our communities during the mass extinction to randomized versions show that this was not the case. The fact that community-level selection was able to exert this effect as such a major event was unfolding implicates it as a significant influence even during mass extinctions.

We envisage the Karoo Basin at the end of the Permian as a geographic landscape with metacommunities (local communities connected by the dispersal of interacting species) distributed across it. As environmental conditions declined, metacommunities would have responded dynamically in ways that varied according to demographic and compositional variation. Species metapopulations would have reacted according to the interaction of trait-based responses to environmental conditions as well as ecologically based responses to the dynamics of other species. Metacommunities in which species having negative impacts on community stability either declined to ecological insignificance or became locally extinct would eventually dominate the landscape because of their greater stability.

Environmental conditions as extreme as those during the PTME have occurred infrequently in Earth's history. If the importance of community stability as an agent of selection was restricted to those times, one would expect to see little relationship between stability and species evolution during the long intervals between global environmental crises. However, the greater stabilities of the pre-PTME paleocommunities and that of the Middle Triassic, compared to models with alternative structures, suggest otherwise. Species composition and guild richnesses changed constantly during the Permian leading up to the PTME, yet transformational ecological innovation was low with few changes to guild structure (figure 13.1), and real paleocommunities were consistently among the most stable models. There are several possible hypotheses for this "stability stasis." First, the stability basin or attractor of the Permian guild configuration might be large and therefore, once established, permitted considerable variation of species composition until the PTME. Second, the established community may have effectively suppressed the establishment of ecologically novel taxa. As noted above, there is evidence that species tend to self-organize into guilds with similar trophic and ecological properties. Species with new or novel ecologies might have found themselves in marginal competitive spaces between established guilds, which would have promoted evolutionary changes in the directions of nearby guilds or extinction. Suppression could also take the form of maintenance of physical environments unsuitable for new species, restricting novel taxa to marginal habitats. Third, low-level perturbations during those times of background extinction might have been sufficient to promote stability as an agent of selection, suppressing rates of evolution for many taxa because of the potential for novelty to decrease stability and consequently the fitnesses of individual taxa while also elevating probabilities of extinction via the sorting of metacommunities as described above.

Major ecosystem transformations could occur, under any of those hypotheses, as the result of destruction of the existing system and environmental restructuring or the removal of constraints on available energy. Evidence in support of the former include the diversification and success of formerly minor clades after mass extinctions, such as elements of the marine modern fauna after the PTME (Sepkoski 1981) or mammals after the end of the Cretaceous (Alroy 1999; Wilson 2013). In the Karoo system, this might be manifested as the radiation of temnospondyl amphibians in the immediate aftermath of the extinction. Some clades might also slowly transform their environments and change the stability landscape to the point where relatively rapid transitions are possible, as has been hypothesized for the slow expansion of angiosperms from marginal habitats in the Mesozoic (Scheffer 2009). Lifting constraints on available energy can result from geophysical and geochemical changes, such as massive injections of nutrients by volcanism or continental erosion or by evolutionary innovations that increase the amount of energy available to the system or rates of energy flow (Bambach 1993; Vermeij 1995). Such innovations include photosynthesis, the exploitation of nutrients trapped in marine and terrestrial sediments, and higher metabolic rates that increase the energetics of biotic interactions (Vermeij 2009).

Only the third hypothesis (selection for community stability), however, offers an explanation for ecologically biased extinction during the PTME and biased turnover at the end of the LAZ community. Stability as an agent of selection during biodiversity crises could explain other observed nonuniformities associated with mass extinctions and recoveries, such as the demise of some clades at the end of the Cretaceous but the survival of others or the absence of significant marine ecological restructuring after the Late Ordovician mass extinction (Droser et al. 2000). At the same time, the results from the initial recovery during the LAZ suggest that by the end of a major mass extinction, the constraint of community selection may be temporarily broken, allowing communities to evolve novel structures.

A key question that remains is what are the long-term (i.e., Phanerozoic-scale) implications of such selection? Have there been long-term trends for increased community stability over time or do ecosystems quickly reach stable states that are difficult to alter outside of major events such as mass extinctions or the elevation of global productivity? Including higher-level community dynamics as an active mechanism in ecosystem evolution, mass extinction, and system restructuring offers a richer framework with which to explain biodiversity during the Phanerozoic and for forecasting the potential futures of current ecosystems. The latter effort is becoming increasingly important as concern grows that the impacts of humans on the biosphere is potentially pushing the global ecosystem toward a planetary-scale critical transition (Barnosky et al. 2012) equivalent to the geological mass extinctions.

Acknowledgments

We thank Charles Marshall for many stimulating "Marshallian" discussions and an anonymous reviewer for several insightful questions. This work was supported by NSF Earth Life Transitions Grant 1336986 (to Peter D. Roopnarine, Kenneth D. Angielczyk, and Christian Sidor).

References

Alroy, John. 1999. "The Fossil Record of North American Mammals: Evidence for a Paleocene Evolutionary Radiation." *Systematic Biology* 48: 107–18.

Angielczyk, Kenneth D., Peter D. Roopnarine, and Steve C. Wang. 2005. "Modeling the Role of Primary Productivity Disruption in End-Permian Extinctions, Karoo Basin, South Africa." In *The Nonmarine Permian*, edited by Spencer G. Lucas and Kate E. Zeigler, 16–23. Bulletin of the New Mexico Museum of Natural History and Science 30.

Angielczyk, Kenneth D., and Bruce S. Rubidge. 2013. "Skeletal Morphology, Phylogenetic Relationships and Stratigraphic Range of *Eosimops newtoni* Broom, 1921, a Pylaecephalid Dicynodont (Therapsida, Anomodontia) from the Middle Permian of South Africa." *Journal of Systematic Palaeontology* 11: 191–231.

Angielczyk, Kenneth D., and Melony L. Walsh. 2008. "Patterns in the Evolution of Nares Size and Secondary Palate Length in Anomodont Therapsids (Synapsida): Implications for Hypoxia as a Cause of End-Permian Tetrapod Extinctions." *Journal Paleontology* 82: 528–42.

Bambach, Richard K. 1993. "Seafood through Time: Changes in Biomass, Energetics, and Productivity in the Marine Ecosystem." *Paleobiology* 19: 372–97.

Botha, Jennifer, and Roger M. H. Smith. 2006. "Rapid Vertebrate Recuperation in the Karoo Basin of South Africa Following the End-Permian Extinction." *Journal of African Earth Sciences* 45: 502–14.

Botha-Brink, Jennifer, and Kenneth D. Angielczyk. 2010. "Do Extraordinarily High Growth Rates in Permo-Triassic Dicynodonts (Therapsida, Anomodontia) Explain Their Success before and after the End-Permian Extinction?" *Zoological Journal of the Linnean Society* 160: 341–65.

Botha-Brink, Jennifer, Adam K. Huttenlocker, and Sean P. Modesto. 2014. "Vertebrate Paleontology of Nooitgedacht 68: A *Lystrosaurus maccaigi*-Rich Permo-Triassic Boundary Locality in South Africa." In *Early Evolutionary History of the Synapsida*, edited by Christian F. Kammerer, Kenneth D. Angielczyk, and Jörg Fröbisch, 289–304. Dordrecht: Springer.

Coney, Louise, W. Uwe Reimold, P. John Hancox, Dieter Mader, Christian Koeberl, Iain McDonald, Ulrich Struck, Vivi Vajda, and Sandra L. Kamo. 2007. "Geochemical and Mineralogical Investigation of the Permian–Triassic Boundary in the Continental Realm of the Southern Karoo Basin, South Africa." *Palaeoworld* 16: 67–104.

Dai, Lei, Daan Vorselen, Kirill S. Korolev, and Jeff Gore. 2012. "Generic Indicators for Loss of Resilience before a Tipping Point Leading to Population Collapse." *Science* 336: 1175–77.

Day, M. O. 2014. "Middle Permian Continental Biodiversity Changes as Reflected in the Beaufort Group of South Africa: A Bio- and Lithostratigraphic Review of the *Eodicynodon*, *Tapinocephalus* and *Pristerognathus* Assemblage Zones." PhD diss., University of the Witwatersrand.

Dineen, Ashley A., Margaret L. Fraiser, and Jinnan Tong. 2015. "Low Functional Evenness in a Post-Extinction Anisian (Middle Triassic) Paleocommunity: A Case Study of the Leidapo Member (Qingyan Formation), South China." *Global and Planetary Change* 133: 79–86.

Dunne, Jennifer A., Conrad C. Labandeira, and Richard J. Williams. 2014. "Highly Resolved Early Eocene Food Webs Show Development of Modern Trophic Structure after the End-Cretaceous Extinction." *Proceedings of the Royal Society of London B: Biological Sciences* 281 (1782): 20133280.

Eldredge, Niles. 2008. "Hierarchies and the Sloshing Bucket: Toward the Unification of Evolutionary Biology." *Evolution: Education and Outreach* 1: 10–15.

Fröbisch, Jörg. 2014. "Synapsid Diversity and the Rock Record in the Permian-Triassic Beaufort Group (Karoo Supergroup), South Africa." In *Early Evolutionary History of the Synapsida*, edited by Christian F. Kammerer, Kenneth D. Angielczyk, and Jörg Fröbisch, 305–19. Dordrecht: Springer.

Gastaldo, Robert A., Rose Adendorff, Marion Bamford, Conrad C. Labandeira, Johann Neveling, and Hallie Sims. 2005. "Taphonomic Trends of Macrofloral

Assemblages across the Permian–Triassic Boundary, Karoo Basin, South Africa." *Palaios* 20: 479–97.

Gastaldo, Robert A., Sandra L. Kamo, Johann Neveling, John W. Geissman, Marion Bamford, and Cindy V. Looy. 2015. "Is the Vertebrate-Defined Permian-Triassic Boundary in the Karoo Basin, South Africa, the Terrestrial Expression of the End-Permian Marine Event?" *Geology* 43, no. 10. doi:10.1130/G37040.1.

Geertsema, H., D. E. Van Dijk, and J. A. Van den Heever. 2002. "Palaeozoic Insects of Southern Africa: A Review." *Palaeontologia Africana* 38: 19–25.

Hancox, P. John, Mikhail A. Shishkin, Bruce S. Rubidge, and James W. Kitching. 1995. "A Threefold Subdivision of the *Cynognathus* Assemblage Zone (Beaufort Group, South Africa) and Its Palaeogeographical Implications." *South African Journal of Science* 91: 143–44.

Hautmann, Michael, Borhan Bagherpour, Morgane Brosse, Åsa Frisk, Richard Hofmann, Aymon Baud, Alexander Nützel, Nicolas Goudemand, and Hugo Bucher. 2015. "Competition in Slow Motion: The Unusual Case of Benthic Marine Communities in the Wake of the End-Permian Mass Extinction." *Palaeontology* 58: 871–901.

Holland, John H. 1995. *Hidden Order: How Adaptation Builds Complexity*. New York: Basic Books.

Holling, Crawford Stanley. 1996. "Engineering Resilience versus Ecological Resilience." In *Engineering within Ecological Constraints*, edited by Peter C. Schulze, 31–43. Washington, DC: National Academy Press.

Holling, Crawford S., and Lance H. Gunderson. 2001. "Resilience and Adaptive Cycles." In *Panarchy: Understanding Transformations in Human and Natural Systems*, edited by Lanche H. Gunderson and Crawford S. Holling, 25–62. Washington, DC: Island Press.

Hubbell, Stephen P. 2001. *The Unified Neutral Theory of Biodiversity and Biogeography (MPB-32)*. Princeton, NJ: Princeton University Press.

Huttenlocker, Adam K. 2014. "Body Size Reductions in Nonmammalian Eutheriodont Therapsids (Synapsida) During the End-Permian Mass Extinction." *PloS ONE* 9 (2): e87553.

Huttenlocker, Adam K., and Jennifer Botha-Brink. 2013. "Body Size and Growth Patterns in the Therocephalian *Moschorhinus kitchingi* (Therapsida: Eutheriodonta) before and after the End-Permian Extinction in South Africa." *Paleobiology* 39: 253–77.

———. 2014. "Bone Microstructure and the Evolution of Growth Patterns in Permo-Triassic Therocephalians (Amniota, Therapsida) of South Africa." *PeerJ* 2: e325.

Irmis, Randall B., and Jessica H. Whiteside. 2012. "Delayed Recovery of Nonmarine Tetrapods after the End-Permian Mass Extinction Tracks Global Carbon Cycle." *Proceedings of the Royal Society of London B: Biological Sciences* 279: 1310–18.

Irmis, Randall B., Jessica H. Whiteside, and Christian F. Kammerer. 2013. "Nonbiotic Controls of Observed Diversity in the Paleontologic Record: An Example from the Permo-Triassic Karoo Basin of South Africa." *Palaeogeography, Palaeoclimatology, Palaeoecology* 372: 62–77.

Levin, Simon A. 1998. "Ecosystems and the Biosphere as Complex Adaptive Systems." *Ecosystems* 1: 431–36.

Lucas, Spencer G. 2010. "The Triassic Timescale Based on Nonmarine Tetrapod Biostratigraphy and Biochronology." In *The Triassic Timescale*, edited by Spencer G. Lucas, 447–500. London: Geological Society, Special Publications 334.

Mitchell, Jonathan S., Peter D. Roopnarine, and K. D. Angielczyk. 2012. "Late Cretaceous Restructuring of Terrestrial Communities Facilitated the End-Cretaceous Mass Extinction in North America." *Proceedings of the National Academy of Sciences of the United States of America* 109: 18857–61.

Moore, John C., and Peter C. de Ruiter. 2012. *Energetic Food Webs: An Analysis of Real and Model Ecosystems.* Oxford: Oxford University Press.

Murray, James D. 1989. *Mathematical Biology.* Berlin-Heidelberg: Springer-Verlag.

Neubert, Michael G., and Hal Caswell. 1997. "Alternatives to Resilience for Measuring the Responses of Ecological Systems to Perturbations." *Ecology* 78: 653–65.

Neveling, Johann. 2004. "Stratigraphic and Sedimentological Investigation of the Contact between the *Lystrosaurus* and the *Cynognathus* Assemblage Zones (Beaufort Group: Karoo Supergroup)." *Council for Geoscience Bulletin* 137: 1–164.

Nicolas, Merrill, and Bruce S. Rubidge. 2009. "Assessing Content and Bias in South African Permo-Triassic Karoo Tetrapod Fossil Collections." *Palaeontologia Africana* 44: 13–20.

———. 2010. "Changes in Permo-Triassic Terrestrial Tetrapod Ecological Representation in the Beaufort Group (Karoo Supergroup) of South Africa." *Lethaia* 43: 45–59.

Pimm, Stuart L. 1982. *Food Webs.* Netherlands: Springer.

Purvis, Andy, John L. Gittleman, Guy Cowlishaw, and Georgina M. Mace. 2000. "Predicting Extinction Risk in Declining Species." *Proceedings of the Royal Society of London B: Biological Sciences* 267: 1947–52.

Retallack, Gregory J., Roger M. H. Smith, and Peter D. Ward. 2003. "Vertebrate Extinction across Permian–Triassic Boundary in Karoo Basin, South Africa." *Geological Society of America Bulletin* 115: 1133–52.

Roopnarine, Peter. D. 2006. "Extinction Cascades and Catastrophe in Ancient Food Webs." *Paleobiology* 32: 1–19.

Roopnarine, Peter D., and Kenneth D. Angielczyk. 2012. "The Evolutionary Palaeoecology of Species and the Tragedy of the Commons." *Biology Letters* 8: 147–50.

———. 2015. "Community Stability and Selective Extinction during the Permian-Triassic Mass Extinction." *Science* 350: 90–93.

Roopnarine, Peter D., Kenneth D. Angielczyk, Steve C. Wang, and Rachel Hertog. 2007. "Food Web Models Explain Instability of Early Triassic Terrestrial Communities." *Proceedings of the Royal Society Series B* 274: 2077–86.

Rubidge, Bruce S., ed. 1995. "Biostratigraphy of the Beaufort Group (Karoo Supergroup)." *South African Committee for Stratigraphy Biostratigraphic Series* 1: 1–46.

Rubidge, Bruce S. 2005. "27th Du Toit Memorial Lecture Re-uniting Lost Continents–Fossil Reptiles from the Ancient Karoo and Their Wanderlust." *South African Journal of Geology* 108: 135–72.

Rubidge, Bruce S., Douglas H. Erwin, Jahandar Ramezani, Samuel A. Bowring, and William J. de Klerk. 2013. "High-Precision Temporal Calibration of Late Permian Vertebrate Biostratigraphy: U-Pb Zircon Constraints from the Karoo Supergroup, South Africa." *Geology* 41: 363–66.

Scheffer, Marten. 2009. *Critical Transitions in Nature and Society.* Princeton, NJ: Princeton University Press.

Scheffer, Marten, Jordi Bascompte, William A. Brock, Victor Brovkin, Stephen R. Carpenter, Vasilis Dakos, Hermann Held, Egbert H. Van Nes, Max Rietkerk, and George Sugihara. 2009. "Early-Warning Signals for Critical Transitions." *Nature* 461: 53–59.

Scheffer, Marten, Stephen R. Carpenter, Timothy M. Lenton, Jordi Bascompte, William Brock, Vasilis Dakos, Johan van de Koppel, Ingrid A. van de Leemput, Simon A. Levin, Egbert H. van Nes, Mercedes Pascual, and John Vandermeer. 2012. "Anticipating Critical Transitions." *Science* 338: 344–48.

Scheffer, Marten, and Egbert H. van Nes. 2006. "Self-Organized Similarity, the Evolutionary Emergence of Groups of Similar Species." *Proceedings of the National Academy of Sciences of the United States of America* 103: 6230–35.

Schwindt, Dylan M., Michael R. Rampino, Maureen B. Steiner, and Yoram Eshet. 2003. "Stratigraphy, Paleomagnetic Results, and Preliminary Palynology across the Permian-Triassic (P-Tr) Boundary at Carlton Heights, Souther Karoo Basin (South Africa)." In *Impact Markers in the Stratigraphic Record*, edited by Christian Koeberl and Francisca C. Martinez Ruiz, 239–314. Berlin: Springer-Verlag.

Sepkoski, J. John, Jr. 1981. "A Factor Analytic Description of the Phanerozoic Marine Fossil Record." *Paleobiology* 7: 36–53.

Sidor, Christian A., Daril A. Vilhena, Kenneth D. Angielczyk, Adam K. Huttenlocker, Sterling J. Nesbitt, Brandon R. Peecook, J. Sébastien Steyer, Roger M. H. Smith, and Linda A. Tsuji. 2013. "Provincialization of Terrestrial Faunas Following the End-Permian Mass Extinction." *Proceedings of the National Academy of Sciences of the United States of America* 110: 8129–33.

Smith, Roger M. H. 1995. "Changing Fluvial Environments across the Permian-Triassic Boundary in the Karoo Basin, South Africa and Possible Causes of Tetrapod Extinctions." *Palaeogeography, Palaeoclimatology, Palaeoecology* 117: 81–104.

Smith, Roger M. H., and Jennifer Botha-Brink. 2014. "Anatomy of a Mass Extinction: Sedimentological and Taphonomic Evidence for Drought-Induced Die-Offs at the Permo-Triassic Boundary in the Main Karoo Basin, South Africa." *Palaeogeography, Palaeoclimatology, Palaeoecology* 396: 99–118.

Smith, Roger, Bruce Rubidge, and Merrill Van der Walt. 2012. "Therapsid Biodiversity Patterns and Paleoenvironments of the Karoo Basin, South Africa." In *Forerunners of Mammals: Radiation, Histology, Biology*, edited by Anusuya Chinsamy-Turan, 223–46. Bloomington: Indiana University Press.

Smith, Roger M. H., and Peter D. Ward. 2001. "Pattern of Vertebrate Extinctions across an Event Bed at the Permian-Triassic Boundary in the Karoo Basin of South Africa." *Geology* 29: 1147–50.

Steiner, Maureen B., Yoram Eshet, Michael R. Rampino, and Dylan M. Schwindt. 2003. "Funal Abundance Spike and the Permian-Triassic Boundary in the Karoo Supergroup (South Africa)." *Palaeogeography, Palaeoclimatology, Palaeoecology* 194: 405–14.

Thompson, John N. 1994. *The Coevolutionary Process*. Chicago: University of Chicago Press.

Van Valen, Leigh. 1973. "A New Evolutionary Law." *Evolutionary Theory* 1: 1–30.

Vermeij, G. J. 1987. *Evolution and Escalation: An Ecological History of Life*. Princeton, NJ: Princeton University Press.

———. 1995. "Economics, Volcanoes, and Phanerozoic Revolutions." *Paleobiology* 21: 125–52.

———. 2009. *Nature: An Economic History*. Princeton, NJ: Princeton University Press.

Ward, Peter D., Jennifer Botha, Roger Buick, Michiel O. De Kock, Douglas H. Erwin, Geoffrey H. Garrison, Joseph L. Kirschvink, and Roger Smith. 2005. "Abrupt and Gradual Extinction among Late Permian Land Vertebrates in the Karoo Basin, South Africa." *Science* 307: 709–14.

Ward, Peter D., David R. Montgomery, and Roger Smith. 2000. "Altered River Morphology in South Africa Related to the Permian-Triassic Extinction." *Science* 289: 1740–43.

Wilson, Gregory P. 2013. "Mammals across the K/Pg Boundary in Northeastern Montana, USA: Dental Morphology and Body-Size Patterns Reveal Extinction Selectivity and Immigrant-Fueled Ecospace Filling." *Paleobiology* 39: 429–69.

CHAPTER 14

Hierarchy Theory in the Anthropocene

Biocultural Homogenization, Urban Ecosystems, and Other Emerging Dynamics

Michael L. McKinney

Introduction

A full understanding of biospheric evolution, past and future, must encompass two largely separate domains: the ecological or economic domain and the genealogical or hereditary domain. In the words of Eldredge (1986, 352), "Evolution is an outcome of interaction between biological entities involved with the two great classes of biological process — (a) matter/energy transfer, and (b) the maintenance, transmission, and modification of genetically based information." Importantly, these two domains interact in complex and often indirect ways. Furthermore, these interactions have not only changed through time, but these changes are clearly accelerating, most recently from human alterations of the biosphere. My goal here is to examine human impacts on the biosphere in the context of these two domains, and especially what human activities imply for future evolution.

The Anthropocene and the Homogeocene

For the vast majority of geological time, natural processes have driven the hierarchical evolution of the biosphere. The geologically rapid emergence of human impacts on the biosphere, sometimes called the "Great Acceleration" (Steffen et al. 2015), is, of course, dramatically altering patterns of extinction, speciation, and ecological functions in ways that we are only beginning to understand. This emerging epoch of geological history is sometimes called the "Anthropocene," roughly defined as a time when many key environmental parameters are outside of their typical Holocene ranges (Corlett 2015). While most of these data focus on physical parameters such as temperature and, especially, chemical patterns, we can expect that biological parameters in the Anthropocene are inevitably also falling outside of their typical Holocene ranges. Recently, it has been argued that either of two key dates could potentially mark the "official" start of the Anthropocene: a carbon dioxide spike in 1610 or a nuclear atmospheric dust spike in 1964 (Lewis and Maslin 2015).

In addition to the Anthropocene, another term used to describe the emerging period of human domination is the Homogeocene (Rosenzweig et al. 2013; Simberloff 2013), also called the "New Pangaea" (Baiser et al. 2012). This refers to the accelerating process of "biotic homogenization" (McKinney and Lockwood 1999), whereby local native species are replaced with widespread nonnative species. Biotic homogenization is one of the major outcomes of human expansion, driven by the two-step process of (1) habitat modifications ("disturbance") from urbanization, farming, logging, and many other activities and (2) transportation of nonnative organisms into an area via massive transportation systems, the globalized economy, and intentional introductions of species for human goals (aesthetics, food, pets, and sport, for example).

Emerging evidence indicates that the Homogeocene is only just beginning. The compositional blending of biological communities is often referred to as "biological homogenization" or BH (McKinney and Lockwood 1999). BH is typically measured using community similarity indices, such as the Jaccard index, to measure a decrease or increase over time in shared species composing two communities (Baiser et al. 2012). There have been many studies, involving many taxa, indicating that most human activities tend to have a homogenizing effect among affected ecosystems, especially in the long term (Dar et al. 2014). Not surprisingly, local areas of intensive environmental modification, such as large cities and commercial farms, tend to be the most homogenized (McKinney and Lockwood 1999; McKinney 2006). Such highly disturbed ecosystems share many widely introduced "weedy" generalist, nonnative species, and they have lost many local native species. However, at larger scales, such as comparisons among the biota of nations between continents, the increase in biotic similarity thus far is a much smaller fraction (Baiser et al. 2014).

On the other hand, it seems quite likely that this incipient homogenization of the Earth's biota will continue at an accelerating pace. Most obviously, this will occur from the projected increase and continued spread of human populations across the planet. Adding to this will be the continued increase in trade and transportation in the globalizing economy. This can only lead to increasing loss of native habitats and species and increasing (intentional and accidental) introduction of nonnative species. In addition, many species already introduced into new regions have yet to attain the full extent of their potential geographic range. Species with this "invasion debt" (Essl et al. 2015) will also contribute to future biotic homogenization.

While we can infer that biotic homogenization will likely increase for many decades and even centuries to come, this raises the crucial question of how thoroughly the biosphere will eventually become homogenized and what impact this would have on global biodiversity. This is of course very speculative, but one estimate based on species-area considerations of a world where all regional barriers among vertebrates are removed indicates that 49 percent of terrestrial vertebrate genera would become extinct (Rosenzweig et al. 2013), perhaps within the next century.

The Genealogical Domain in the Homogeocene

One of the most significant effects of the intermixing of biotas is the blending of formerly isolated gene pools. While there is much attention on the ecological impacts of species introductions, discussed below, it is clear that biotic homogenization often involves substantial genetic intermixing. Specifically, this is the hybridization of species or otherwise genetically distinct and genetically separated populations. Hybridization between closely related species is a natural process that has typically affected 10 percent of animals and 25 percent of plants (Mallet 2008).

With the advent of anthropogenic mixing of biotas, the occurrence and extent of hybridization has increased dramatically, with potentially profound implications for conservation of species and future evolution of the biosphere. Of immediate concern is that genetic "pollution" or "swamping" by widespread invasive species is causing the extinction of many closely related endemic species. A review by Rhymer and Simberloff

(1996) documented many examples where endangered and threatened animals and plants were losing their genetic identity from extensive interbreeding with a closely related introduced species. Some of the more notable examples would include the loss of several endemic duck species from hybridization with the range-expanding mallard duck and the loss of several wild felid and canid species from interbreeding with domestic cats and dogs, respectively (Rhymer and Simberloff 1996). Vuillaume and others (2015) describe an excellent recent example of the gradual genetic disappearance of the endangered island endemic *Iguana delicatissima* by interbreeding with its invasive and widespread relative, *Iguana iguana*.

Aside from outright elimination (extirpation) of local native endemic species, another long-term impact of hybridization on the biosphere is the loss of fitness of many native populations. In some cases, the influx of genetic diversity via the natural process of hybridization will increase the fitness of the local population and even encourage adaptation and evolution (Arnold 2006). However, this does not seem to be the case with much of the hybridization occurring today, wherein hybridizing local native populations appear to become less well adapted to their local environments (Rhymer and Simberloff 1996).

What is the long-term effect on the biosphere of extensive anthropogenic genetic intermixing of biotas? In the short term, this may often increase local genetic diversity but at the cost of reducing global genetic diversity. This pattern has been most commonly discussed at the species level, where it has been noted that species introductions often increase local diversity while eroding global diversity (Sax and Gaines 2003). However, the long-term effects of intermixing on biospheric evolution may be more sanguine. It is clear that many species are adapting rather rapidly to human-modified habitats throughout the globe (Palumbi 2001).

Given long periods of time for local and hybridized populations to become adapted to their habitats and time for continued gene flow into those habitats, it seems likely that some redifferentiation and speciation will eventually occur. This has been the explicit expectation of some biologists, who predict that a homogenized biosphere will reach a new (if lower than current) equilibrium of species diversity in the next few centuries or millennia, a new equilibrium that is driven in part by adaptation to new habitats and climatic regimes (Rosenzweig et al. 2013; McKinney 2014).

The Ecological Domain in the Homogeocene

There is increasing recognition among conservation biologists that "novel ecosystems" not only are inevitable but must be studied as functional systems. Novel ecosystems are generally defined as biological communities in which the species composition and/or function have been completely transformed from the historic (natural) system (Hobbs et al. 2009). Examples would include most of the highly altered ecosystems found in urban and suburban areas and most agricultural systems. These tend to be dominated by invasive and other nonnative species that were not part of the natural system and were either intentionally cultivated by humans or have colonized disturbed habitats on their own. It is also possible to describe a "hybrid" ecosystem that is transitioning to the novel state, defined as one that retains characteristics of the historic system but whose composition or function now lies outside the historic range of variability (Hobbs et al. 2009).

Novel ecosystems have major implications for hierarchy theory and for practical goals of biodiversity conservation. For example, there is much discussion over the functionality of novel ecosystems. Do these rapidly assembled and very dynamic (with high species turnover) ecosystems have comparable ecosystem functions (and services for humans) to natural systems? By "ecosystem services," I refer to the many contributions that ecosystems provide for human welfare: clean water, waste decomposition, climate regulation, and pollination, to name just a few. It has been argued that, in fact, novel ecosystems do have comparable functionality and service capacities, and they might therefore be used to replace natural ecosystems lost by human impacts (Hobbs et al. 2006). In addition, many restoration ecologists have suggested that novel ecosystems can never be fully restored: the abiotic and/or biotic parameters of the system have been altered so much by human impacts that they are irreversible. Therefore, efforts at total restoration are doomed to fail, and we should focus on more realistic restoration goals, such as restoring ecosystem function and structure (Hobbs et al. 2009). In fact, many anthropogenically created hybrid and novel ecosystems have greater species diversity, most commonly measured as species richness, than the native ecosystems that they have replaced (Kowarik 2011). Given that ecosystem function often increases with species diversity, one could argue that hybrid and novel ecosystems could often have greater functional properties than those they replace.

The increasingly widespread replacement of native ecosystems with hybrid or novel ecosystems has an interesting, and perhaps alarming, secondary consequence: compositional and functional homogenization of the biosphere. Because these anthropogenically altered ecosystems are composed of the relatively small subset of species that can disperse into and adapt to human habitats (especially cities, farms, and so on), these hybrid and novel ecosystems tend to be both compositionally and functionally similar.

Of particular relevance to the ecological domain is the increasing number of studies finding that humans cause "functional homogenization" patterns among ecosystems (Clavel et al. 2011). Studies of several taxa, including fishes (Villeger et al. 2014) and birds (Godet et al. 2015), have shown that functional homogenization is driven by the replacement of specialist species with generalist species, resulting in simplified ecosystems that have functional similarities shared among disturbed habitats. This not only includes such disturbances as farming, urbanization, and logging but may also be driven on a global scale by climate change, which is apparently increasing functional homogenization in birds (Davey et al. 2012). Another interesting dimension of functional homogenization is that it is often decoupled from taxonomic homogenization—that is, changes toward a shared species composition (Villeger et al. 2014).

Cities as Primary Engines of Change in the Ecological Domain

In 1900, only 13 percent of humanity inhabited urban areas. Now more than half of the human population lives in an urban environment, and that proportion is predicted to rise to nearly 70 percent by 2050 (Seto et al. 2013). Unfortunately for conservation goals, most of this growth is occurring in the remaining biodiversity-rich areas of Latin America, Asia, and Africa. This massive episode of urbanization obviously has many profound implications for the dynamics of the biosphere, and it is useful to view them through the lens of the ecological and genealogical domains.

In terms of the ecological domain, urban habitats create some of the most novel ecosystems on the planet. (In this chapter, "urban" is loosely used to refer to any habitat associated with significant human settlements including urban core, suburban, and periurban habitats.) There are two basic reasons for this. One reason is dispersal: as transportation hubs, human settlements are foci for an unending stream of intentional and unintentional biological introductions (von der Lippe and Kowarik 2008). As a result, the high "propagule pressure" of individuals, seeds, spores, and other reproductive agents virtually guarantees that a large number of nonnative species find their way into urban habitats.

A second reason for the extreme ecosystem novelty is the habitat itself: urban habitats are among the extremely altered ("disturbed") environments on the planet. We often forget that in the process of creating a habitat to meet the relatively narrow needs and preferences of humans, we drastically modify the physical parameters of the local natural environment. As Marzluff (2008) has discussed, compared to most other human activities such as logging, farming, and mining, urbanization is far greater in terms of both temporal duration (often lasting centuries) and magnitude of modification from the natural baseline environment. As a result, in heavily urbanized areas, relatively few species have features that can be evolutionarily exapted (i.e., functionally co-opted; see Gould and Vrba 1982), allowing them to thrive in such extreme habitats. Such species, sometimes called "urban exploiters" (McKinney 2002), tend to be the same exotic commensals that have accompanied human expansion for many years. Consequently, the majority of species in biological communities of downtown (urban core) areas of most cities typically consist of nonnative species (reviewed in McKinney 2006). Familiar animal examples would include the rock dove, house sparrow, house mouse, and brown rat. Similarly, the gingko tree, tree of heaven, dandelion, and clover are probably familiar to most urban visitors in many parts of the world.

Driven by this twofold dynamic of propagule influx plus intensive habitat creation, these urban ecosystems are notable for their extreme novelty and their spatial uniformity. Their uniformity emerges from the similar conditions found in nearly all highly urbanized environments, especially the abundance of impervious surfaces and the heat island effect. This physical uniformity, combined with the relatively small subset of species that occupy such habitats, has promoted biotic homogenization among large cities (McKinney 2006) with some evidence of functional homogenization (Pickett et al. 2011).

The extreme novelty of urban ecosystems, I would argue, is one of the most interesting and also neglected aspects of this dynamic. Compared to natural ecosystems, urban systems are often very complex, with greater species diversity. For example, large urban gardens have more species than any habitat on Earth, even more than tropical forests (Thompson et al. 2003). In addition, urban ecosystems are assembled from species that have little historical connection to the local ecology, and they are part of an extremely dynamic and frequently fluctuating environment that requires rapid adaptation and changing interactions with other species. As a result, urban ecosystems are complex but very dynamic and loosely connected and often characterized as "amplified" in having very high inputs of energy and materials from anthropogenic activities (Adler and Tanner 2013).

In addition to their high amplification, loose connectivity, and (often) high complexity, urban ecosystems can have one other trait that renders them truly novel compared to natural systems: in some cases, they are functionally supplemented by human technology. Urban habitats are of course largely engineered to accommodate human needs. A recent development has been the increasing use of "green infrastructure" to more efficiently provide ecosystem functions ("services") to meet human needs such as water purification and stormwater regulation. Some of these efforts are based on so-called engineered ecosystems that are explicitly designed to combine biological communities such as wetlands with technological systems such as pumps, pipes, and other hydraulic equipment (Adler and Tanner 2013).

These combined biological-technological ecosystems are clearly extremely novel in every sense of the word, not only combining a diverse array of previously unassembled species, but also including human technology into its functional dynamics. For lack of a better name, I have suggested these be called *cyborg ecosystems* (McKinney, in prep.) because the cyborg concept describes a biological organism that is functionally enhanced by technology. This, in fact, describes these urban ecosystems except that it is biological communities, rather than a single organism, whose function is technologically enhanced.

Cities as Primary Engines of Change in the Genealogical Domain

The "Great Acceleration" of environmental changes wrought by humans in the Anthropocene has also caused an acceleration of evolutionary changes. While there has (justifiably) been much discussion of anthropogenic increase in extinction rates, it is also true that many species are adapting to human changes at an accelerating rate. Indeed, humans have been called the "world's greatest evolutionary force," reflecting three fundamental ways in which we cause evolutionary change (Palumbi 2001): (1) rapid evolution of resistance to antibiotics, pesticides, and other attempts by humans to control other species, (2) rapid evolution of many invasive species introduced into new habitats by humans, and (3) rapid evolution of heavily harvested species, such as many fish, game, and plant species to help them adapt to the novel human "superpredation."

While these three forces occur across the globe, they tend to be concentrated in urban areas and are greatly enhanced by the presence of high human abundance. It is, therefore, not surprising that increasing numbers of studies are documenting rapid evolutionary changes in many taxa in urban habitats. Much of this evolution begins with behavioral and physiological adaptations. Nocturnal lighting in cities, for example, has drastically altered behavior in many animals from insects to birds, perhaps most commonly by extending foraging activity to a longer day (Ditchkoff et al. 2006). Noise is another interesting selective element, causing urban birds to sing more at night than their rural conspecifics (Fuller et al. 2007). Urban plants extend their growing season due to higher temperatures in urban habitats (Zhang et al. 2004).

Subsequently, genetic evolutionary changes also occur rapidly from the intense and dynamic urban selective pressures. Urban house finches in Phoenix, Arizona, for example, are showing a genetic adaptation for increased beak size and bite force as an adaptation to the abundance of sunflower seeds at feeders (Badyaev et al. 2008). In contrast, house finch populations in rural areas that consume much smaller grass and cactus seeds retain the smaller beaks. A more general evolutionary adaptation in urban birds is an apparent tendency for them to evolve smaller body sizes relative to non-urban populations, perhaps as an adaptation to poor food quality or urban predator avoidance (Chamberlain et al. 2009).

The implications of all this for the planet's genealogical domain are potentially quite profound. On the one hand, there are very powerful selective forces acting on each urban species, causing it to rapidly adapt to its own local habitat. At first glance, we might argue that urbanization could therefore lead to rapid speciation and the production of many branches on the evolutionary tree of life. And this would be true if cities were isolated from each other and each city represented a relatively unique array of different habitats and challenges. However, as we have seen, in fact the opposite is true: cities are a powerful homogenizing force in at least two basic ways. One, as centers of species importation, they serve as foci of genetic mingling. Two, as they are built to meet the relatively predictable needs of humans, cities tend to share many physical characteristics (heat islands, impervious surfaces, fragmented green spaces, many predatory cats, and so on; McKinney 2006). As a result, the selective forces on species in all cities are often quite similar, especially compared to the surrounding natural areas. Evidence for these convergent selective pressures is illustrated the "urban bird syndrome," which is the documented observation that successful urban birds in all cities tend to share a suite of traits: omnivorous diet, relatively large brain size, sedentary, tree or rock nesting, and high fecundity (Moller 2009).

Furthermore, there is emerging evidence that these homogenizing tendencies are indeed becoming manifest at the phylogenetic level. The most successful urban birds, for example, are disproportionately concentrated in just two families: the Corvidae (crows, jays, magpies, ravens) and the Paridae (tits and chickadees). Similarly, an analysis of plants in German cities indicates that successful urban plant species tends to come from a relatively few number of closely related groups—that is, "phylogenetic clustering" (Knapp et al. 2008).

Cities as Primary Engines of Change in the Human Genealogical Domain

Due to its major role in the human environment, urbanization will undoubtedly influence future human biological and cultural evolution. Eldredge and Grene (1992) have made the key observation that social systems in animals tend to "reintegrate" the genealogical and ecological domains in units above the level of the individual organism. More specifically, they state, "Social systems patently fit in at the local level, roughly at the level of demes and avatars. Social systems, moreover, appear to represent a distinct reversal of the ever-widening gulf between economic and reproductive activity as one goes from organism to deme/avatar to species/ecosystem" (Eldredge and Grene 1992, 131–32).

In the case of humans, on which I will focus here, this reintegration can be readily seen in our evolutionarily older social systems, such as at the tribal level, whereby the tribe clearly influenced both genealogical (reproductive) and economic (ecological) processes of our species. In this case, I use "tribe" in the traditional anthropological definition of a relatively small band or collection of bands of hunters and gatherers, which formed the basic social unit for much of human evolutionary history. Tribal membership played a large role in marriage and family and, as a hunting and gathering unit, also influenced the acquisition of matter and energy resources. It is, therefore, especially interesting in the context of this chapter to note the role of cities, as our much newer and larger social system, in their influence on the genealogical and economic domains.

Turning first to the genealogical domain, I would suggest that the most obvious and profound consequence of cities has been the accelerating genetic intermixing of human populations. For hundreds of thousands of years, limitations on human dispersal meant that the local human (and protohuman) social group was a much more isolated and coherent spatiotemporal unit than we see today—so much so that increasing numbers of evolutionary biologists are making the case for "group selection" in hominid evolution (Puurtinen and Mappes 2009; Wilson 2012). In this view, natural selection has operated to favor not only individuals but also the survival of certain social groups (e.g., bands or tribes of hunters and gatherers) over others. If true, then natural selection operating on the group level (e.g., selection of groups with more or less aggressive tendencies or more stable social structures) could have influenced the reproductive success of the group as well as the individuals and genes within those groups (Wilson 2012).

The rise of cities, however, has greatly altered this dynamic: cities have increasingly become focal points of trade and, especially, immigration from many areas. This trend has of course only become more prominent in recent years with "globalization" increasing exponentially (Friedman 2008). The subsequent accelerating genetic mixing of human populations in cities must certainly be reducing, if not eliminating, any natural selection of the local social system (e.g., tribe, village) that once operated in these areas. Instead, it would seem that the primary driver of human gene pool composition in cities would be the huge influx of new genes from other geographic areas.

This brings us back to the earlier discussion of the future of the Homogeocene: how far will anthropogenic blending of the biosphere go? In this case, how thoroughly mixed will the gene pool of humanity become? I noted above that emerging evidence indicates that biological communities are, thus far, still far from being fully homogenized. The same is clearly true for humans, where even many urban areas retain many distinctive local genetic elements. But as the globalization process continues exponentially, we can only speculate how long this will remain true.

Cities as Primary Engines of Change in the Human Ecological Domain

Cities have clearly an enormous influence on how human social groupings have rapidly influenced the flow of matter and energy through natural ecosystems. It is now estimated that humans appropriate about 28 percent of available photosynthetic production (Ma et al. 2012), and most of this is either directly or indirectly appropriated by humans in urban habitats. Indeed, viewed from the perspective of increasingly efficient use of material and energy flow, the evolution of increasingly larger (more populated) cities seems virtually inevitable. This is demonstrated by an emerging literature of urban scaling, documenting that many social, economic, and physical variables (including matter and energy consumption) show relatively robust patterns of change with increasing city size. For example, both per-capita wealth and rate of technological innovation consistently increase by about 15 percent as cities double in size (Bettencourt et al. 2007).

More important, these patterns indicate that, as human settlements grow in size (population), they become more efficient in "metabolizing" matter and energy resources, providing more resources (and wealth) per capita. Data from hundreds of cities of different sizes show, for example, that infrastructure efficiency (e.g., roads, energy transmission) and income (GDP) scale in a predictable way with population size (Bettencourt et al. 2007; Bettencourt 2013). Importantly, these increases are "superlinear" in that they increase at a faster (disproportionate) rate relative to population increase. This pattern is quite robust with these (and related variables) increasing systematically on a per capita basis by about 15 percent with every doubling of a city's population, regardless the city's initial size. Of equal interest is that cultural variables such as intellectual and technological innovation (e.g., new patents, inventions) also scale superlinearly with population size (Bettencourt et al. 2007). Social life, such as the number of social contacts per capita, also increases in a superlinear pattern (Schlapfer et al. 2014).

A key implication of these urban scaling results is a sense of "inevitability" about the evolution of dense human aggregations. There are so many physical advantages (infrastructure, resource gain and usage), economic advantages (specialization of labor), and cultural advantages (innovation, social life) that the growth of cities across evolutionary time would seem to be strongly selected for, perhaps at both the level of the individual organism and the group (city). Although this is speculative, this inference of an evolutionary universality is supported by the finding that urban scaling patterns are found in the archaeological record as well, and these patterns seem to parallel those found in the historical record (Ortman et al. 2014).

Cities as Primary Engines of Biocultural Homogenization

In addition to the homogenization of biological composition and function noted above, cities are increasingly homogenizing human societies. As with the loss of global biodiversity, this will inevitably lead to the loss of cultural diversity. Indeed, there are so many parallels between biological and cultural homogenization that Rozzi (2012) has introduced the term *biocultural homogenization* as a unifying dynamic describing much of what may occur in the Anthropocene. Specifically, this dynamic leads to the disruption of coevolutionary interrelationships between cultures and their land and massive replacements of native biota and cultures by a few cosmopolitan species, languages, and cultures. For example, of the nearly seven thousand languages currently spoken, for the first time in human history, more than half of the world's population speaks just seven of them (Rozzi 2013). This growing disproportion is a direct consequence of the homogenizing effects of colonization events.

I have already addressed the role of cities in biological homogenization, but it is clear that cities also play a large role in cultural homogenization. Rozzi (2013) regards the migration of rural human populations into cities as one of the primary drivers of biocultural homogenization. In promoting similar habitats, urban settlements promote not only similar biological communities but also similarity of culture, encouraging the growth of a "cosmopolitan culture." Obvious examples of material culture would include the rapid spread of housing styles, clothing fashions, and restaurant chains in many cities of the world. This includes many relatively isolated indigenous cultures (e.g., American Indian tribes in South America) that are increasingly adopting elements of the cosmopolitan global culture. Ritzer (1993) refers to the homogenization of cultures through global domination by multinational corporations as "McDonaldization."

Biocultural homogenization clearly represents a drastic alteration in the evolution of life. In addition to the dramatic changes in both the genealogical and ecological domains described above, we can add a similarly drastic change in an emerging "cultural" domain, representing the domain of cultural ideas. This is also largely tied to urbanization and would include the spread of not only material culture but also the culture of ideas: the spread of a few values, beliefs, and many other "memes" (Buskes 2013) that are replacing those of long-standing indigenous cultures (Maffi and Woodley 2010). While such intangibles are difficult to measure, the rapid and measurable spread of a few languages at the expense of many others, noted above, serves to illustrate this point.

In the context of hierarchy theory, the overarching theme of this chapter is the blending, at many scales, of the biological and cultural units of our biosphere and global human society. Driven by the inexorable spread of humans into all corners of the globe, we see, in the genealogical realm, the intermixing of units at several scales of the hereditary hierarchy: genes, populations, species, and perhaps even monophyletic taxa. Similarly, in the ecological realm, we see intermixing of units at several scales of the hierarchy to produce new combinations of avatars and, of special importance for ecological functions and services, "novel ecosystems." Given that diversity is the raw material of evolution, it is difficult to see how the anthropogenically driven reduction of diversity of genealogical units, ecological units, and cultural units (social groups) into a far fewer number of units portends well for the future evolution of the biosphere (Maffi and Woodley 2010). While it seems unlikely that humans will succeed in completely homogenizing the biosphere, hierarchy theory would suggest that the overall diversity of life on this planet will ultimately lose much of its uniqueness at several scales of the genealogical and ecological domains.

References

Adler, Frederick, and Colby J. Tanner. 2013. *Urban Ecosystems*. Cambridge: Cambridge University Press.

Arnold, Michael L. 2006. *Evolution through Genetic Exchange*. Oxford: Oxford University Press.

Badyaev, Alexander V., Rebecca L. Young, Kevin P. Oh, and Clayton Addison. 2008. "Evolution on a Local Scale: Developmental, Functional, and Genetic Bases of Divergence in Bill Form and Associated Changes in Song Structure between Adjacent Habitats." *Evolution* 62: 1951–64.

Baiser, Benjamin, Julian D. Olden, Sydne Record, Julie L. Lockwood, and Michael L. McKinney. 2012. "Pattern and Process of Biotic Homogenization in the New Pangaea." *Proceedings of the Royal Society B* 279: 4772–77.

Bettencourt, Luis M. A. 2013. "The Origins of Scaling in Cities." *Science* 340: 1438–41.

Bettencourt, Luis M. A., José Lobo, Dirk Helbing, Christian Kuhnert, and Geoffrey B. West. 2007. "Growth, Innovation, Scaling, and the Pace of Life in Cities." *Proceedings of the National Academy of Science of the United States of America* 104: 7301–6.

Buskes, C. 2013. "Darwinism Extended: A Survey of How the Idea of Cultural Evolution Evolved." *Philosophia* 41: 661–91.

Chamberlain, Daniel E., Arthur R. Cannon, Michael P. Toms, and Kevin J. Gaston. 2009. "Avian Productivity in Urban Landscapes: A Review and Meta-analysis." *Ibis* 151: 1–18.

Clavel, Joanne, Romain Julliard, and Vincent Devictor. 2011. "Worldwide Decline of Specialist Species: Toward a Global Functional Homogenization?" *Frontiers in Ecology and the Environment* 9: 222–28.

Corlett, Richard T. 2015. "The Anthropocene Concept in Ecology and Conservation." *Trends in Ecology and Evolution* 30: 36–41.

Dar, Pervaiz A., and Zafar A. Reshi. 2014. "Components, Processes and Consequences of Biotic Homogenization: A Review." *Contemporary Problems of Ecology* 7: 123–36.

Davey, Catherine M., Dan E. Chamberlin, and Stuart E. Newson. 2012. "Rise of the Generalists: Evidence for Climate Driven Homogenization in Avian Communities." *Global Ecology and Biogeography* 21: 568–78.

Ditchkoff, Stephen S., Steven T. Saalfeld, and Catherine J. Gibson. 2006. "Animal Behavior in Urban Ecosystems." *Urban Ecosystems* 9: 5–12.

Eldredge, Niles. 1986. "Information, Economics, and Evolution." *Annual Review of Ecology and Systematics* 17: 351–69.

Eldredge, Niles, and Marjorie Grene. 1992. *Interactions: The Biological Context of Social Systems*. New York: Columbia University Press.

Essl, Franz, Stefan Dullinger, Wolfgang Rabitsch, Philip E. Hulme, Petr Pysek, John R. Wilson, and David M. Richardson. 2015. "Delayed Biodiversity Change: No Time to Waste." *Trends in Ecology and Evolution* 30: 375–78.

Friedman, Thomas. 2008. *Hot, Flat, and Crowded: Why We Need a Green Revolution—and How It Can Renew America*. New York: Farrar, Straus and Giroux.

Fuller, Richard A., Philip H. Warren, and Kevin J. Gaston. 2007. "Daytime Noise Predicts Nocturnal Singing in Urban Robins." *Biology Letters* 3: 368–70.

Godet, Laurent, Pierre Gauzere, Frederic Jiguet, and Vincent Devictor. 2015. "Dissociating Several Forms of Commonness in Birds Sheds New Light on Biotic Homogenization." *Global Ecology and Biogeography* 25: 416–26.

Hobbs, R. J., James Aronson, Jill S. Baron, Peter Bridgewater, and Viki A. Cramer. 2006. "Novel Ecosystems: Theoretical and Management Aspects of the New Ecological World Order." *Global Ecology and Biogeography* 15: 1–7.

Hobbs, Richard J., Eric Higgs, and James A. Harris. 2009. "Novel Ecosystems: Implications for Conservation and Restoration." *Trends in Ecology and Evolution* 24: 599–605.

Knapp, Sonja, Ingolf Kuehn, Oliver Schweiger, and Stefan Klotz. 2008. "Challenging Urban Species Diversity: Contrasting Phylogenetic Patterns across Plant Functional Groups in Germany." *Ecology Letters* 10: 1054–64.

Kowarik, Ingo. 2011. "Novel Urban Ecosystems, Biodiversity, and Conservation." *Environmental Pollution* 159: 1974–83.

Lewis, Simon L., and Mark A. Maslin. 2015. "Defining the Anthropocene." *Nature* 519: 171–80.

Ma, Ta, and Chenghu H. Zhou. 2012. "Simulating and Estimating Tempo-Spatial Patterns in Global Human Appropriation of Net Primary Production (HANPP): A Consumption-Based Approach." *Ecological Indicators* 23: 660–67.

Maffi, Luisa, and Ellen Woodley, eds. 2010. *Biocultural Diversity Conservation*. Washington, DC: Earthscan.

Mallet, James. 2008. "Hybridization, Ecological Races and the Nature of Species: Empirical Evidence for the Ease of Speciation." *Philosophical Transactions of the Royal Society B* 363: 2971–86.

Marzluff, John M. 2008. "Island Biogeography in an Urbanizing World." *Urban Ecology* 8: 355–71.

McKinney, Michael L. 2002. "Urbanization, Biodiversity, and Conservation." *Bioscience* 52: 883–90.

———. 2006. "Urbanization as a Major Cause of Biotic Homogenization." *Biological Conservation* 127: 247–60.

———. 2014. "Impacts of Global Warming, Habitat Loss, and Homogenization on Global Biodiversity." *Evolutionary Ecology Research* 16: 285–89.

McKinney, Michael L., and Julie L. Lockwood. 1999. "Biotic Homogenization: A Few Winners Replacing Many Losers in the Next Mass Extinction." *Trends in Ecology and Evolution* 14: 450–53.

Moller, Anders P. 2009. "Successful City Dwellers: A Comparative Study of the Ecological Characteristics of Urban Birds in the Western Palearctic." *Oecologia* 159: 849–58.

Ortman, Scott G., Andrew H. F. Cabaniss, Jennie O. Sturm, and Luis M. A. Bettencourt. 2014. "The Prehistory of Urban Scaling." *PLoS ONE* 9: e87902. doi:10.1371/journal.pone.0087902.

Palumbi, Stephen R. 2001. "Humans as the World's Greatest Evolutionary Force." *Science* 293: 1786–90.

Pickett, Stewart T., Mary Cadenasso, Jean M. Grove, Christopher G. Boone, and Peter M. Groffman. 2011. "Urban Ecological Systems: Scientific Foundations and a Decade of Progress." *Journal of Environmental Management* 92: 331–62.

Puurtinen, Mikael, and Tapio Mappes. 2009. "Between-Group Competition and Human Cooperation." *Proceedings of the Royal Society B* 276: 355–60.

Rhymer, J. M., and Daniel Simberloff. 1996. "Extinction by Hybridization and Introgression." *Annual Review of Ecology and Systematics* 27: 83–109.

Ritzer, George. 1993. *The McDonaldization of Society*. London: Sage.

Rosenzweig, Michael L., Vanessa Buzzard, John Donoghue, Gavin Lehr, and Natasha Mazumdar. 2013. "Patterns in the Diversity of the World's Land Vertebrate Genera." *Evolutionary Ecology Research* 15: 86–882.

Rozzi, Ricardo. 2012. "Biocultural Ethics: The Vital Links between the Inhabitants, Their Habits and Regional Habitats." *Environmental Ethics* 34: 27–50.

———. 2013. "Biocultural Ethics: From Biocultural Homogenization toward Biocultural Conservation." In *Linking Ecology and Ethics for a Changing World: Values, Philosophy, and Action*, edited by Ricardo Rozzi, 9–31. Dordrecht, Netherlands: Springer.

Sax, Dov F., and Stephen D. Gaines. 2003. "Species Diversity: From Global Decreases to Local Increases." *Trends in Ecology and Evolution* 18: 561–66.

Schlapfer, Markus, Luis M. A. Bettencourt, Sebastian Grauwin, Mathias Raschke, Rob Claxton, Zbigniew Smoreda, Geoffrey B. West, and Carlo Ratti. 2014. "The Scaling of Human Interactions with City Size." *Journal of the Royal Society Interface* 11: 20130789. http://dx.doi.org/10.1098/rsif.2013.0789.

Seto, Karen C., Burak Guneralp, and Lucy R. Hutyra. 2013. "Global Forecasts of Urban Expansion to 2030 and Direct Impacts on Biodiversity and Carbon

Pools." *Proceedings of the National Academy of Science of the United States of America* 109: 16083–88.

Simberloff, Daniel. 2013. "Introduced Species, Homogenizing Biotas and Cultures." In *Linking Ecology and Ethics for a Changing World: Values, Philosophy, and Action*, edited by Ricardo Rozzi, 33–48. Dordrecht, Netherlands: Springer.

Steffen, Will, Wendy Broadgate, Lisa Deutsch, Owen Gaffney, and Corneila Ludwig. 2015. "The Trajectory of the Anthropocene: The Great Acceleration." *Anthropocene Review* 2: 81–98.

Thompson, K., Kevin C. Austin, Richard M. Smith, Philip H. Warren, Penny G. Angold, and Kevin J. Gaston. 2003. "Urban Domestic Gardens (I): Putting Small-Scale Plant Diversity in Context." *Journal of Vegetation Science* 14: 71–78.

Villeger, Sebastien, Gael Grenouillet, and Sebastien Brosse. 2014. "Functional Homogenization Exceeds Taxonomic Homogenization among European Fish Assemblages." *Global Ecology and Biogeography* 23: 1450–60.

von der Lippe, Moritz, and Ingo Kowarik. 2008. "Do Cities Export Biodiversity? Traffic as Dispersal Vector across Urban-Rural Gradients." *Diversity and Distributions* 14: 18–25.

Vuillaume, Barbara, Victorien Vallette, Olivier Lepais, Frederic Grandjean, and Michel Breuil. 2015. "Genetic Evidence of Hybridization between the Endangered Native Species Iguana Delicatissima and the Invasive Iguana Iguana in the Lesser Antilles: Management Implications." *PLoS ONE*. doi:10.1371/journal.pone.0127575.

Wilson, Edward O. 2012. *The Social Conquest of Earth*. New York: W. W. Norton.

Zhang, Xiaoyang Y., Mark A. Friedl, Crystal B. Schaaf, Alan H. Strahler, and Annemarie Schneider. 2004. "The Footprint of Urban Climates on Vegetation Phenology." *Geophysical Research Letters* 31: L12209.

Conclusion

Hierarchy Theory and the
Extended Synthesis Debate

Telmo Pievani

Summing Up: Hierarchy Theory in a Nutshell

As several of the chapters in this volume show, the so-called hierarchy theory of evolution is an ecological and multilevel approach to potentially any evolutionary phenomena. According to this view, the micro- and the macrodimensions of the evolutionary patterns and processes are conceptually separated but empirically intertwined levels. The former pertain to any change inside the biological populations, the latter pertain to the birth, death, persistence, and branching of species and higher taxa (Eldredge and Lieberman 2014). The vertical dimension of evolution (i.e. changes due to the continuous and differential gene transmission) is strictly interrelated to the horizontal sphere of climatic and geographical factors, able to produce episodic ecological changes that, in turn, affect genetic and genealogical relations among organisms, populations, and species.

According to Eldredge's "sloshing bucket" model, environmental events may have very different magnitudes—namely, localized or subregional ecosystem disturbances, regional disturbances, and global environmental disruptions (the case study proposed in this volume by Roopnarine and Angielczyk explains how a paleocommunity responded, in ecological time, after particular types and magnitudes of disturbances). The greater the magnitude, the greater ecosystems change (from rapid recoveries to turnover pulses of species to mass extinctions at the maximum, with subsequent new radiations of larger groups); the greater the loss of higher taxa, the more different will be the newly evolved taxa. We may thus consider evolutionary units as belonging to a multilevel hierarchy: the microevolutionary levels of genes and their organismic environments; the intermediary, basic level of organisms; and the macroevolutionary levels of groups and species (as for the plurality of definitions of "levels of organization," see Umerez in this volume).

At the highest level, ecosystems (containing abiotic elements) show typical hierarchical and organizational features (see Cooper et al. in this volume). On the opposite side, there are hierarchical layers also below the organismic level: according to Gregory et al. in this volume, an evolutionary analysis of the transposable elements in the human genome demonstrates the necessity of a multilevel approach to explain genome features and their variation all across the phylogenetic tree of life. Transposable elements are biological entities that reproduce, compete, and evolve at their focal level, with cross-level effects. Caianiello's chapter in this volume outlines a tentative hierarchy theory at the organismic and suborganismic level, elucidating the genetic and epigenetic causality of phenotypic determination at the organismic level.

Each level should be considered autonomous (with its own patterns and focal "individual" entities—about the emergence of novel "individuals" in a hierarchical perspective, see Pavličev et al. in this volume) and at the same time interdependent with the others (cross-level patterns and processes). An event produced at the highest level can have repercussions downward at lower levels, or vice versa, a microevolutionary transformation can be propagated upward to higher levels. Among the downward causations, according to the quite original view proposed by Daniel W. McShea in this volume, even an apparent teleology is a byproduct of an inherent hierarchical organization of living systems.

The two "walls" of the bucket metaphorically represent the two different hierarchies described in this volume. Each hierarchy goes from microevolutionary to macroevolutionary levels and vice versa. The genealogical hierarchy, concerning reproduction or replication, involves genetically based information systems: the microevolutionary level of genes is part of the upper level of organisms, which are nested into local breeding populations and species. The ecological hierarchy is about matter-energy transfer systems: organisms are parts of local conspecific populations seen in this case as "economic" entities acting for physical survival in their ecological niches. Local ecosystems are parts of regional ecosystems, up to embrace the whole biosphere.

Evolution is a process occurring at different levels, from genes to ecosystems. At all scales, changes in ecological dynamics affect the information stored in the genealogical hierarchy and vice versa. Lower level phenomena can penetrate the higher levels, but higher-level events set the stage for the agency of lower-level processes. In other words, hierarchy theory is a research program aiming at the unification of micro- and macroevolution: a possible solution for a very old problem in evolutionary debates since Darwin.

Hierarchy theory is still a Darwinian research program for two reasons: (1) because the original structure of Darwinian evolutionary theory was much more ecologically centered and pluralistic as for the causes and patterns of evolutionary change than the gene-centered neo-Darwinian reloads in the twentieth century and (2) because natural selection (among individuals) stands at the center of the sloshing bucket. Organisms are simultaneously systems and lineages (see Caponi in this volume) and simultaneously part of the two different interacting hierarchies: as reproducing "packages" of genetic information (replicators), they are part of the genealogical hierarchy; as matter-energy transfer systems (interactors), they are part of the ecological hierarchy, and their business is to survive. Thus the process of natural selection at the level of organisms incorporates the two dimensions (ecological pressures and genetic variations), and it is the bottom of the "bucket" in Eldredge's metaphor. Genetic variations and ecological pressures are two sides of the same coin. Hierarchy theory is a nonreductionist Darwinian research program.

As Niles Eldredge pointed out in the historical and introductory chapter of this volume, the first pillar of the approach is the idea that "biological nature is complexly organized into economic and genealogical hierarchies, and [we] have come to ask how those twin hierarchies interact to produce evolutionary history." The second pillar is that in the history of life, "large-scale environmental disruptions produce mass extinctions, with proportionally great concomitant evolutionary reactions—and with smaller environmental perturbations producing correspondingly lesser effects."

A Practical Application to a Specific Disciplinary Field: Paleoanthropology

Hierarchy theory is a unifying frame in which life emerges from a complex architecture of intersecting hierarchies of levels: the genealogical hierarchy of reproduction, the economic hierarchy of survival and finding resources, and even the outer hierarchy of physical structures of the Earth's crust. The evolution of organisms that reproduce, the evolution of ecosystems, and the evolution of the planet are inextricably intertwined and interdependent: the fluctuations of one reverberate proportionally on the other. They form an integrated functional network.

Yet it must be said that the mainstream of evolutionary biologists still retains an antihierarchical attitude, even among the supporters of the "extended evolutionary synthesis." Probably all of them associate hierarchy theory to another version of the rather inconclusive theories of complexity or to vaguely holistic approaches. As a theoretical attempt, hierarchy theory seems too large and too abstract, mixing together everything under the sky of evolution. However, as many chapters in this volume (written by scientists working in the field) explain, such a multilevel approach illustrates the need for a global and extended coherence in a rapidly expanding field. It has a heuristic and a theoretical relevance.

Starting with the first—heuristic and methodology—hierarchy theory could be a useful framework in order to understand the rapid advancements in specific evolutionary disciplines. We take human evolution as a practical example. An amount of recent findings in paleoanthropology are stressing climate instability and ecological disturbances as key factors affecting the highly branching hominin phylogeny, from early hominins to the appearance of cognitively modern humans (for a recent review, see Parravicini and Pievani 2016). Allopatric speciations due to geographic displacements, turnover pulses of species, adaptive radiations, mosaic evolution of traits in several coeval species, bursts of behavioral innovations, and serial processes of dispersal out of Africa are some of the macroevolutionary patterns emerging from the field. In this enlarged picture, human evolution is seen as an integration of different levels of evolutionary change, from local adaptations in limited ecological niches to dispersal phenotypes able to colonize huge, unprecedented ranges of ecosystems.

Human evolution has been the outcome of global ecological changes that transformed first our African cradle and then the Old World and the New World. The interactions among different levels and between the biotic and abiotic factors shaped the main features of human evolution. Genus *Homo* is the descendant of such a complex intertwining of different levels of evolutionary change. It is the consequence of an explosion of punctuated equilibria and turnover pulses in early Pleistocene, which are in turn a side effect of complex environmental changes caused by the effects of the ice ages in Africa. The same environmental conditions might have promoted our attitude to dispersal until the final wave of cognitively modern humans. We may be the effect of a sequence of large climatic and ecological disturbances that moved the water in the African "sloshing bucket" of evolution.

This shift in the general perspective about human phylogeny has been due to dramatic technological advances. New tools of integrated analysis and the extensive use of new dating techniques are available. The paleontological and archaeological record has been largely expanded. The "mosaic" anatomy of the earliest hominins (with unique combinations of retained and derived traits) suggests the idea of a series of adaptive postural "experiments" right around the putative time, when hominin lineage branched from chimpanzee lineage (five to six million years ago). Climate and environmental changes transformed the eastern and southern African environment from a relatively flat, homogeneous region covered with tropical mixed forest to a heterogeneous region, with high mountains and a mosaic of habitats ranging from cloud forests and closed woodlands to grasslands and deserts. In such a macroevolutionary process, we see the matching between a mosaic of unstable environments (the ecological hierarchy) and a mosaic of transitional forms with different adaptations in the bipedalism trend (the genealogical hierarchy).

The impressive radiation of hominin forms in Africa between three and two million years ago (australopithecines, genus *Paranthropus*, genus *Homo*) and the concomitant climatic instability and habitat fragmentation strongly suggest a high incidence of geographic speciations and a punctuational pattern of change. This branching picture of human phylogeny is very far from the anagenetic view held by the fathers of modern synthesis's phyletic gradualism and much more similar to the evolutionary history observed in other coeval mammalian taxa, revealing its higher parsimony.

Other subexamples show the same ecological pattern. When genus *Homo* emerged around 2.8 million years ago, bipedalism became first prevalent and then obligate. This evolutionary transition was carried out by a few species morphologically instable or by a plurality of separated species, each one with a specific set of traits. It follows that also the transition from a smaller hominin, more adapted to an arboreal lifestyle, to an obligate bipedal reveals the pattern of a "mosaic" evolution. Climate instability and habitat fragmentation were the macroevolutionary challenges. A plurality of adaptive strategies, sharing the same flexibility, was the answer.

Global ecological phenomena triggered severe ice ages, with intense and continuous fluctuations between glacial and interglacial phases. As a consequence of these global changes, East African climate became even drier because of the long-term aridity trend, while in the short term it became more instable with the alternation of high and low climate variability and cycles of moisture and aridity. Due to habitat fragmentation in African environments, many peripheral populations became isolated, providing the best conditions for the expected occurrence of extinction events, allopatric speciations, and adaptive radiations in a punctuational evolutionary mode. Climate instability, turnover pulses, and a plurality of genera feature the still unclear birth of genus *Homo* (still less clear after the discovery of *Homo naledi*, another mosaic of traits, in South Africa).

The still cryptic interspecific trend in genus *Homo* encephalization could be an example of the macroevolutionary pattern that Allmon in this volume defines as the result of lineage splitting, not only of within-lineage transformation. New, interesting ecological hypotheses have been proposed in order to account for the repeated hominin dispersals out of Africa as well. Again, microevolutionary factors (adaptations to local selective pressures) match with macroevolutionary patterns and processes. During the Pleistocene and its ice ages, the territories were plenty of mobile physical barriers. Geographical expansion through highly fragmented and diversified habitats produced a predictable fragmentation of populations. Vicariance and dispersal of human populations out of Africa, thus, may have triggered pulses of allopatric speciations, now in Eurasia too. This could be the reason for the remarkable morphological variation showed by the fossil record assigned to genus *Homo* and found in different parts of the Old World after 1.8 million years ago until very recent times with the cohabitations of *Homo sapiens*, *Homo neanderthalensis*, Denisovans, and *Homo floresiensis*.

Also the speciation of *Homo sapiens* in Africa, around two hundred years ago, has been affected by climate instability and dry phases. The birth of cognitively modern humans, however, seems to be more recent, as if anatomy and behavior were temporally disjointed. Only around sixty to eighty thousand years ago, robust findings point to an unprecedented appearance of symbolic expressions throughout the areas of distribution of *Homo sapiens*. Biogeography and migrations become crucial also in this case. Recent evidence corroborates the idea that the bursts of behavioral and cognitive innovation in South Africa (seventy thousand to eighty thousand years ago) are related to a growing population of *Homo sapiens* that was expanding out of Africa, carrying the symbolic and linguistic capacities showed by the members of this group later in Europe and Far East—again, small adaptive changes in fragmented populations and regional ecological disturbances due to global changes. In a pluralistic approach based on the twin hierarchies, gradualistic changes and punctuational patterns, genetic and epigenetic variation, adaptive transformations and plasticity, and biological evolution and cultural evolution can be seen as no longer in contrast. Old dichotomies crumble.

As we see in this example, the theory of evolution has a highly heterogeneous empirical basis, ranging from molecules to fossils, from changes occurring in a few minutes in bacteria to changes lasting for geological eras. Hierarchy theory predicts the growing role of interdisciplinary, integrated models. The recent advancements in paleoanthropology integrate archaeological, paleoclimatological, molecular, and demographic data in order to explain the biological and cultural evolution of cognitively modern humans. The consilience of very different data—biogeography, molecular biology, paleoclimatology, geophysics, paleoecology, and archaeology—highlights the most important patterns in hominin phylogeny today.

The multilevel approach to evolution provides a useful framework for bringing together and unifying the complex interplays observed among patterns and processes that belong to different evolutionary hierarchies (ecological and genealogical) structured along different levels. Microevolutionary explanations that appeal to processes occurring within local populations—that is, adaptation through natural selection and genetic drift—are fundamental, but they have to be integrated with macroevolutionary patterns in a pluralistic perspective, giving due relevance to the ecological and geophysical patterns and processes in human evolution. The physical and ecological world is not a mere backdrop of evolution: it changes, and such changes had profound effects on human evolution as well.

Extended Syntheses: A Matter of Focus

We anticipated that hierarchy theory could have also a theoretical relevance. In the last two decades, a lot of accumulating discoveries show that not all the evolutionary game is genetic and selective. Phenotypic and developmental plasticity (morphologies and behaviors that vary under changing environmental circumstances without genetic modifications) is a powerful and widespread adaptive strategy that can even cause the diversification of species. Furthermore, the material on which the selective processes act is not only given by small, continuous, and spontaneous genetic mutations. Rather, selection has to find trade-offs with internal constraints and channelings of development that do not always have a negative and limiting power but can positively orient evolution, generating crucial innovations.

According to a growing number of scholars, these findings should lead to an extension of the explanatory structure of the standard neo-Darwinian theory of evolution based on genetic variation and natural selection (Pigliucci and Müller 2010). The promoters of the extended evolutionary synthesis stress that the processes by which organisms grow and develop are active causes of evolution and speciation. While the views of the opposing side (the so-called standard theory) are polemically branded as "gene-centered" and too narrow (Laland et al. 2014), it is interesting to note that so far, the extended synthesis is strongly focused on organism as the fulcrum of evolutionary change. Development (so, the organismic level) is the influential internal "environment" of any variation. Due to epigenetic changes and inheritance and through the influence of developmental constraints biasing the emergence of novel phenotypic traits, the external environment can interact with organisms' traits in a constructive way. However, even in the extended synthesis, the larger ecological scenario gains its relevance only through the mediation of lower causal levels (the organism in this case).

Considering what is happening in the field, the extended synthesis is already there, we need just to recognize it. Nevertheless, the so-called extended evolutionary synthesis does not have a coherent structure yet: it still seems a summation of observed processes, very frequent and deserving great attention, even though sometimes dramatized, without a common theoretical thread. The only glue that apparently emerges is the opposition to genetic reductionism, but the fight against the straw man of gene-centrism and the focus on organisms seem not enough to launch a post-Darwinian coherent theory of evolution.

For reasons of focusing, the extended synthesis is not yet a multilevel explanation of the evolutionary phenomena because it lacks a theoretical frame for the interconnections between different levels of change (ecological and genealogical). Let us take the example of another pillar of the extended synthesis: niche construction. The point is that organisms (the main focus) are not passive entities, malleable at will by selection. The metabolic and behavioral activities of biological populations (to build a termite mound, to erect a dam on the river, or to pursue a cultural and technological advancement) change the ecological niches, thus influencing the environmental resources and the selective pressures that in turn retroact on organisms themselves. Such a process produces a crucial feedback and recursive phenomenon, called "niche construction" (Odling-Smee et al. 2003). Organisms actively change their environment, and the environments selectively change organisms.

Then the organism is an active player who codirects its own evolution, systematically changing the environments and so influencing the frame of selective pressures. Any biological population inherits from the previous generation not only a package of genes but also an amended ecological niche. If we see the process through the lens of hierarchy theory, niche construction becomes another example of water sloshing in Eldredge's bucket. Selective pressures come from the ecological hierarchy, affecting populations of organisms in their differential survival. But organisms can actively transform their environments for adaptive reasons and so construct new ecological niches that will be the frame of selective pressures for the next generations. The feedback and recursive processes occur at different levels: the niche construction of a species affects the ecological niche of other species; a limited niche construction process can be extended, influencing a larger ecological surrounding.

Such a "multilevel niche construction" is an excellent example of the recursive relationships between the twin hierarchies, with ecological and macroevolutionary patterns as crucial drivers of evolution. In this sense, hierarchy theory is another kind of "extended synthesis," a multilevel extended synthesis, still strongly Darwinian in its core (the bottom of the bucket). The dynamics of growth and extension of the theory of evolution is based on processes of theoretical revision and empirical enlargement of an elastic set of explanations already supported but constantly needing adjustments and integrations (Pievani 2012). New problems and apparent exceptions can be solved and understood through integrative explanations, different modulations of the empirical domain of already established patterns, and new quantitative calculations of the relative frequency of a pattern with respect to another.

In the case of punctuated equilibria theory (Eldredge and Gould 1972), after decades of debates, the emerging consensus around the mechanisms of speciation is that we need a multiplicity of processes and modes of birth of new species (punctuated in some ecological circumstances and gradual in others), a multiplicity of possible rates of speciation, and a multiplicity of levels of change (from an ecological and a genealogical point of view) to be considered (Coyne and Orr 2004). Thus the main methodological stance today is a calculation of the relative frequencies of one speciational pattern (punctuationism) with respect to another (gradualism and trends), using for instance meta-analyses on molecular and integrated phylogenies (like in Pagel et al. 2006). A radical alternative among incompatible patterns (stasis, slow phyletic gradualism, abrupt punctuations) is no longer the case (about the causes of evolutionary and ecological stasis, see Brett et al. in this volume).

This is what we know as a "pluralistic" approach to biological evolution, considering that punctuated equilibria (being a matter of both "tempo" and "mode" of evolution—see Allmon in this volume) has been the access door to hierarchy theory. We see here a passage from an exclusive pattern (previous phyletic gradualism) to a plurality of patterns about the rates of speciation, being species a fundamental unit in macroevolutionary patterns. Summing up, hierarchy theory is a different extended synthesis able to cover all the levels that make the evolutionary game so complex, from genes to organisms to species and to the largest ecological scenarios. Its theoretical potential lies here.

Future Directions

The strong methodological assumption that every macroevolutionary phenomenon should be reduced to the uniform accumulation of microevolutionary processes has been definitely challenged. Adopting hierarchy theory as a theoretical tool of assimilation and accommodation in different evolutionary fields, we could better appreciate the multiple patterns involved in the most promising lines of evolutionary research today. The fieldwork of Peter and Rosemary Grant on the Galápagos finches— mixing genes and developmental and ecological conditions and involving all the core patterns of evolution (mutation, natural selection, drift, migration, hybridization, speciation, biological and cultural evolution, regional ecological disturbances)—is an outstanding case study of this extended toolbox (Grant and Grant 2008).

As for another example, regarding transgenerational epigenetic inheritance, rather than relying on the unlikely return of Lamarckism, any Darwinian pluralist should agree with Robert J. Schmitz: "These results provide strong evidence that epialleles contribute to the heritability of complex traits and therefore provide a substrate on which Darwinian evolution may act" (Schmitz 2014). Genetic and epigenetic inheritable variability is the complex substrate for Darwinian evolution to work. In epistemological terms, we see a diversification of the variational materials on which the selective processes may act (and not a revolutionary neo-Lamarckian overthrow). Here and in many other examples, the structure of the current evolutionary research program is continuously evolving toward a more pluralistic version (Ayala and Arp 2010), and hierarchy theory is a promising candidate for such unification.

Biological evolution has expressed some recurrent "lawlike" patterns, somehow ordered configurations, or repeated schemes of historical events, but each time its specific historical outputs are unpredictable and unique (for a reconciliation of nomothetic and idiographic sciences in evolutionary biology according to a hierarchical perspective, see Lieberman in this volume). In the dialectic between the arrow of history and the cycle of recurrent regularities, when a pattern is assumed from data, it selects the pertinent further data and influences the scientific questions (i.e. it has a heuristic agency). Thus as a tool of scientific discovery and explanation, the pattern has both an epistemological status (it is in our minds) and an ontological status (it raises from objective data out there; Eldredge 1999). The task facing evolutionists is seeking recurrent patterns within a multiplicity of interconnected evolutionary lines whose trajectories appear highly unforeseeable. The goal is a more realistic comprehension of the evolutionary transformations and processes. In this sense, hierarchy theory is a metapattern gathering all the main patterns of evolutionary change (in this volume, William Miller III sketches a "metatheory" of evolution— namely, a "macroevolutionary consonance theory"—including also the macroecological aspect of evolution).

Today cosmology is somehow an "evolutionary" science. In the last decades, geology, thanks to the theory of plate tectonics, has become an important "evolutionary" discipline, and it is alongside biology and ecology in the reconstruction of the natural history of life on Earth. Ecologists discover historical patterns of coevolution between species and their environmental niches. Some patterns of change and repeated schemes of historical events seem to emerge from extremely diversified fields of research, even in the context of cultural and technological evolution. Even historians of science sight in the collective adventure of human knowledge a few "evolutionary" regularities. In the varied scientific disciplines involved in the study of history, a common sensibility for explanatory patterns of an evolutionary type is dawning, along with the possibility to apply more generally (even though prudently) the twin hierarchy of ecological (contextual) factors and internal factors.

As another future direction—following the idea that human social systems reintegrate the genealogical and the ecological domains at a level above the individuals—we could explore the possibility that the interactive distinction between the economic, material, physical, and ecological dimension of evolutionary processes (the history of "matter in motion" and energy transferring) and the gene-molecular dimension of the transmission of biological information from one generation to the next applies also to nonstrictly biological evolutionary processes (i.e., social, cultural, and technological evolution). As an example, McKinney in this volume applies the hierarchy approach to describe the multidimensional reduction of diversity in the biosphere during the Anthropocene, focusing on urbanization and urban scaling.

This is an exciting time for evolutionary biologists, as their discipline is facing a vivid debate on its epistemological status. The parties involved are not simply dicotomically lined up. A plurality of contributions is working for the enrichment of the evolutionary research program (Pievani 2015). Human evolution, again, with its sets of open problems still looking for proper evolutionary explanations (the interspecific encephalization trend, the evolution of language, the appearance of behavioral modernity, etc.), stands as an ideal arena on which novel models, theories, and hypotheses can be advanced and whose validity can be tested. Hierarchy theory is able to display a sound coupling between acknowledged sets of heterogeneous data coming from paleoanthropological studies and heuristic and theoretical models, stressing the relevance of a multilevel analysis. Moreover, human evolution has to be understood in light of the complex interplay between biological and cultural evolution. This could be another exciting line of future researches in hierarchy theory.

The most recent archeological findings (D'Errico and Stringer 2011; D'Errico and Banks 2013) are informing us that *Homo sapiens* was not the only *Homo* species capable of behaviors whose complexity might have played a still underestimated evolutionary role. The emergence and persistence of material cumulative cultures (the capacity for accumulating modifications over time) represent an outstanding case of niche construction activity being able to change the ecological and social environment, systematically biasing selective pressures. The role of ecological and biogeographical factors in the evolution of material cultures in genus *Homo* is an area waiting for a systematic study, especially focusing on the different regional trajectories and on the potential interplay among cultural adaptations, brain plasticity, environmental changes, and ecological and demographical factors. Such a feedback and recursive biocultural process needs to be appreciated in its multilevel deployment: effects and by-products are recognizable at the individual, group, and ecosystem levels but also at the behavioral, cognitive, neural, and genetic levels. Gene-culture coevolutionary explanations are among the most promising research paths in this field (Laland et al. 2010), and, although this is still poorly appreciated, their explanatory structure is intrinsically a multilevel one.

Evolutionary biology itself is an evolving scientific discipline, demanding for pluralistic explanatory models. Key concepts, advanced by the extended synthesis supporters, such as reciprocal causation (Laland et al. 2015), or catching the constructive relationship between the ecological environment and organisms' behavior and development, could perfectly match with the multilevel framework proposed by hierarchy theory. The fruitfulness of such an enterprise would be twofold: (1) providing the coherent and unified structure that the pluralistic debate on the need for an updating of the evolutionary research program is looking for, and (2) addressing a set of specific unsolved or open problems (such as the ones mentioned above) with a proper set of extended explanatory tools, being able to put them under a new light in order to increase the empirical content of the set of explanations in respect of new observations.

New discoveries pile up. Models are changing. New concepts gain attention. The theory of evolution is evolving. The wish is that hierarchy theory could be another brick in this field under construction. To the young generations of evolutionary biologists, the task is shaping the future landscape of this fascinating interdisciplinary enterprise.

References

Ayala, Francisco J., and Robert Arp, eds. 2010. *Contemporary Debates in Philosophy of Biology*. New York: Wiley.

Coyne, Jerry A., and H. Allen Orr. 2004. *Speciation*. Sunderland, MA: Sinauer Associates.

D'Errico, Francesco, and William E. Banks. 2013. "Identifying Mechanisms behind Middle Paleolithic and Middle Stone Age Cultural Trajectories." *Current Anthropology* 54 (S8): S371–S387.

D'Errico, Francesco, and Chris B. Stringer. 2011. "Evolution, Revolution or Saltation Scenario for the Emergence of Modern Cultures?" *Philosophical Transactions of the Royal Society B* 366: 1060–69. doi:10.1098/rstb.2010.0340.

Eldredge, Niles. 1999. *The Pattern of Evolution*. New York: W. H. Freeman.

Eldredge, Niles, and Stephen J. Gould. 1972. "Punctuated Equilibria: An Alternative to Phyletic Gradualism." In *Models in Paleobiology*, edited by Thomas J. M. Schopf, 82–115. San Francisco: Freeman, Cooper.

Eldredge, Niles, and Bruce S. Lieberman. 2014. "What Is Punctuated Equilibrium?

What Is Macroevolution? A Response to Pennell et al." *Trends in Ecology and Evolution* 29: 185–86.

Grant, Peter R., and B. Rosemary Grant. 2008. *How and Why Species Multiply.* Princeton: Princeton University Press.

Herschel, John. 1838. Letter to Charles Lyell, in Charles Babbage, *The Ninth Bridgewater Treatise*, 2nd ed. London.

Laland, Kevin N., John Odling-Smee, and Sean Myles. 2010. "How Culture Shaped the Human Genome: Bringing Genetics and the Human Sciences Together." *Nature Reviews Genetics* 11: 137–48.

Laland, Kevin N., Tobias Uller, Marcus W. Feldman, Kim Sterelny, Gerd B. Müller, Armin Moczek, Eva Jablonka, and John Odling-Smee. 2015. "The Extended Evolutionary Synthesis: Its Structure, Assumptions and Predictions." *Proceedings of the Royal Society B* 282: 20151019.

Laland, Kevin, Tobias Uller, Marc Feldman, Kim Sterelny, Gerd B. Müller, Armin Moczek, Eva Jablonka, John Odling-Smee, Gregory A. Wray, Hopi E. Hoekstra, Douglas J. Futuyma, Richard E. Lenski, Trudy F. C. Mackay, Dolph Schluter, and Joan E. Strassmann. 2014. "Does Evolutionary Theory Need a Rethink?" *Nature* 514: 161–64.

Odling-Smee, John, Kevin N. Laland, and Marcus W. Feldman. 2003. *Niche Construction.* Princeton, NJ: Princeton University Press.

Pagel, Mark, Chris Venditti, and Andrew Meade. 2006. "Large Punctuational Contribution of Speciation to Evolutionary Divergence at the Molecular Level." *Science* 314: 119–21.

Parravicini, Andrea, and Telmo Pievani. 2016. "Multi-level Human Evolution: Ecological Patterns in Hominin Phylogeny." *Journal of Anthropological Sciences* 94: 1–16. doi:10.4436/JASS.94026.

Pievani, Telmo. 2012. "An Evolving Research Programme: The Structure of Evolutionary Theory from a Lakatosian Perspective." In *The Theory of Evolution and Its Impact*, edited by Aldo Fasolo, 211–28. New York: Springer.

———. 2015. "How to Rethink Evolutionary Theory: A Plurality of Evolutionary Patterns." *Evolutionary Biology*. doi:10.1007/s11692-015-9338-3.

Pigliucci, Massimo, and Gerd B. Müller, eds. 2010. *Evolution: The Extended Synthesis.* Boston: MIT Press.

Schmitz, Robert J. 2014. "The Secret Garden—Epigenetic Alleles Underlie Complex Traits." *Science* 343: 1082–83.

Contributors

About the Editors

Niles Eldredge has been, since 1969, paleontologist on the curatorial staff of the American Museum of Natural History in New York, where he is now curator emeritus in the Division of Paleontology. His specialty is the evolution of trilobites—a group of extinct arthropods that lived between 535 and 245 million years ago. Throughout his career, he has used repeated patterns in the history of life to refine ideas on how the evolutionary process actually works. The theory of "punctuated equilibria," developed with Stephen Jay Gould in 1972, was an early milestone. Eldredge went on to develop a hierarchical vision of evolutionary and ecological systems, and in his book *The Pattern of Evolution* (1999), he unfolds a comprehensive theory (the "sloshing bucket") that specifies in detail how environmental change governs the evolutionary process. Eldredge was curator-in-chief of the American Museum's Hall of Biodiversity (May 1998) and has written several books on the subject—most recently *Life in the Balance* (1998). He has also combated the creationist movement through lectures, articles, and books—including *The Triumph of Evolution . . . and the Failure of Creationism* (2000). An amateur jazz trumpeter and avid collector of nineteenth-century cornets, Eldredge has turned his evolutionary approach to cornet history—and to the comparison of patterns and processes of material cultural and biological evolution. A critic of gene-centered theories of evolution, Eldredge's *Why We Do It* (2004) presents an alternative account to the gene-based notions of "evolutionary psychology" to explain why human beings behave as they do. His most recent book is *Eternal Ephemera: Adaptation and the Origin of Species from the Nineteenth Century through Punctuated Equilibria and Beyond* (2015). http://www.nileseldredge.com.

Telmo Pievani is full professor in the Department of Biology, University of Padua, where he covers the first Italian Chair of Philosophy of Biological Sciences. He also teaches anthropology. He is the author of two hundred publications, including *Introduction to Philosophy of Biology* (2005; Portuguese edition, 2010), *The Theory of Evolution* (2006; new edition, 2010); *Creation without God* (2006; Spanish edition, 2009), *Born to Believe* (with V. Girotto and G. Vallortigara, 2008), *The Unexpected Life* (2011), *Homo sapiens: The Great History of Human Diversity* (with L. L. Cavalli Sforza, 2011), *Evolved and Abandoned: Sex, Politics, Morals: Does Darwin Explain Everything?* (2014). He is a fellow of the Istituto Veneto di Scienze, Lettere ed Arti, Class of Sciences, Venice; Turin Academy of Sciences, Class of Biological Sciences; Italian Society of Evolutionary Biology; "Umberto Veronesi" Foundation for the Progress of Sciences, Milan; Istituto Italiano di Antropologia, Steering Board, Rome. He is a member of the editorial board of *Evolution: Education and Outreach*, *Evolutionary Biology*, *Rend. Lincei Sc. Fis. Nat.*, and *Le Scienze*, the Italian edition of *Scientific American*. Engaged in several projects regarding communication of science in Italy (e.g., with Luigi Luca Cavalli Sforza, curator of the international exhibition *Homo sapiens*: The Great History of Human Diversity), he is a fellow of the Scientific Board of Genoa Science Festival (Secretariat, 2003–2010). He is director of Pikaia, the Italian website dedicated to evolution and philosophy of biology. http://www.telmopievani.com/en.

Emanuele Serrelli is a researcher in the philosophy of science, specialized in philosophy of biology and evolutionary theory but vocated to interdisciplinary projects that span the whole range of disciplines and fields of knowledge. Currently fellow at the University of Milano-Bicocca, Italy, CISEPS Center for Interdisciplinary Studies in Economics, Psychology and Social Sciences, where he leads the "Cultural Evolution" research program, he also works as a freelance research consultant for several universities around the world. He is a member of the editorial board at *Frontiers in Ecology and Evolution: Evolutionary Developmental Biology*. His recent international works include *Macroevolution: Explanation, Interpretation, and Evidence* (edited with N. Gontier, 2015) and *Understanding Cultural Traits. A Multidisciplinary Perspective on Cultural Diversity* (edited with F. Panebianco, 2016), both published by Springer.

Ilya Tëmkin is an interdisciplinary scientist who studies how evolution works in nature and in human culture in the general framework of the hierarchy theory. An expert on bivalve mollusks, Dr. Tëmkin analyzes the relative roles that ecology, history, and individual development play in diversification and the evolution of organic form. As a specialist on the history of musical instruments (and a passionate musician), he explores the question of the extent to which the mechanisms of information transmission and historical change in human culture mirror evolutionary changes in living systems using musical instrument design as an example. Currently, Dr. Tëmkin is assistant professor of biology at Northern Virginia Community College and a research associate at the National Museum of Natural History, Smithsonian Institution.

About the Authors

Warren D. Allmon is director of the Paleontological Research Institution (PRI) in Ithaca, New York, and Hunter R. Rawlings III Professor of Paleontology in the Department of Earth and Atmospheric Sciences at Cornell University. He is the author of more than 250 technical and popular publications. He is a fellow of the Geological Society of America and the Paleontological Society and the recipient of the 2004 Award for Outstanding Contribution to Public Understanding of Geoscience from the American Geological Institute.

Kenneth D. Angielczyk is associate curator of paleomammalogy at the Field Museum of Natural History in Chicago. He is a vertebrate paleontologist interested in understanding the broad implications of the paleobiology and paleoecology of nonmammalian synapsids and in trophic network-based approaches to paleoecology. He is coeditor (with C. F. Kammerer and J. Fröbisch) of *Early Evolutionary History of the Synapsida* (2014), as well as the author of a number of papers published in journals, including *Science, Nature, PNAS, Biology Letters*, and *Proceedings of the Royal Society B*.

Carlton E. Brett is University Distinguished Research Professor in the Department of Geology, University of Cincinnati, Cincinnati, Ohio. His research lies at the interface between paleontology and sedimentary geology. He is pursuing studies relating regional and global changes in climate, sea level, and the carbon cycle to episodes of biotic change (bioevents) and extinction as well as time-specific facies. Among the highlights of his career are the receipt the Paleontological Society's Schuchert Award; election as a fellow of the Paleontological Society and of the Geological Society of America; receipt in 2008 of the Digby McClaren Medal for Lifetime Achievement in Stratigraphic Paleontology; and receipt in 2005 of an Alexander von Humboldt Research Prize. He is the 2012 recipient of the Raymond C. Moore Medal of SEPM for excellence in paleontological research and the 2013 recipient of the AAPG's Grover Murray Outstanding Geological Educator Award.

Silvia Caianiello is senior researcher at the Istituto per la Storia del Pensiero Filosofico e Scientifico Moderno (ISPF) of the Italian National Research Council (CNR) in Naples. Her research interests and experiences range from history of European philosophy to history and philosophy of life sciences. She has conducted extensive research on the correlation between representations of time and epistemological approaches to history from the eighteenth to the twentieth century and has authored several papers on the interactions between human and life sciences in the nineteenth century. Currently her main research field is in history and epistemology of life sciences, with particular focus on evolutionary theory and evo-devo ("L'interno della selezione," in *Confini aperti in biologia*, edited by B. Continenza et al., 2013; "Les modules de la variation. L'évodévo ou la nouvelle genèse des formes," in *Critique* 2011; "Adaptive vs. Epigenetic Landscape: A Visual Chapter in the History of Evolution and Development," in *Graphing Genes, Cells and Embryos: Cultures of Seeing 3D and Beyond*, edited by S. Brauckmann et al., Max Planck Institute Pre-Print Series, 2009). In 2007/2008 she was a visiting scholar at the Max Planck Institute for History of Science in Berlin, Germany. She is a member of ISPHSSB and associated with "Res viva" (Italian Interuniversity Research Centre on Epistemology and History of Life Sciences).

Gustavo Caponi is professor of philosophy and history of biology at the Federal University of Santa Catarina, Department of Philosophy, Florianópolis, Brazil. He has published more than one hundred works (most of them on issues of philosophy and history of biology) in several anthologies and international journals. He is also the author of six books on philosophical and historical issues related to biology: *Cuvier: un fisiólogo de museo* (México, 2008); *Buffon* (México, 2010); *La segunda agenda darwiniana: contribución preliminar a una historia del programa adaptacionista* (México, 2011); *Função e desenho na Biologia Contemporânea* (São Paulo, 2012); *Réquiem por el centauro: aproximación epistemológica a la Biología Evolucionaria del Desarrollo* (México, 2012); *and Leyes sin causa y causas sin ley en la Explicación Biológica* (Bogotá, 2014).

Gregory J. Cooper is professor of philosophy in the Philosophy Department of the Washington and Lee University. His research interests include the history and philosophy of ecology, the subject of his book *The Science of the Struggle for Existence: On the Foundations of Ecology* (2003) and the philosophy of biology more generally. He also works in the general area of environmental philosophy, including topics such as environmental ethics, ecosystem health and integrity, environmental economics, and environmental aesthetics. His work in applied ethics goes beyond environmental ethics to include medical ethics and legal ethics. Finally, he is currently engaged in research on the ecology of the freshwater fishes of the Amazon basin. His publications include books, book chapters, encyclopedia entries, and papers published in major international journals of philosophy.

Charbel N. El Hani is associate professor at the Institute of Biology, Federal University of Bahia, Brazil, where he coordinates the History, Philosophy, and Biology Teaching Laboratory. He is currently book review editor for *Science and Education*. His areas of research are science education research, philosophy of biology, and evolutionary biology. He worked as a postdoctoral researcher at the Center for Philosophy of Nature and Science Studies, University of Copenhagen, Denmark. He coordinates the Pronex project Integrating Levels of Organization in Ecological Predictive Models: Contributions from Epistemology, Modeling and Empirical Investigation (INOMEP). He coordinates the science popularization initiative Cafe Scientifique Salvador (http://cafecientificossa.blog spot.com). He is vice president of the Brazilian Association for History and Philosophy of Biology (ABFHIB). He is the author or editor of nine books and has published more than one hundred scientific papers, many of them in international journals.

Tyler A. Elliott completed his BSc in molecular biology and genetics at the University of Guelph, where he also went on to earn his MSc and is nearing completion of his PhD. He is primarily interested in transposable element evolution, in particular the development of a conceptual framework based on an element-level perspective and large-scale comparisons of transposable element abundance and diversity across genomes.

T. Ryan Gregory earned his BSc in biology from McMaster University and his PhD in evolutionary biology from the University of Guelph, where he is now associate professor. His primary research interests include large-scale genome evolution, biodiversity, and macroevolution. He has been the recipient of several prestigious scholarships, fellowships, and awards, including the 2003 NSERC Howard Alper Postdoctoral Prize, a 2006 American Society of Naturalists Young Investigator Prize, the 2007 Canadian Society of Zoologists Bob Boutilier New Investigator Award, and the 2010 Genetics Society of Canada Robert H. Haynes Young Scientist Award. He is currently editor-in-chief of the journal *Evolution: Education and Outreach.*

Bruce S. Lieberman is professor in the Department of Ecology and Evolutionary Biology at the University of Kansas and senior curator at KU Biodiversity Institute. He is a paleontologist and evolutionary biologist interested in macroevolution. He has published articles in major scientific journals, including *TREE, Proceedings of the Royal Society, Proceedings of the Academy of Sciences,* and *Paleobiology.* His research uses the fossil record of arthropods and other taxa to reconstruct evolutionary patterns in order to gain insight into the nature of the evolutionary process. He is particularly interested in research approaches that use biogeography, phylogenetics, and ecological niche modeling. He has considered various topics in macroevolutionary theory, including levels of selection; punctuated equilibria and stasis; evolutionary and adaptive radiations; the role abiotic as opposed to biotic factors play in governing macroevolution; and the effects of past, present, and future climate change on ecology and evolution.

Stefan Linquist holds a BA in philosophy from Simon Fraser University, an MSc in biology from Binghamton University, and a PhD in philosophy from Duke University. He is currently associate professor in the Department of Philosophy at the University of Guelph. His research covers a range of topics, including environmental science, biodiversity, evolution, ecology, and genomics.

Michael L. McKinney is professor in the Department of Earth and Planetary Sciences and director of the Environmental Studies Program at the University of Tennessee. He teaches environmental geology, restoration ecology, and an introductory course in environmental science and serves as faculty advisor to the student environmental club. He has published several books, including *Environmental Science* (6th ed.) and *Biotic Homogenization* (with Julie Lockwood). His research interests have generally focused on biological issues. He started out in paleobiology, in which he still has an active interest. In this area, he has published several papers on extinction and evolution as seen in the fossil record. However, in recent years he has published increasing numbers of articles on topics relating to modern biodiversity issues, especially the effects of urbanization on biodiversity and how human activities are homogenizing the biosphere. He has served on the editorial boards of *Evolutionary Ecology Research*, *Animal Conservation*, *Urban Naturalist*, and *Frontiers in Ecology and Society*.

Daniel W. McShea is professor of biology in the Department of Biology, Duke University, Durham. His major papers have been in the field of paleobiology, with a focus on large-scale trends in the history of life, especially documenting and investigating the causes of the (putative) trend in the complexity of organisms. A significant part of this work involves operationalizing certain concepts, such as complexity and hierarchy, as well as clarifying conceptual issues related to trends at larger scales. His more recent work is in the philosophy of biology, especially on the problems of goal directedness in biological systems and in machines. He is coauthor (with Robert Brandon) of *Biology's First Law* (2010) and (with Alex Rosenberg) of *Philosophy of Biology: A Contemporary Introduction* (2007). He serves on the editorial board of *Biology and Philosophy* and is codirector of Duke's Center for the Philosophy of Biology.

Arnold I. Miller is professor of geology and senior associate dean in the McMicken College of Arts and Sciences Department of Geology at University of Cincinnati. He is an evolutionary paleobiologist and paleoecologist, with research and teaching interests in biodiversity throughout geological time and in the present day. Current projects include the investigation of geographic and environmental selectivity during global mass extinctions and major diversification events, assessment of anthropogenic impacts on shallow-water molluscan communities as recorded in skeletal accumulations, numerical modeling of time-averaged fossil assemblages, and assessments of the distributions of animals and plants along present and past environmental gradients. The author of many important publications, his articles have been published in major international scientific journals, including *Science*, *Paleobiology*, and *Paleontology*. He is also the author (with M. Foote) of *Principle of Paleontology* (2007, Chinese translation, 2013).

William Miller III is professor of geology at the Humboldt State University (Arcata, California) and adjunct research professor of geology at Appalachian State University (Boone, NC). His research interests include paleocommunity temporal dynamics; late Cenozoic molluscs from the Atlantic Coastal Plain; deep-marine trace fossils and giant foraminiferans from Western North America, Ecuador, and Italy; ichnotaxonomic theory; trace fossil evidence of earliest animals; species delineation problems and speciation theory; application of hierarchy theory to problems of macroevolution; and the connections between evolutionary patterns detected in the fossil record and development of large-scale ecologic systems. He is the author or editor of more than 100 publications on these topics, spanning more than three decades of work.

Nei Freitas Nunes-Neto is professor of philosophy at the Institute of Biology, Federal University of Bahia. He obtained a PhD in ecology at the Federal University of Bahia (2013) and worked as a visiting researcher at the IAS Research Group—Centre for Life, Mind and Society—associated with the Department of Logic and Philosophy of Science, at the University of Basque Country, Spain (2015). His main research interests are the functional and teleological discourse in current ecology and environmental sciences; the complex relationships established between ecology, evolution, and ethics, and the design of strategies for a science teaching based on socioscientific issues and aiming at sociopolitical actions.

Andrea Parravicini is a postdoctoral research fellow in philosophy of biology at the Department of Biology of the University of Padua. He is the author of *Darwin's Mind: Philosophy and Evolution* (2009) and *Thought in Evolution: Chauncey Wright between Darwinism and Pragmatism* (2012), as well as some articles and book chapters on topics in the philosophy and history of evolutionary biology in the history of American thought (in particular the so-called American pragmatism) and in philosophy of paleoanthropology. He is a member of the editorial board of *Noema— Online Journal of Philosophy*. He is a member of the cultural association Pragma for the study of the American pragmatism, of the Italian Society for Evolutionary Biology, and he collaborates with the interuniversity research center Pragmatismo, Costruzione dei Saperi e Formazione. Currently, his studies focus on the conceptual framework and the historical arrangement of the hierarchical perspective in the evolutionary theory, and he coordinates the Templeton project Hierarchy Group: Approaching

Complex Systems in Evolutionary Biology, directed by Niles Eldredge and Telmo Pievani (www.hierarchygroup.com).

Mihaela Pavličev is assistant professor in the University of Cincinnati Department of Pediatrics, Cincinnati Children's Hospital. She leads the Pavlicev lab, interested in evolutionary systems biology. Current research projects revolve around questions related to contribution of multiple traits and multiple genes to complex adaptation, focusing on the properties that organisms must possess in order to function, persist, and evolve. The aim of the lab is to understand the genetic basis of phenotypic traits, asking how genetic change leads to variation among individuals. She has published articles in major international scientific journals, including *TREE*, *Evolutionary Biology*, *Proc. Roy. Soc.*, *Nature*, *Evolution*, and *Genetics*.

Richard O. Prum is William Robertson Coe Professor of Ornithology at the Department of Ecology and Evolutionary at the Yale University. He is also curator of ornithology and head curator of vertebrate zoology at the Peabody Museum of Natural History, and he is director of the Franke Program in Science and the Humanities, which is a new initiative at Yale that aims to foster communication, mutual understanding, collaborative research, and teaching among diverse scientific and humanistic disciplines. He is an evolutionary ornithologist with broad interests in avian biology. He has done research on diverse topics, including avian phylogenetics, behavioral evolution, feather evolution and development, sexual selection and mate choice, sexual conflict, aesthetic evolution, avian color vision, structural color, carotenoid pigmentation, evolution of avian plumage coloration, historical biogeography, avian mimicry, and the theropod dinosaur origin of birds. He has conducted field work throughout the Neotropics and in Madagascar and has studied fossil theropods in China. http://prumlab.yale.edu.

Peter D. Roopnarine is curator invertebrate zoology and geology at the Institute for Biodiversity Science and Sustainability, California Academy of Sciences. His research focus is evolutionary ecology, with a strong emphasis on paleontology and deep time perspectives. Most of his research these days centers around global change biology and how we can further develop our understanding of Earth's past ecosystems to better forecast our future. His articles have been published in major international journals including *Science*, *PNAS*, *Nature*, and *Paleobiology*.

Gary Tomlinson is John Hay Whitney Professor of Music and Humanities and Director of the Whitney Humanities Center at Yale University. He is a musicologist and cultural theorist known for his interdisciplinary breadth, and his work has ranged from the philosophy of history to opera, from critical theory to Aztec song. His latest book, *A Million Years of Music: The Emergence of Human Modernity* (2015), takes the evolutionary coalescence of human musical capacities as the anchor point for an investigation into the formation of our modernity. He is now elaborating the ideas there in a new book on the biocutural evolution of humans over the last 100,000 years. Tomlinson has garnered prizes from several organizations, was elected to the American Academy of Arts and Sciences, and has received a Guggenheim Fellowship and a MacArthur Award.

Jon Umerez is a research fellow in the Department of Logic and Philosophy of Science at the University of the Basque Country (UPV/EHU). Previously he held a Ramón y Cajal research contract and its extension (2001–2011). He graduated from the Universidad Complutense in Madrid (1985) and worked for some years as a secondary school philosophy teacher before and after getting his doctoral degree at the University of the Basque Country (1994). He has also been a visiting research fellow at the University of Binghamton in New York (1994–1995, 2009). His research interests are in philosophy of biology, philosophy and history of science, complex systems and artificial life, and biology and society and the public understanding of science. He has published papers in important international journals, including *Biology and Philosophy*, *Biosemantics*, *Artificial Life*, *BioSystems*, *Brain and Cognition*.

Günter P. Wagner is Alison Richard Professor of Ecology and Evolutionary Biology, Department of Obstetrics, Gynecology and Reproductive Sciences, Yale University, and he is adjunct professor of obstetrics and gynecology, Wayne State University. His research interests include homology/character identity/novelty, digit homology from a developmental point of view, the role of transcription factor protein change in the evolution of transcriptional control, and the evolution of gene regulatory networks. He is the author of several books, book chapters, and articles published by important international publishers and scientific journals.

Andrew A. Zaffos is research associate at the University of Wisconsin–Madison. He is primarily interested in the extinction and diversification of marine organisms and how patterns of stratigraphic turnover have controlled diversity throughout the Phanerozoic by restructuring habitats and/or creating gaps in the fossil record. He is also interested in changes in the geographic and environmental distribution of organisms, either regionally along onshore-offshore transects of the continental shelf or globally along continental shorelines. He is a member of the University of Wisconsin's Macrostratigraphy Lab under Shanan E. Peters. In particular, he works on group projects utilizing the Macrostrat Database, Paleobiology Database, and the GeoDeepDive research and development teams.

Index

Page numbers in italic refer to figures, tables, and notes.

abiotic factors, 131–32, 227–29
Abouheif, Ehab, *167*
Acadian Orogeny, 288
acyclical graph, 20
adaptation, 37, 53, 87, 93, 96–99, 106, 162,
    164, 186–87, 206, 227, 231–33, 235, 243,
    246–47, 308, 337, 341–42, 354–57, 362,
    365; group-level, 176
adaptationism, 189
adaptive cycle, 322, 324
adaptive zone, 231
Africa: climate, 356; out of, 354, 356; South,
    241, 308, 356
allopatric speciation, 12–13, 33, 37, 39, 42,
    263–64, *273*, *274*, 354, 356
Alon, Uri, *166*
Altenberg, Lee, *164*
anagenesis, 240, 260, 262–63, 265, 269–73,
    296
ancestors/descendants, 2, 4–5, 7, 9–11, 53,
    57, 86, 127–28, 227, 229, 246, 261, 270,
    272, 354
Anelosimus studiosus, 179
*Animal Dispersion in Relation to Social
    Behavior* (Wynne-Edwards), 176
Anthropocene, 241, 334–35, 341, 346, 362
apomorhy, 57–59
Appalachian Basin, 39, 292, 298–99
Aquarius remigis, 179
arms race, evolutionary, 231
assemblage zone: Beaufort Group, *312*;
    Cynognathus (CAZ), *312*, 314–15, 318,
    319, 321, *322*, 323; *Dicynodon* (DAZ),

312, 314, 315, *319*; *Lystrosaurus* (LAZ),
    312, *312*, *319*, *322*; *Tropidostoma*, *319*
asymmetry, 22, 27, 55, 154
attractor, 163, 315–16, 318, 326
Auletta, Gennaro, 155
avatar, 51–52, 54–55, *60–61*, 103, 133, 191,
    193, 231, 283, 298–300, 308, 343, 347

Bahia Blanca (Argentina), 8
Beagle, HMS, 7–10
Bechtel, William, 74
benthic assemblages, 283
biocoenoses, 133, 283, 298, 300
biocultural, 202, 205, 211–20, 344, 346, 362
biogeography, 228, 230, 356–57, 370, 374
biome, definition of, 229. *See also* Clements,
    Frederic E.
biosphere, 63, 103, 192, 203, 233, 241, 328,
    334–39, 344, 347, 352, 362, 371
Biston betularia, 49
bivalves, 267, 292
Boltzmann, Ludwig, 228
Borrello, Mark, 176
Bothrops alternatus (yarará), 57, 58
Bouchard, Frédéric, 113–14
boundaries (ecological), 20, 78, 90, 144
boundary conditions, 22–23, 129, 151, 156
brachiopods, 292, 298
Brett, Carlton E., 288
Broad, Charles D., 68
Brocchi, Gianbattista, 2–11, 14; Brocchi's
    analogy, 2, 6–8, 10; hierarchical view of,
    5–8; species as individuals, 2, 5

Bucephalus, 47, 48, 55, 56
buffering, 110, 132–33
Buffon, Georges-Louis Leclerc de, 2
Bunge, Mario, 67

Campbell, Donald T., *81*, 92
Cape Verde islands, 8
carbonates, 292
cascading effects, 22, 132, 301
cascading extinctions on graphs (CEG), 318
causation: cross-level, 131, 190, 352; downward, 22, 24, 64–65, 75, 76, 89, 130, 160, 352; future, 92, 98; upward, 22, 24, 65, 76, 90, 130, 151–52, 188, 352
C. elegans, 155
cellular level, 24, 128
Centerfield Member, 289, 290, *293*
Centronella, 292
character identity networks (ChINs), 203
chromosomes, 59–60, 137, 146–47
Cimitaria, 292
Cincinnatian faunas, 297
Cincinnati Arch, 283
cladogenesis, 263, 269, 273
Clements, Frederic E., 228–29. *See also* biome, definition of
closure: causal, 75, 77; of constraints, 117–18, 120; organizational, 115–17
clustering coefficient, 21
coadaptation, 162
coevolution, 23, 215, 231, 325, 346, 361, 363
coextinction, 231
cohesion, 51–52, 152, 164,
common descent, 53, 59
community (ecological), 131–33, 228, 230–31, 233
complexity, 19, 21–22, 31, 63, 65, 67, 76–77, 78–80, 91, 93, 129, 131–32, 138, 151–52, 193, 202–8, 214, 216, 219, 227, 229, 233–34, 341, 354, 362, 371
composition/compositional hierarchy, 19, *20*, 23, 24, 27, 65, 69–75, 77, 80–81, 87, 128, 131, 160, 183; components (homogeneous or heterogeneous), 73–75, 77, 80
consonance theory, 239, 243, 247, 252, 361
constraint, 22–23, 75–81, 99, 115, 117–18, 120, 128–29, 151, 153–54, 228, 246, 322, 324–25, 328; developmental, 96, 358; energetic, 310, 324; holonomic and nonholonomic, 79, *82*; interlevel, 27, 78;

internal, 94; lower-level, 156; organizational, 154; upper-level, 90
consumers, 105, 109, 118, 131–32, 234
contingency/contingent, 26, 29, 30–42, 53, 203–4, 218, 220, 230, 326
control, 19–20, 22, 24, 70, 74–75, 77–79, *82*, 88, 129–31, 155, 160, 231; hierarchy and, 19, *20*, 88, 97, 118, 130, 160
convergence, 30, 34, 96, 245
coral, 292, 300
cospeciation, 231
Craver, Carl F., 69, 108, 109
cross-level: by-products, 188–89; causation, 189–90; interactions, 151, 160
Cummins, Robert, 108, 119
Cuvier, George, 4–5, 7
cyclicity, 39
Cystiphylloides, 292

Damuth, John, *60*, 184
Darwin, Charles, 4, 6–14, 53, 86–87, 100, 174–75, 177; altruism and, 175; *The Descent of Man*, 175; *February 1835*, 10; *Geological Diary*, 8; gradualism of, 4, 11–12 (*see also* gradualism); group selection and, 195, 227, 231–33, 268, 270, 353 (*see also* group selection); hierarchical thinking and, 8–12 (*see also* hierarchy); natural selection and, 129, 174, 189, 193–94; *On the Origin of Species*, 7, 10, 12; *Ornithological Notes*, 9; "per saltum," 10; *Red Notebook*, 6, 10; *Transmutation Notebooks B–E*, 6, 11, 12, 13
Darwin, Erasmus, 2
Darwinism, *195*; levels of selection and, 182–83; ultra-, 155
Davidson, Eric H., 158–59, *167*, 209
Dawkins, Richard, 31, 176, 180, 181, 192, 193, *195*; *The Selfish Gene*, 176
DCA axis 1, 291–92
DCA axis 2, 291–92
Deacon, Terrence, 135, 203, 204, 211, 220
degree distribution, 21
demes, 54, 56, 61, 72, 103, 175, 183, 193, 245–47, 343
Denisovans, 211, 356
Descent of Man, The (Darwin), 175
detrended correspondence analysis, 289
development (ontogenetic), 8, 27, 78, 87, 89, 96, 100, 134, 138, 140–41, 153, 156–62,

164–67, 208–11, 215–19; developmentalist challenge, 154, 166; developmental
constraints, 96; developmental plasticity,
357; developmental system, 154, 163, 165;
developmental system drift, 162; developmental system theory, 166; ecology
and, 231; evolution and, 58; hierarchy
and, 14. *See also* evolutionary developmental biology (evo-devo)
discreteness, 20, 21
disruption, 14, 22, 127, 131–32, 191, 254,
298–301, 318, 320, 346, 351, 353
disturbance/perturbation, 14, 21–22, 39,
105, 112–13, 127, 130, 132–33, 155, 167,
191, 217, 241, 249, 250–52, 254, 298–99,
301, 308–9, 313, 315–22, 325, 327, 335,
339, 351, 354–55, 357, 360
DNA: junk, 138, 144, 148; selfish, 142, 176
Dobson, Cristopher M., 77
Dobzhansky, Theodosius, 12
Dollo, Louis A. M. J., law of, 30
Drude, Carl G. O., 228
Dussault, Antoine C., 113, 114

ecological evolutionary developmental biology (eco-evo-devo), 165
ecological evolutionary unit (EE unit), 293;
subunit, 293, 295
ecological locking, 300
ecology, 28, 65, 67, 76, 95, 103–7, 110, 119,
121, 165, 194, 229, 231, 233–36, 252, 340,
361, 367, 370–75; behavioral, 177, 247;
coining the term, 227 (*see also* Haeckel,
Ernst); community, 110, 230, 231, 246;
connections, 230; ecosystem, 21, 27, 28,
51, 52, 54, 55, 63, 66, 72, 103–7, 109–14,
118–21, 136, 160, 165, 191, 192, 194, 195,
215, 219, 229, 231, 233, 234, 239, 241, 246,
247, 248, 250, 251, 253, 254, 255, 298, 300,
301, 308, 310, 322, 327, 328, 334, 335, 338–
41, 343, 345, 347, 351, 352, 353, 354, 362,
369, 374 (*see also* Tansley, Arthur); evolutionary, 104, 227, 232–33, 247; functional,
104–7, 113–14, 230, 232–33; history of,
227–28; paleo-, 357, 368; philosophy of,
369; population, 181, 247; urban, 241
economic entities, 247, 252, 298, 334, 343,
352
economy of nature, 7, 103, 123, 227. *See also*
Haeckel, Ernst

Edinburgh, 6–7
Eifelian Stage, 285
Eldredge, Niles, 6, 29–30, 35–41, 51, 60, 109,
152, 190–94, 235, 241, 247, 249, 251–52,
263–64, 267, 270, 296, 298, 308, 334, 343,
351, 353, 359, 365; *Eternal Ephemera*,
1, 365. *See also* punctuated equilibria;
sloshing bucket model
Elonichthys, 323
emergence, 135–36, 159, 177, 186, 202–5,
211, 218, 220. *See also* traits
ENCODE project, 138–39, 144
energy, 13, 21, 23, 34, 109, 127–33, 192–93,
228, 298, 310, 318, 320, 327, 343, 345;
evolution and, 132, 235; information
and, 127, 132; matter and, 13, 21, 23, 51,
103, 105, 117–18, 192–93, 228, 234–36,
307, 344; transfer, 334, 352–53, 362–63;
trophic, 131–32, 231
environmental tolerance, 141, 240, 250, 282,
284, 289, 296, 298–99
Eochonetes, 298
epiboles, 295, 299, 301
epicycle, biocultural, 203, 212–15, 218
epistemology, 19, 36, 367–70
equilibrium, 22–23, 131–32, 134, 228, 234,
316–18, 337; far from, 74, 131; mutational, 140
Equus caballus, 47, 48, 55, 56
Erwin, Douglas H., 158–59
Escherichia coli, 179
Eternal Ephemera (Eldredge), 1, 365
eukariotes, 24, 146
eusocial insects, 24, 174, 178, 187
evolutionary developmental biology (evo-
devo), 152, 157, 159, 160
evolutionary individuals, 151–53, 160, 163–64
evolutionary theory, 1, 5–6, 8, 14, 26, 137, 177,
185, 231–32, 235–36, 243, 244, 247, 353
evolvability, 134, 146–47, 162
exaptation, 188, 190, 261, 340
expectations, 120, 243, 250
extended synthesis, 351, 358–60, 363
extinction, 2, 3, 5, 7–8, 14, 31, 39, 40, 49,
52, 53, 130, 133, 147, 178, 236, 248, 249,
251, 252, 311, 312, 315, 318–20, 323–28,
335–36, 356, 368, 371, 375; "law of" (Van
Valen), 230–31; mass, 14, 33, 39, 40, 189–
91, 241, 245, 251, 254, 307–9, 311, 312,
319, 321, 322, 326–28, 351, 353, 372 (*see*

extinction (*cont.*)
also Permian-Triassic mass extinction);
rates, 40, 96, 286, 341; secondary, 309,
312, 315, 318, 319, 320, *321, 322*
extrapolationism, 152

Falkland fox, 9
Favosites, 292
February 1835 (Darwin), 10
feedback, 21, 23, 136, 156–57, 166, 203,
209–15, 218, 220, 232, 309, 312, 359, 362;
niche construction and, 212–13; positive,
209–10, 220
Feibleman, James K., 68
Feldman, Marcus W., 231
Finger Lakes region, 285
Fisher, Ronald, 165, 175
fitness, 111–12, 129, 132–33, 174, 176,
178–79, 183–85, 188, 231, 325, 327, 337;
absolute, 178; collective (fitness II), 184;
inclusive, 176, 181, 190; particle's (fitness I), 184–85; relative, 178
Flack, Jessica C., 166
food web, 21, 131, 234, 309–10, 312–14,
316–20
fossil record, 31, 39, 230, 244–45, 247, 249–
51, 255, 261, 266–67, 269–70, 272, 282,
297, 313, 356
fractals, 22–23
freedom: degrees of, 78, *82*, 117–18; lower-
level, 87, 91, 98–99
functional organization, 112–13, 177–78, 187
Fusco, Giuseppe, 158, 160

Galápagos, 10, 11; finches, 360; mocking-
bird, 9
Gallus gallus domesticus, 57
Gaussian curve, 284
genealogy, 24, 37, 42, 60, 127–28, 129–32,
229–30, 235
generative entrenchment, 164
genetic drift, 49, 95, 98–99, 129, 144, 162,
228, 283, 357, 360
genetics, 24, 232; developmental, 160; population, 60, 177, 182, 189, 232, 366
genome, 134, 137–48, 154, 157–58, 192, 204
genome evolution, 139–40, 143, 148
Geological Diary (Darwin), 8
geological time, 132, *312,* 323, 334, 372
Gerhart, John, 166
germline, 24, 103, 207

Ghiselin, Michael, 14, 56–57, 164
Gilpin, Michael, 177
Givetian Stage, 285, 299, 300
goal directedness, 92, 100
goniatites, 292
Gould, Stephen J., 13, 29–42, 152, 160, 189–
93, 249, 260, 263–64, 267, 270, 296, 340
gradients: environmental, 283, *284, 287,*
289–92, 295, 297–98, 300, 372; faunal,
283; food-concentration, 88–91, 93,
97–98
gradualism, 4, 12–14, 269, 355, 359–60
Grammysoidea, 292
Grant, Peter and Rosemary, 360. *See also*
Galápagos: finches
Grant, Robert, 7–8
Grene, Marjorie, 41, 157, 241, 343
Griesemer, James, 159–60
group selection, 129, 174–95, 344; altruism
and, 175, 181; among-group selection,
176, 178–79, 181, 187; cultural, 177;
David Sloan Wilson and, 177, 179, 182,
186–88, 190, 193; Edward O. Wilson
and, experimental evidence of, 177;
homind evolution and, 344; inclusive
fitness and, 181; kin selection and, 129,
180, 194; rejection of, 177, 179; selfishness and, 176, 179, 180–81, 183–85, 187;
species selection and, 184, 191; within-
group selection, 176, 178–79, 181; Wynne
Edwards and, 176
guild, 309–10, 312–14, 317–18, 320–27; structure, 241, 309–10

Haeckel, Ernst, 227. *See also* ecology
Haldane, John B. S., 175
"halo" pattern, 9–10
Hamilton, William, 176, 179, 181
Hamilton Group, 240, 283–86, 288, 290, *291,*
292, 295, 297, 299, 300, 301
Heliophyllum, 292
heritability/heredity, 23, 159–60, 161, 165,
183, 188, 203–4, 228, 360
Herschel, John, 7
hierarchy: of control, 30; definition of, 19;
dual hierarchy, 24, 235–36, 298; ecological (economic) hierarchy, 3, 13–14, 24,
28, 39, 48, 54–55, 59, 87, 103–9, 127, 129–
32, 156, 165, 190, 192–93, 235, 283, 308,
352–54, 355, 359; genealogical hierarchy,
3, 13–14, 24, 26–27, 34, 39, 48, 54–55,

59–61, 76, 88, 103, 105–6, 108–9, 112,
127–32, 153, 165, 191–94, 229–31, 235,
241, 282, 298, 308, 334, 336, 339, 342–44,
346–47, 351–61; nested compositional,
19, 20, 23, 24, 27, 65, 128, 131, 132, 183,
235; theory, 6, 12, 14, 17, 19, 23–24, 26,
28, 48, 68, 76, 135, 151–53, 156, 164–66,
190, 194, 236, 239–41, 243–44, 248, 256,
298, 308, 334, 338, 347, 351–54, 357,
359–63, 366–67
holistic, 229, 354
Homo: *floresiensis*, 211, 356; genus, 50,
   355, 356, 362; *heidelbergensis*, 211, 213;
   naledi, 356; *neanderthalensis* (Neander-
   thal), 211, 216, 356; *sapiens*, 50, 56, 205,
   206, 211, 213, 216, 356, 362
homogenization, 241, 334–36, 339–40, 346
Homogeocene, 241, 334, 335, 336, 338, 344
homology, 37–38, 375; autonomy, 162–63;
   character identity and character state,
   162; as continuity of information, 160;
   core character identity network, 162;
   definition of, 210; epistemological and
   ontological relevance, 161; heritability,
   161; hierarchy and, 153, 160–61, 163; homologues as attractors, 163; homologues
   as natural kinds, 164; individualization,
   160–63; mechanistic basis, 153, 160–61;
   organizational homology concept, 163;
   sequence, 261
Horner, John, 6
Hull, David, 177
Hutchinson, George E., 230
Hutton, James, 6, 32, 33, 36
Huxley, Julian, 255

idiographic approach, 26, 29, 361. *See also*
   contingency/contingent
inclusion, 19–20, 22–23, 51, 65, 72, 128–29,
   131, 189, 229, 231, 250
incorporation, 26, 51, 77, 109, 244
individuality, 159, 163–65, 186, 191, 193, 205,
   208, 210, 218
information, 13, 23, 127–32, 153–56, 159–
   60, 162, 165, 166, 192, 211–12; analog
   coding, 155; biological, 156, 146, 362;
   coding, 129; context-dependency, 154–
   55; cultural, 218; definition, 127–28, 130;
   energy and, 127, 132; evolution and,
   128; extinction and, 130; flow of, 13, 23,
   127, 130; genetic, 59–60, 193, 235, 334,

352–53; genealogical hierarchy and,
   13, 165, 192, 193; loss, 130; molecular
   syntax vs molecular semantics, 154–55,
   162; packaging, 153, 159; processing,
   128–29, 155; receiver, 154; selection, 154;
   semiotics, 130; social, 215; statistical,
   154; theory, 130, 154–55; transmission,
   23, 103, 127; Wiener's semantic, 154
integration, 26–27, 50–52, 70–71, 73–74, 80,
   125, 163–64, 244
interactor, 23–24, 129–30, 159–60, 191–93,
   252, 254, 353
invasion (biological), 234, 241, 251–52, 297,
   301, 336
isolation: geographic, 11–12, 14, 35, 49;
   reproductive, 266
isomorphism 22, 152

Jablonski, David, 185
Jameson, Robert, 6–7

Kant, Immanuel, 31, 99
Karoo: basin, 241, 308, 311, 315, 326; pa-
   leocommunities, 309, 310, 312, 313, 314,
   317, 318, 322, 324, 325; species, 313
Kestrosaurus, 323
Kimura, Mooto, 231
kin selection, 129, 176, 179, 180, 182, 194–95
Kirschner, Marc, 166
Korn, Robert W., 165

Lack, David, 176
Laland, Kevin, 231
Lamarck, Jean Baptiste, 1, 4–7, 11–12; Lamarckism, 360–61; *Système des Animaux
   sans Vertèbres*, 1
Laubichler, Manfred, 154–55
law of extinction, 230
levels: focal, 20, 22, 23, 27, 91, 128–29, 184–
   85, 187, 189, 243, 246, 352; hierarchical,
   19–23, 27, 34, 63, 76, 118, 128–29, 132,
   151, 183, 186, 190, 246, 250, 273, 308–9;
   higher, 20, 22, 27, 32, 39, 64, 69, 74–75,
   78, 81, 89, 91, 93–94, 100, 128–29, 132,
   146, 151, 153, 155–56, 159, 166, 175–76,
   183, 185–87, 190–92, 204–5, 240, 245–46,
   300, 319, 328, 352–53; lower, 19–20, 22,
   23, 27, 32, 64, 69, 72, 74–75, 78–79, 81,
   87, 89–91, 94, 96, 98–99, 108–9, 128–29,
   131, 151–52, 155–56, 160–61, 166, 183–86,
   189, 192, 321, 352–53; of organization,

levels (*cont.*)
  21–22, 63–67, 70–71, 75, 77–79, *82*, 100,
  141, 146, 155, 157, 161, 182, 219, 232,
  244–45, 250, 290, 352; population level,
  20, 21, 24, 98, 128, 131–32, 190, 193, 231–
  34; of selection, 64, *148*, 153, 175, 178,
  180, 182, 183, 185, 186, 187, 188, 189,
  195, 370
Lewes, George H., 68
Lewontin, Richard, 164, 183
Lieberman, Bruce S., 39, 40
lineages, 235
Lloyd, Elyzabeth, 177, 182
Lotka, Alfred J., 228
Love, Alan, 77
Ludlowville formation, 286, 289, *293. See
  also* Hamilton Group
Luscina megarhynchos, 53
Lyell, Charles, 4, 6, 7, 10, 32
Lynnean hierarchy, 1, 3, 5
Lystrosaurus, 323

macroevolution, 26, 29, 34–35, 38–42, 49,
  64, 96, 135, 152–53, 158–59, 161, 165,
  190, 192, 194, 225, 230, 243–56, 260, 262,
  270–73, 352–61
macrostate, 128
major transition, 100, 135, 159, 186–88,
  204–5
Mammalia, 50
Margulis, Lynn, 186
Marine Biology Laboratory (MBL) group,
  31, 41
mathematics, 228, 232, 234
matter, 127–32, 157, 192–93, 228
Maynard Smith, John, 159–60, 176
Mayr, Ernst, 12, 35, 108
MBL group. *See* Marine Biology Laboratory group.
mechanistic/neomechanistic, 27–28, 34, 64,
  66–67, 69, 73, 96, 99, 108, 112, 153, 157–
  58, 161–63, 165, 208, 210
Megatherium, 8
Melott, Adrian L., 39, 40
metacommunities, 326–27
metastability, 22
microevolution, 34, 41, 49, 53, 152, 158–59,
  192, 239, 244–48, 250, 255, 265, 273, 352,
  356, 360
microstates, 128

migration, 175, 248, 299, 346, 356, 360
Milankovitch cycles, 286
Mill, John S., 68
Minelli, Alessandro, 161
Modern Synthesis, 151–53, 175–76, 228, 235
modularity, 158–59, 164
monophyly, 40, 50, 53, 103, 251, 347
Morange, Michel, 158
Moreno, Alvaro, 114–15, 117
Morgan, C. Lloyd, 68
morphospecies, 272
Moscow Formation, 286, *293. See also*
  Hamilton Group
Mossio, Matteo, 114–15, 116, 117
mudstone, 286, 288
Müller, Gerd B., 161, 163
Muller, Richard A., 39.
multilevel: approach, 134, 139, 141, 145–48,
  152–53, 156–57, 351; selection (MLS),
  135, 152, 174, 177, 179–91, 194
mutual dependence, 118, 120

natural kinds, 47, 50, 164
natural selection, 11, 30, 37, 53–54, 59, *81*,
  87, 94, 99–100, 112–14, 116, 119, 129,
  131–32, 142, 145–46, 149, 166, 174–76,
  178–80, 183, 185, 188–89, 191, 193–94,
  196, 204, 206, 208, 212, 215, 228, 232,
  235, 237, 243, 246–47, 254, 261, 344, 353,
  357–58, 360
nautiloids, 292
Needham, Joseph, 68, 79
nesting, 19–23, 88, 128–32, 183, 235, 343
networks: complex biological, 21–22, *23*
  132–33, 219, 248, 375; core, 209–10,
  212, 215; developmental gene regulatory (dGRN), 158, 162; ecological, 131;
  epigenetic character identity, 203, 209;
  interactors (ecologic), 251–55, 300;
  metabolic, 131; trophic, 313
New Edinburgh Philosophical Journal, 6
Newman, Stuart A., 163
Newton, Isaac, 99–100
Newtonian: naturalism, 4–5; physics, 31, 33,
  41–42; worldview, 1
New York: outcrop belt, 285–86, 292; State,
  284–86, 288, 299
niche: conservatism, 284–86, 289–90, 293–
  94, 296–301; evolution, 282; fundamental, 296–97; modification, 282–83, 285,

293–96, 299, 301; parameterization, 290;
realized, 295, 297, 301
niche construction, 113, 231–32; cumulative
culture and, 362; extended synthesis
and, 358; feedback and, 212; hierarchy
theory and, 359; node, 21
nomothetic, 26, 29–42, 361
nonequilibrium, 22, 131
nontransitivity, 21
Notaelurodon (*Promoschorhynchus*), 323
novelty (evolutionary), 74, 132, 146, 158,
164, 218, 327, 340
Novikoff, Alex B., 68
Nowak, Martin, 182

Oatka Creek formation (upper Marcellus),
286, 288. *See also* Hamilton Group
Odling-Smee, John, 231
Ohta, Tomoko, 231
Okasha, Samir, 64, 152, 182–84, 188–89
O'Malley, Maureen A., 157
On the Origin of Species (Darwin), 7, 10, 12
ontology, 12, 19, 23, 36, 67, 75, 156
Ophidia, 57–58
ordination, 289–92, 295; space, 292
Ordovician, Late Ordovician, 283, 297, 328
Orians, Gordon H., 232–33
origination, 31, 39–40, 152, 161, 163, 165,
286
Ornithological Notes (Darwin), 9
Owen, Richard, 10, 210

paleobiology, 26, 230, 250, 261–62, 301, 368,
370–75
paleocommunity, 241, 308–19, 322–26, 351
paleoecology, 357, 368
Paleozoic Era, 38, 97, 297
Pancaldi, Giuliano, 2
Panthera leo, 47, 48, 49, 50, 52, 53, 60; *persica*, 47, 49, 50, 52, 60
Panthera pardus, 47, 53, 60
Paranthropus, 355
Paris basin, 5
Patagonian wilds, 9, 14
path length, 21
Pattee, Howard H., 78–79
pattern, 2, 5, 7, 9–11, 26, 29–42, 74, 76, 78,
87, 89–90, 103, 113, 127–29, 132, 135,
152, 155, 165, 174, 181, 186, 189, 191–92,
194, 220, 225, 229–34, 243–56, 261–66,

269–70, 272, 283–85, 294, 297, 300–301,
310–11, 314, 318, 326, 335, 339, 345, 351–
61; processes and, 29–42
peak abundance, 240, 284, 289–95
Pearsons's correlation coefficient, 293
Pentamerella, 292
Permian-Triassic mass extinction, 307, 308,
311
phenotype, 94–97, 135, 153–58, 161, 164,
166, 184, 208–10, 215–18, 234, 266, 268,
270, 354; genotype–phenotype mapping,
157
phylogenetics, 35, 37, 42, 370
phylogeny, 37, 53, 60, 203, 354–55, 357
plant biology, 227–28
plesiomorphy, 57–59
Polanyi, Michael, 78–79
Popper, Karl, 244, 249, 255
population structure, 128, 179–82, 185, 194
predator–prey relationship, 132
Price, George, 177, 182, 184; equation,
181–82
process: evolutionary, 1, 4, 5, 12, 31, 36–37,
42, 47, 114, 127, 141, 151, 175, 179, 203,
219, 232–34, 243, 247, 250, 298, 310, 362,
365; macroevolutionary, 250; microevolutionary, 152, 192, 244, 355, 360. *See
also* macroevolution
producer, 105–6, 109–10, 118, 159–60, 165,
234, 307, 310, 313
proliferation, 48, 214–16, 299
protein evolution, 231
Pseudoatrypa, 292
Pseudomonas fluorescens, 179
punctuated equilibria (or punctuated
equilibrium), 12, 26, 30, 33, 35, 38, 39,
42, 190, 192, 235, 239, 249, 250, 254, 263,
267, 268, 270, 272, 354, 359, 360, 365. *See
also* Eldredge, Niles; Gould, Stephen J.

random walk, 31, 240, 265–66, 269, 271, 273
rank, 19–20
rate: birth, 129; death, 129; extinction, 96,
286, 309, 341
Raup, David M., 29, 39
reciprocal averaging, 289
Red Notebook (Darwin), 6, 10
Red Queen hypothesis, 132, 231, 325
regulation, 21, 27, 71, 74, 80, 82, 131
replacement of species, 7, 9

replicator, 23–24, 127, 129–30, 159–60, 182, 192–93, 353
reversibility, 21, 23, 129–30, 163
Richmondian invasion, 297–98
Rio Negro, The, 9
robustness, 21, 136, 163, 166, 203, 212, 215–17, 220
Rohde, Robert A., 39
Rosenzweig, Michael L., 233
Ruse, Michael, 195

Saborido, Cristian, 114–15
Salthe, Stanley N., 50, 69, 71, 82, 91, 108–9, 156
Scheffer, Marten, 316
Schimper, Andreas F. W., 228
Schmitz, Oswald, 219
Schmitz, Robert J., 360
sea-level cycles, 286
sedimentary cycles, 283
Selfish Gene, The (Dawkins), 176
selfishness, 175–76, 178–81, 183, 184–85, 187
self-organization, 99, 322
self-similarity, 22
semiotics, 130
Sepkoski, J. John, 39
shale, 285, 286, 288
Shannon, Claude, 154
Siliciclastic, 286, 287, 291, 292, 300
Simon, Herbert, 69, 92
Simpson, George Gaylord, 5, 30, 262–66, 268–70; Tempo and Mode in Evolution, 260, 262
simulation, 234, 289
Skaneateles Formation, 286, 289, 293. See also Hamilton Group
sloshing bucket model, 3, 14, 39, 42, 191, 193, 235, 247, 251–53, 298–301, 308, 351, 353, 355, 365. See also Eldredge, Niles
Sober, Elliott, 177, 195; Unto Others, 177
social networks, 21, 128
sorting: notion of, 152, 189; processes, 76, 78, 106, 128–29, 132, 135, 190, 327; species, 14, 49
Sowerbyella, 298
speciation: 128, 131–32, 147, 158, 188, 192, 228
species: concept, 231; drift, 129; as individuals, 37; invasive, 336, 341
stability, 2, 10, 12–13, 22, 74, 87, 91, 95, 98, 114, 128, 131–33, 154, 158, 161–62, 203,

215, 218, 220, 235, 347–252, 285, 300, 301, 307–28
stasis, 12–13, 132, 235, 251, 254–55, 262, 265–66, 268–73, 284, 297, 326, 360; coordinated, 235, 243, 247, 249; ecological, 282, 296
stratigraphic, 283, 284, 285, 247, 250, 251, 253, 283, 288, 289, 292, 293, 294, 295, 299, 312, 313, 314, 322, 323, 368, 375
struggle for life, 54, 227
succession, ecological, 228
system (definition of), 19, 128
Système des Animaux sans Vertèbres (Lamarck), 1
systems biology, 157, 373
Szathmáry, Eörs, 159

Tansley, Arthur, 229. See also biome, definition of
Tapirus indicus, 52
tectophase, 288
teleology, 27, 86, 88–89, 91–99, 107–20, 352
Tellinopsis, 292
tempo and mode, 30, 260, 262, 264–66, 269–73
Tempo and Mode in Evolution (Simpson), 260, 262
thermodynamics, 228, 236
Tigrisuchus (Moschorhinus), 323
time's arrow, 32–41
time's cycle, 32–36, 39–41
traits: aggregate, 20, 183, 184, 195; emergent, 20, 183, 184, 195. See also emergence
Transmutation Notebooks B–E (Darwin), 6, 11, 12, 13
transposable elements, 134, 139–41, 143, 145
trophic: energy, 231; interactions, 105, 131, 234, 310–11, 324, 327, 368; webs, 20–21
Tully Formation, 285
turnover pulse, 26, 30, 38–39, 42, 191, 247, 249, 251, 253, 351, 354, 356. See also Vrba, Elisabeth
Tuscany, 2
type II error, 293, 295

unimodal response curves, 283, 289
Union Spring formation (lower Marcellus), 286. See also Hamilton Group
units of selection, 160, 177, 181, 182, 183, 186, 191

unity of type, 53
Unto Others (Sober and Wilson), 177

Van Valen, Leigh, 230–31, 235
Venn diagram, 20
Vertebrata, 47, 50, 56, 57
volatility, 40
Vrba, Elisabeth, 29–30, 38–39, 152, 188–89,
    247, 251–52. *See also* turnover pulse

Wade, Michael, 177
Wagner, Günter P., 58, 153–56, 160–66
Warming, Eugenius, 228

Watsonisuchus, 323
Whewell, William, 2
Wiener, Norbert, 155
Wiley, Edward O., 29–30, 34, 37–38, 42
Williams, George C., 176, 179, 181, *195*
Wilson, David Sloan, 177, 179–80, 182, 186–
    88, 190, 193; *Unto Others*, 177
Wilson, Edward O., 179–80
Wimsatt, William, 69, 74, 93
Wright, Sewall, 175
Wynne Edwards, Vero C., 176, 195; *Animal
    Dispersion in Relation to Social Behav-
    ior*, 176. *See also* group selection